The Complete
BASIC BUILDER

Blue Circle
Mortar & Rendering

The Complete BASIC BUILDER: Techniques, Projects and Materials

Edited by Mike Lawrence

ORBIS · LONDON

Acknowledgements
Photographers: Jon Bouchier, Simon Butcher, Paul Forrester, Simon Gear, Jem Grischotti, Barry Jell, Keith Morris, Karen Norquay, Roger Tuff.

Artists: Roger Courthold Associates, Bernard Fallon, Nick Farmer, Trevor Lawrence, Linden Artists, David Pope, Mike Saunders, Ed Stuart, Craig Warwick, Brian Watson.

This edition published 1985 by
Orbis Publishing Limited, London
under licence from
Whinfrey Strachan Limited
315 Oxford Street
London W1R 1AJ

Printed in Yugoslavia
ISBN: 0-85613-832-0

CONTENTS

1 Bricks and brickwork
Basic bricklaying techniques 8
Forming corners and piers 14
Laying special bonds 19

2 Using concrete
Mixing and laying concrete 25
Basic concreting techniques 26
Foundations for garden walls 30
Laying concrete slabs 35
Laying ready-mixed concrete 40

3 Paths, steps and patios
Laying paths with paving slabs 46
Laying paths with bricks 50
Building free-standing steps 54
Creating built-in steps 58
Laying crazy paving 61
Building a paved patio 64
Laying block paving 68

4 Walls and walling
Building a screen block wall 74
Building retaining walls 78
Building arches in brickwork 83
Rendering exterior walls 88

5 Alterations
Making a new doorway I – the lintel 95
Making a new doorway II – the frame 100
Installing patio doors I – preparation 104
Installing patio doors II – installation 108
Fitting a new window I – the opening 113
Fitting a new window II – the frame 118
Fitting a new fireback 123
Removing a fireplace 127
Replacing an old ceiling 132
Block partition walls 137
Building a stud partition wall 142

6 Plastering walls and ceilings
Basic plastering techniques 148
Plastering angles and reveals 154
Patching plaster ceilings 158
Plastering plasterboard 162

7 Materials, tools, rules and regulations
Bricks 168
Cement, sand and aggregate 170
Concrete paving slabs 172
Garden walling blocks 174
Ironmongery for building and repair work 176
Platform towers 178
Safety gear 180
Masonry tools 182
Building rules and regulations 185
Index 190

CHAPTER 1

BRICKS AND BRICKWORK

Successful bricklaying relies on three easily-mastered elements. The first is mixing mortar properly – like a cook's recipe, sloppy measuring and mixing spells disaster, while a correctly-mixed mortar will be as strong and long lasting as the bricks or blocks it bonds together. The second is laying a brick squarely and level. The third – and arguably the hardest – is laying all the other bricks square and true with their neighbours, yet provided you can lay one properly and you're prepared to check your work regularly as you proceed, a perfect wall will be the end result.

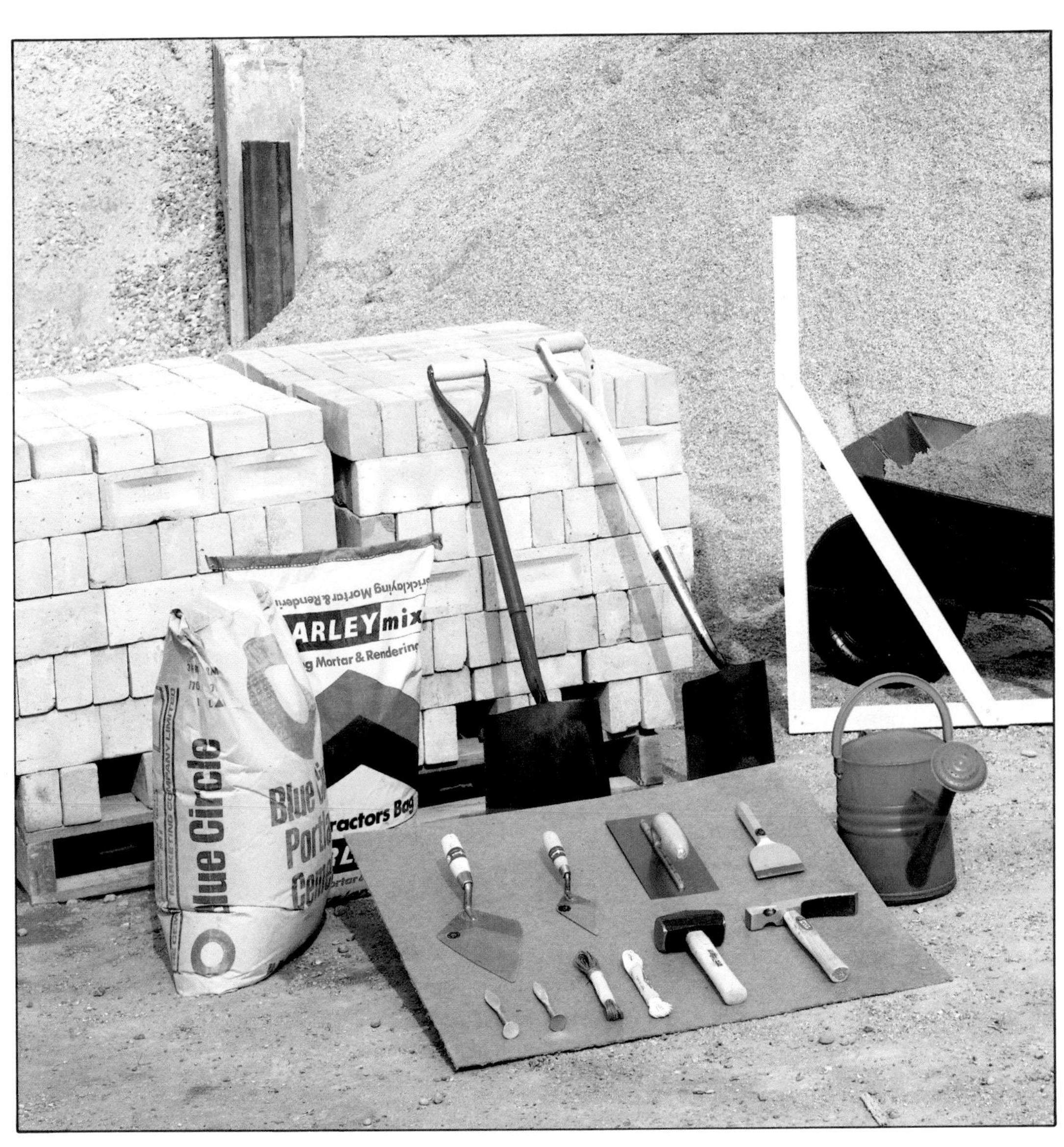

BASIC BRICKLAYING TECHNIQUES

There's a lot in bricklaying that you only really pick up with practice. But there are some rules which can guide you every step of the way. Understanding bricks themselves, the right way to mix mortar and how to use the trowel correctly will help you achieve a result to be proud of.

Brickwork is made up of two things: the bricks, and the mortar which forms the joints. Building a wall that's going to last needs careful attention to both, and the first thing is to choose the right bricks for the job.

Know your bricks

There are three groups of clay bricks:

1 Common bricks have no special finish because they are made to be used where they will not be seen or be subjected to major stress or load. They are mostly used in situations where they will be covered by paint, plaster, cladding, rendering etc. They are a relatively inexpensive brick and are usually a rather patchy pink in colour.

2 Facing bricks come in a variety of colours and textures for they are made to be displayed indoors and out. Also called *stocks,* they are capable of bearing heavy loads. If classed as *ordinary quality,* it means they can be used for most projects, but in very exposed conditions outdoors will need to be protected by a damp proof course at ground level (either a course of engineering bricks, see below, or a layer of bituminous felt) or with a coping above to prevent the bricks becoming saturated with rain. Without this protection they are liable to be affected by frost which would cause disintegration. *Special quality* facing bricks are suitable for use in exposed places or where great strength is needed, eg, for paving, retaining walls, garden walls and steps.

3 Engineering bricks are smooth and dense, designed to be used where strength and low water absorption is essential – for example in foundation courses (thus providing a damp proof course for a wall or planter) and load bearing walls.

Another type of bricks completely are the *calcium silicate bricks.* These are flint/lime bricks which are whitish when steam-hardened in an autoclave (they aren't fired like clay), but these are available in many colours because they take pigment well. They absorb moisture easily, so must never be laid with a mortar that doesn't contain a plasticiser. They can be used in just the same way as clay bricks. Like engineering bricks, they are also more regular in shape and vary less in size than ordinary bricks.

Brick types

Bricks also vary in their character as well as their composition: they may be solid, perforated or hollow, but most fall into the solid category. Even bricks with small or large holes in them (these are also known as cellular) are classed as solid so long as the perforations do not exceed 25% of the total volume. The same is true of bricks with a shallow or deep indentation known as a *frog.* As well as making the bricks lighter, perforations and frogs give bricks a better key (ie, the mortar is better able to bond them together).

Bricks are measured in two ways: when they come from the works the actual size is 215mm long, 102.5mm wide and 65mm deep; the format size, however, is the one used for calculating the number of bricks you need. This needs an allowance of about 10mm added to each of these dimensions for the mortar joints – ie, 225mm long, 113mm wide and 75mm deep. Bricks are also made in special shapes and sizes for particular uses (copings, bullnose and angles are some examples). For information about these see pages 168 -169.

Storing bricks

As all bricks (except engineering bricks) are porous, they should be stacked on a level area away from damp, otherwise long after you've used them the mineral salts inside the clay will stain the surface with an unsightly powdery white deposit (known as *efflorescence*). In the garden, put bricks on planks or a metal sheet and cover them with plastic sheeting. Apart from anything else, bricks which are saturated with water (as opposed to just being wet) are hard to lay and will prevent a satisfactory bond between bricks and mortar.

Mortar for bricklaying

Cement and sand made into mortar with water will set quickly, but is liable to create a crack between the mortar and the brick if it shrinks during drying. The ideal mortar, in fact, doesn't set too quickly, doesn't shrink much and can take up settling movements without cracking. There are two ways of making a mortar like this:

- The first way is by adding hydrated lime to the mix. This makes the mortar more workable and smooth (or 'buttery' as the experts say).
- The second is by adding a plasticiser – a proprietary liquid or powder. Air bubbles are formed which provide spaces for the water to expand into, thus preventing cracks.
- Basic to mortar is cement. This acts as the adhesive, binding the particles of sand together. Ordinary portland cement is the one most commonly used.
- Fine sand is used for mortar to give it its correct strength. Use clean builder's sand (also known as 'soft' sand) which does not contain clay, earth or soluble salts (these can lead to efflorescence).

Buying the materials

Cement is usually sold in 50kg (112lb) bags, although you may also find smaller sizes. Sand is sold by the cu metre (1⅓ cu yd) and in parts of a cu metre. To give you a sense of scale, a cu metre of sand weighs about 1,500kg (1½ tons) — a very large heap. Both are usually bought from builders merchants, where you can also buy lime or

MIXING MORTAR

1 Unless you're using dry ready-mix (most suitable for small jobs) carefully proportion 1 part cement to 6 parts builders sand.

2 Thoroughly mix the cement into the sand so that you end up with an entirely consistent colour. Turn the mix over at least three times.

3 Adding a little plasticiser to the water (the amount will be specified on the container) will make the mortar easier to work with.

4 Form a crater and pour in half the water. In total you'll need about the same amount of water (by volume) as cement — but add the rest gradually.

5 Mix in the dry mortar from the inside walls of the crater. As the water is absorbed, add a little more. Turn the whole mix over several times.

6 The final mix should look like this. Check it by stepping the shovel back — the ridges should be firm and smooth, holding the impression of the shovel.

READY REFERENCE

BRICKLAYING GLOSSARY

Frog: the identation in the top of a brick. Generally bricks are laid frog up except on the top course of a free-standing wall.

Header face: the small end of the brick. Bricks laid so this end shows on the face of the wall are called headers.

Stretcher face: the long side of a brick. Bricks laid lengthways are called stretchers.

Bed face: the underside of the brick which is set into a mortar bed.

Bonds: the patterns of bricklaying to give walls strength. Vertical joints (called perpends) in one course must be overlapped by bricks in the next so the joint doesn't run through.

Course: name for a single row of bricks. A course of stretchers is known as ½ brick thick, a course of headers 1 brick thick.

Joints: the average of 10mm of mortar between bricks.

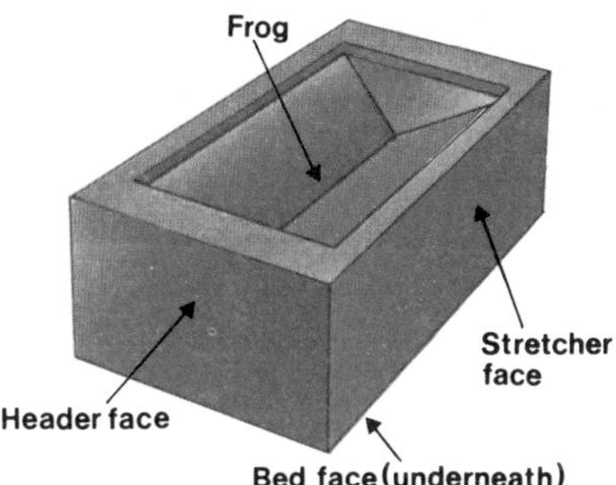

BRICK SIZES

A standard brick is 215mm (8½in) long, 102.5mm (4in) wide and 65mm (2⅝in) deep. For estimating purposes the 'format size' is used instead. This includes 10mm in each dimension to allow for one mortar joint. *Format size* is 225 x 113 x 75mm (8⅞ x 4⅜ x 3in).

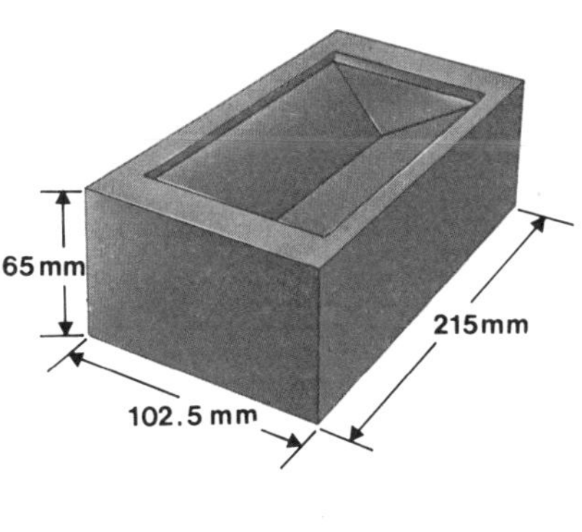

USING THE TROWEL

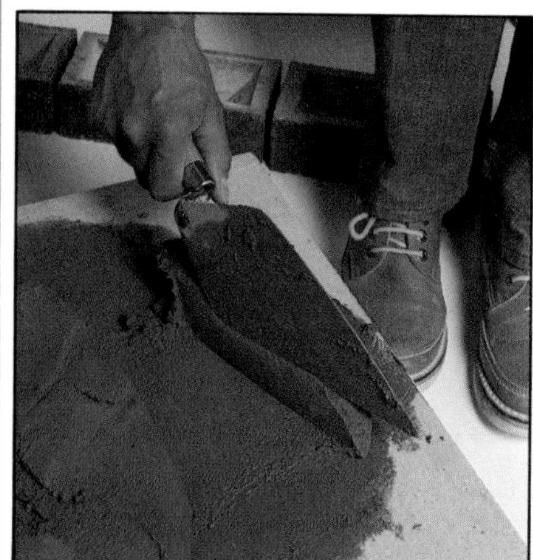

1 *Use the trowel to chop off a section of mortar (about the same size as the trowel) and separate it from the rest with a clean slicing action.*

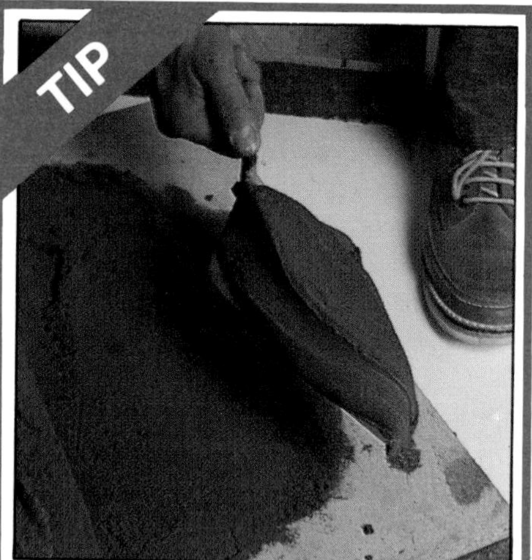

2 *Shape the back of the mortar slice into a curve — so that it's pear-shaped. Sweep the trowel underneath to lift the mortar off the board.*

3 *Slide the trowel sharply backwards to lay the mortar in a 'sausage' shape. Spread it out by stepping the tip of the trowel down the middle.*

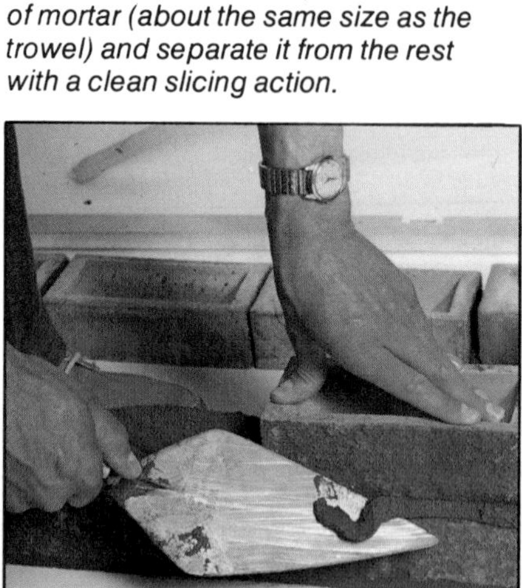

4 *When you've laid a brick in position, remove the mortar that squeezes out of the joint by sweeping the trowel upwards with its edge just scraping the brick.*

5 *To create the vertical joints between bricks, you 'butter' one end before you lay it. Scrape the mortar on by sliding the trowel backwards.*

6 *Hold the brick upright and scrape down all four edges. Finally spread out the mortar evenly to a thickness of 10mm-12mm (about ½in).*

proprietary plasticisers. Alternatively you can buy special masonry cement which has a plasticiser already in it and only needs to be mixed with sand and water.

Proprietary plasticisers are available in 5kg containers and you will have more than you need if you're only doing a small job — only a capful or two for each bucket of cement. But always follow manufacturer's instructions for use. Hydrated lime is a powder bought in 25kg bags.

Dry ready-mix mortars are also available with all the necessary ingredients ready mixed — so you just add water. Although more expensive than buying the sand and cement separately, it's a convenient way of buying for small projects. Bags usually come in 10kg, 25kg, 40kg and 50kg sizes. Alternatively, you can buy bags in which the cement is packaged separately from the sand.

Remember that it's always better to have a little more than you need — so be generous in estimating (see *Ready Reference).* Also make sure that any surplus cement or dry-mix is well sealed. This is vital to prevent it going off.

Rules to remember

- When filling the cement bucket (proportions are by volume, not weight), tap it frequently to disperse any trapped air.
- Mortar that has begun to set is no use. Any not used within 2 hours of the wetting of the cement should be discarded – if used it would dry too quickly and would not give the required strength to the brickwork.
- The sand and cement have to be thoroughly mixed before any water is added. Turn mixture over and over with the shovel until the pile is a consistent colour all through. The same rule applies to dry-mix mortar.
- When mixing in water, make crater on top of the pile, add some water and bring dry materials from sides to centre. Turn over whole pile several times, make another crater and repeat until mixture has a consistency which will hold the impression of the fingers when squeezed, or the impression of the trowel point.
- As builders' sand is rarely dry it is not possible to know how much water will be

needed to achieve the right consistency. Using a small container such as an empty tin will give you more control than using a bucket – and add water bit by bit.

The vital bricklaying tool

The trowel is the tool which makes the job, and no other tool can be substituted for it. A bricklayer's trowel is heavier and less flexible than any other trowel, and can be used to pick up and smooth down a required amount of mortar. Brick trowels can be bought in various blades sizes (from 225 to 350mm, or 9in to 14in) but the easiest to handle is the 250mm/10in one.

Brick trowels are roughly diamond shaped with a sharp point at the end opposite to the handle. The left side has a straight edge for scooping up mortar; the right side has a slight curve used for cleaning up the edges of bricks and for tapping the brick down into the mortar to level it. These are reversed in left-handed trowels.

Professional bricklayers use the curved edge of the trowel to cut bricks, but a more accurate and cleaner cut can be made with a brick hammer and bolster chisel. The trowel has a wooden handle raised slightly above the diamond, and at an angle to it to prevent you brushing your knuckles on the bricks as you are working. Getting the feel of the trowel and handling it properly is the key to good brickwork.

The trowel must be manipulated so that the mortar is scooped up in what's called a 'pear' or 'sausage' shape (see left) and placed on the bricks. This action is one that needs a lot of practice, for mortar that isn't compact is hard to manoeuvre and won't go where you want it to.

Practice routines

Make up a small amount of mortar (or 1 part of lime to 6 of sand, plus water to make it pliable) and practise combining it with bricks before you undertake a bricklaying project. The bricks can be scraped off within 2 hours (before the mortar sets). You have longer with

BRICK JOINTS

1 The simplest brick joint is a 'struck' joint — do the vertical joints first, drawing the trowel upwards or downwards with a firm action.

2 Next do the horizontal joints — use the full length of the trowel and drawing it firmly backwards with a sliding action.

Types of pointing

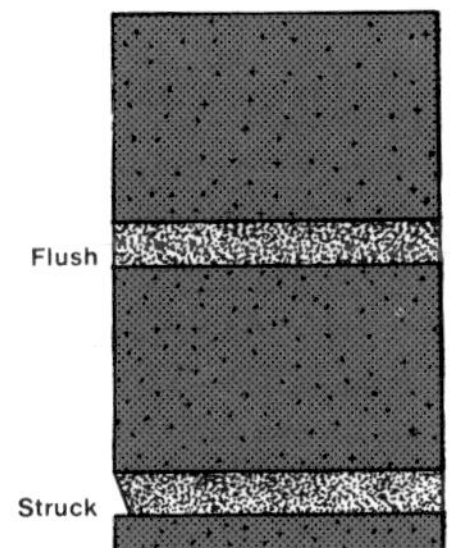

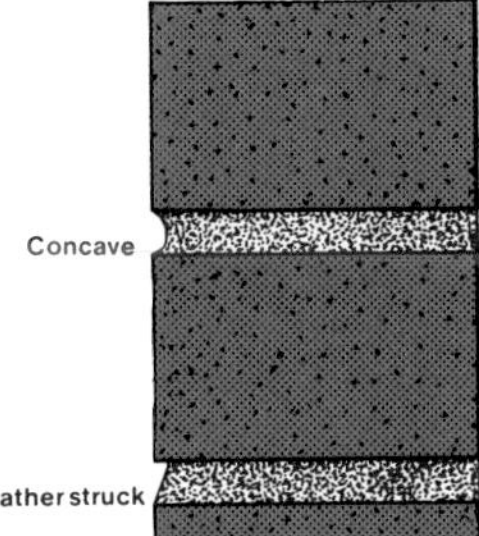

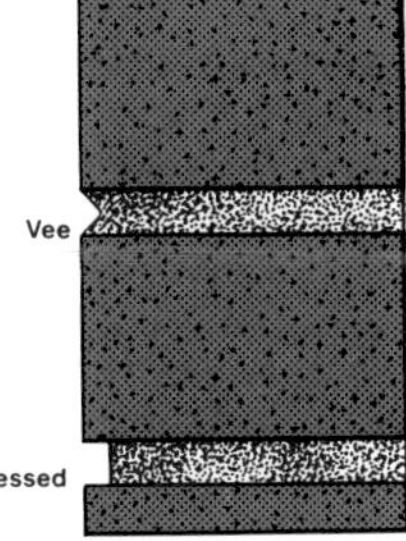

There are lots of ways of finishing off the joints in brickwork. Above is a selection of six of the most common. The flush joint is finished flush with the brick surface, while struck, weatherstruck and vee joints are all formed with the point of the trowel. Concave or rounded pointing is formed by running the edge of a bucket handle or a piece of hose pipe along the joints, while recessed pointing is pressed back with a piece of wood planed to the same size as the brick joints.

READY REFERENCE

HOW MANY BRICKS?

For a straight run of walling, remember
- One brick length + one joint = 225mm
- One brick height + one joint = 75mm

So for each metre of wall length you need 4.5 bricks; for each metre of wall height you need 13.3 bricks.

Example: a wall 4 m long and 2 m high needs 4 x 4.5 = 18 bricks in each horizontal course, and a total of 2 x 13.3 = 27 courses (to the nearest whole number). 18 x 27 = 486 bricks in total.

On more complex structures, dry-lay the first course of bricks and count how many are required. Then multiply this number by 13.3 for each metre of height required.

HOW MUCH MORTAR?

For small jobs: buy dry ready-mixed mortar, available in 10kg, 25kg and 50kg bags. A 50kg bag makes enough mortar to lay 50-60 bricks.

For larger jobs: it's more economical to buy materials separately and mix them up yourself. Here's how they are sold:
- Cement — usually in 50kg bags
- Lime — usually in 25kg bags
- Sand — by the 1/4, 1/2 or whole cu m.
- Plasticisers — usually in 5kg containers

MORTAR MIX

Measure ingredients by volume, using a stout bucket or similar container. For general brickwork above ground level, mix:

1 bucket ordinary Portland cement
1 bucket hydrated lime
6 buckets soft sand

- Replace lime with proprietary liquid plasticiser, added according to manufacturer's instructions, if preferred.
- One 50kg bag of cement mixed in the above proportions will give enough mortar to lay about 450 bricks.

TIPS: MORTAR MIXING

The amount of water you add to give the right consistency depends on how damp the sand is. As a rough guide allow 1 bucket of water for each bucket of cement. Remember:
- always add water gradually, never all at once, mixing after each addition
- always use mortar within 2 hours of mixing; it cannot be reconstituted once it has begun to set.

LAYING AND LEVELLING

1 Check layout by 'dry laying' the bricks, setting them a finger-width apart. Use string and pegs as a guide, fixing the ends with bricks, as shown.

2 Use your gauge rod — see **8** *— to check that a corner is square. Measure 3 marks along one side,4 along the other; if square, the diagonal will be 5 marks.*

3 With the line and pegs still in position, lay the mortar on the base by drawing the trowel sharply backwards. Lay enough for at least 2 bricks at a time.

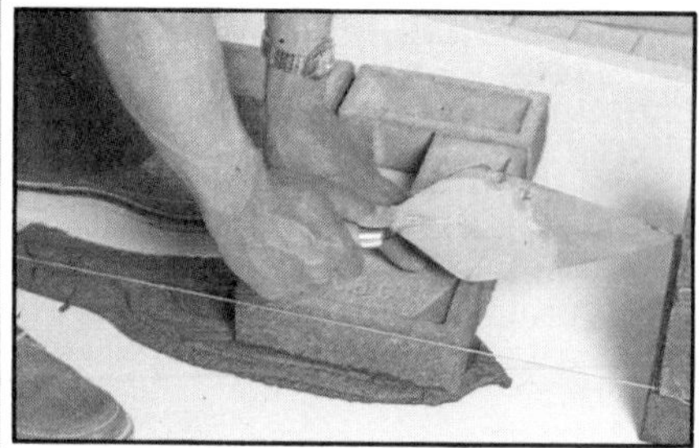

4 Tap the first brick into position, using the string as your guide. The mortar should make a joint 10mm (just under ½in) thick.

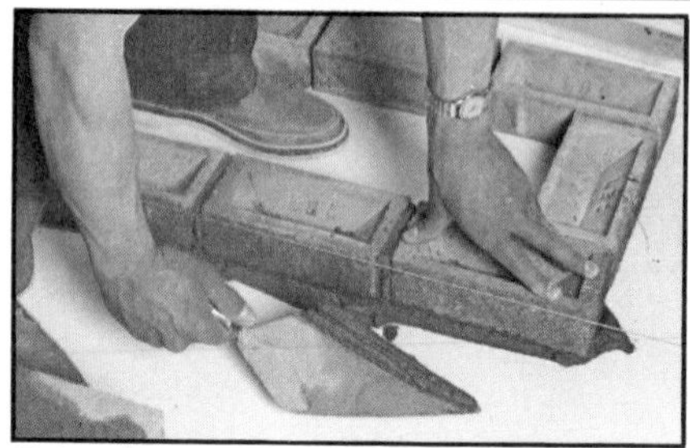

5 Lay the next 4 or 5 bricks, still following the line, making sure all mortar joints are the same thickness. Carefully scrape off the excess mortar.

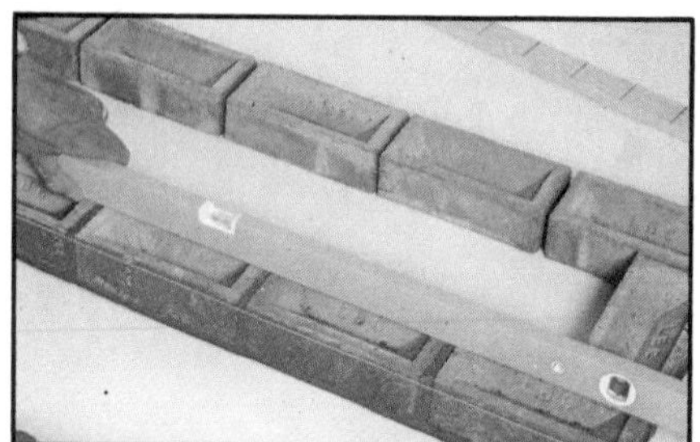

6 Use the spirit level to check that the bricks are sitting perfectly level. If one is too high, tap it down. If too low, remove it and add mortar.

7 Each brick for the next course should straddle two on the first course. This creates the 'stretcher bond', evenly distributing the weight of bricks.

8 Check that the courses are rising correctly with a 'gauge rod'. The rod is marked at 75mm intervals — brick height plus a 10mm mortar joint.

9 With string and pegs removed, check each brick with the spirit level as you lay it. Tap the bricks gently with the trowel handle.

10 With each new course, check the corner with the gauge rod. With the first brick correctly positioned, other bricks are aligned with it.

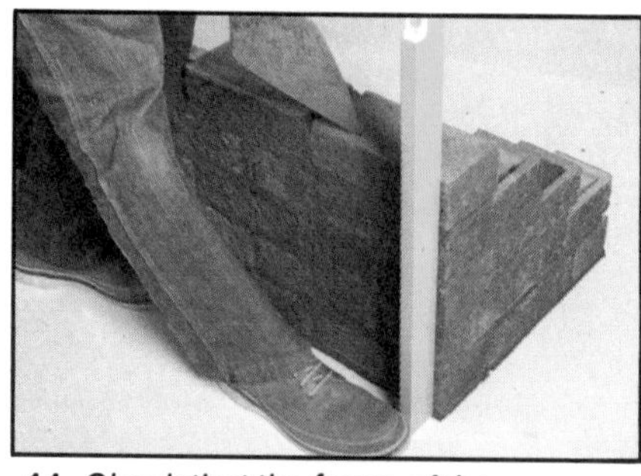

11 Check that the faces of the bricks are vertical and aligned with each other. If not tap bricks back into position with the trowel.

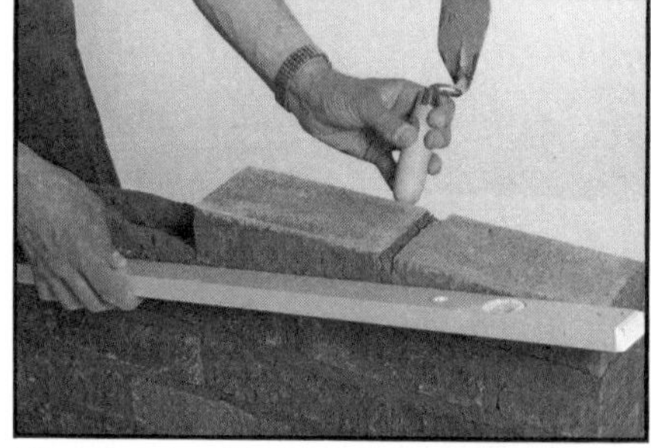

12 Also check diagonally across the face of the bricks. Lay the bricks frog (the indentation in the top of the brick) down only on the final course.

lime mortar. The 'sausage' or 'pear' is the basic shape of mortar lifted onto the trowel. The following sequence is worth practising over and over until it becomes easy to do. Chop down into the mortar and draw a slice of it towards the edge of the board. Move the trowel to and fro, along the length of the slice, pressing the body of the trowel on to the mortar till you have shaped the back of the slice into a curve – the mortar should be smooth and have no cracks.

Now sweep the trowel underneath the curved slice and load it on to the trowel, it will either look like a sausage or a pear, hence the name. Put it back on the spot board, shape it again, then sweep it up ready for placing. This amount of mortar should give you a 10mm thick bed for two stretchers. Hold the trowel parallel to the course, then, as you draw it back towards you, lift and jerk it slightly so the mortar rolls off gradually in a smooth elongated sausage. Press the mortar along the middle with the point of the trowel so a furrow is made in the mortar. When you place a brick on it a small amount of mortar should ooze out.

Joints in brickwork

Bricks are laid with both horizontal and vertical joints to keep the bricks apart. After the excess mortar has been removed from the face of the bricks (and behind), the joints can be finished in various ways – for an attractive effect as well as for protection against the weather. Coloured mortar is a specially prepared dry mix to which only water needs to be added. Pigment can be bought to colour your own mix of mortar but it can be difficult to obtain the same colour for each batch.

Making a cross joint

Sometimes 'buttering' is used to describe the technique of coating the end of a brick with mortar to form the vertical joint. Sweep up enough mortar to cover about a third of the trowel. Now sharply flick the trowel so the mortar lifts up, then falls back onto the trowel, (this squashes out air and makes the mortar 'sticky'). Hold the brick at a slight angle, then scrape the trowel against the bottom edge. Use the trowel point to flatten and level the mortar on the header – it should be 10mm thick.

Cleaning off

The other important trowel action is removing excess mortar from the side of the bricks as you lay them. Cut the mortar off cleanly by firmly lifting the trowel upwards (if you do it horizontally it will smudge the bricks). This leaves a flush joint.

Tricks or bad habits?

Bricklayers will often add a few squirts of washing-up liquid to the water when mixing mortar. This on-site plasticiser, used instead of lime or a proprietary plasticiser, is not added in any precise manner. Although it might make the mortar more pliable it could also weaken it. And how much is a squirt anyway? If you want a pliable mix, buy a proprietary plasticiser additive, or ready-mix with it already added.

- The shovel is frequently used as a measuring stick when mixing mortar, and there's no doubt it's an easy way of proportioning the ingredients. It can, however, give wildly inaccurate results. A mound of powdery cement won't sit on a shovel in the same way as sand will. Measurement should always be by volume. A bucket is ideal for most quantities – although if you're only making a very small amount use a small metal container instead.
- The curved edge of the bricklayer's trowel will effectively cut bricks when wielded by a professional. Apart from doing a great deal of damage to the trowel (the edge of which is needed for the upward sweep required to remove mortar from brickwork), it is easier to cut bricks on a sandy or soft ground with a bolster chisel and a hammer.

Protecting brickwork from damage

As soon as you have finished bricklaying, and you've cleaned off and finished all the joints, it's worth taking a few simple precautions to protect your work until the mortar has set and it is able to take care of itself.

The biggest enemy is rain. A heavy downpour could wash mortar out of freshly-pointed joints – which you would then have to re-point – and stain the face of the brickwork. Such stains are particularly difficult to remove except by hosing and scrubbing. Furthermore, if your brickwork is set on a hard surround – a patio, for example, or alongside a path – rain could splash up from the surface onto your brickwork, again causing staining and erosion of mortar joints at or near ground level.

So on small projects it's a good idea to cover your work, at least for 24 hours or so, until the mortar has had time to set to something like its final hardness. Drape polythene or similar water-proof sheeting over the brickwork, anchoring it on top with several loose bricks, and drawing the sides of the sheeting away from the face of the brickwork before anchoring them at ground level a foot or so away from the wall. In windy weather, lay a continuous line of bricks, or use lengths of timber, to prevent the wind from whipping underneath the sheeting.

Remember that until the mortar has hardened any knocks will displace bricks and break mortar joints. Corners are particularly prone to knocks and accidental collisions. So it's well worth erecting some kind of simple barricade in front of the new work for a day or two.

READY REFERENCE

BRICKLAYING TOOLS

Bricklayers' trowel: these come in various sizes and shapes; a blade 250mm (10in) long is easiest to manipulate.
Builders' spirit level: 1 metre long (3ft) so it can check levels across several bricks. It has two bubbles so you can check both horizontal and vertical levels. Alternatively, use a shorter spirit level laid across a timber straight edge.
Gauge rod: to check that successive courses are evenly spaced. Not a tool you buy, but one you can easily make yourself. Use a piece of 50 x 25mm (2 x 1in) timber about 1 m long, marked with saw cuts at 86mm intervals to match the height of one brick plus one mortar joint.

Line and pins: to help mark out building lines and guide you as you lay the bricks. The pins are shaped for wedging into brickwork joints or into the ground.

SIMPLE WALL FOUNDATIONS

All brick walls need foundations. They are usually made of concrete, but if you're building a single brick wall not more than 4-6 courses high you can use bricks.
- make a trench one brick-length wide and as deep as a brick
- sprinkle base with 12mm (½in) of sand
- lay bricks, frog down, side by side in the trench; tamp down very firmly
- fill the gaps with mortar

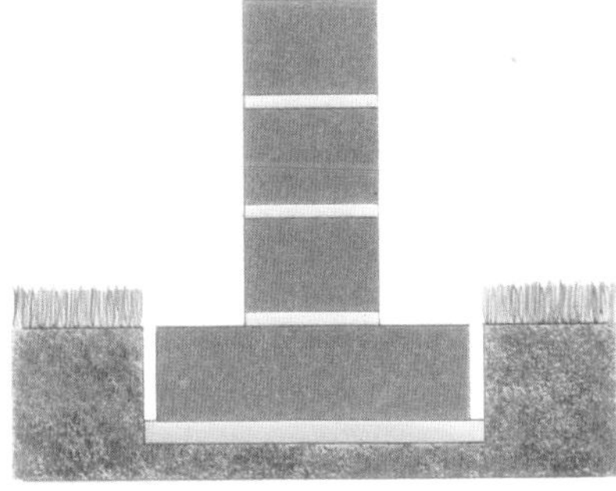

For concrete foundations see pages 30 -34.

FORMING CORNERS & PIERS

The techniques involved in making a brick wall turn a corner or to finish with a pier require an understanding of bonding and how cut bricks might have to be used to keep a design symmetrical.

Building walls isn't simply a matter of arranging bricks in straight lines. You may have to include corners and, when you come to the end of the wall, it must be finished off properly. The techniques for doing this effectively are relatively easy once you know the basis of brick bonding.

Brick bonds are crucial to bricklaying; simply stacking bricks one above the other without any kind of interlocking would neither distribute the weight of the wall evenly nor provide the wall with any kind of strength, however strong the mortar between the bricks. And because the joints line up they would provide a perfect channel for water to get in and wash out the mortar.

The simplest way of bonding is to overlap the bricks, with no vertical joints continuing through adjacent courses. This kind of bonding can create numerous different patterns — some very simple, such as the stretcher bond used on pages 8 -13, others much more complicated and requiring advance planning.

Exactly the same principle applies whether you're building a wall a half-brick thick (a single line of bricks) or one that needs to be one brick thick (two adjacent lines of bricks or one line laid header on). The difference is that instead of only overlapping the bricks lengthways as in a *stretcher bond* you can also overlap them widthways. With the *header bond*, for instance, all of the bricks are arranged header on to the face of the wall — and again the vertical joints only line up in alternate courses. In effect, the bricks overlap by half their width.

With any bonding pattern, there may be a need for cut bricks to maintain the bond. This may happen at the end of a wall built in stretcher bond where half bricks (called ½ bats) are needed in alternate courses. It may also occur where a new wall is being tied in to an existing wall (see below).

Similarly, with a wall built in header bond the ends need two three-quarter bricks (called ¾ bats) laid side by side in alternate courses to maintain the symmetry, the overlap and wall thickness. With other types of bond, the number and variety of cut bricks increases. The *English bond*, for instance, alternates a course of bricks laid stretcher face on with a course header face on to make a one-brick thick wall — and it needs a brick cut in half lengthways (called a queen closer) in each header course or two ¾ bats laid side by side in the stretcher course.

Corners in brickwork

When it comes to turning a corner in brickwork (known as a *quoin*) the importance of correct bonding is even more apparent. Without it, you'd be building two walls which weren't interlocked and so lacking in real strength. In a half-brick thick wall in stretcher bond the corner is easy to make. Instead of cutting ½ bats for alternate courses, a whole brick is placed header face on at right angles to the front face of the wall.

The necessary 'tying in' of bricks with other bonding patterns, however, usually requires additional cut bricks and careful planning. In effect, the bond may change when you turn a corner. In header bond for example, which has alternate courses starting with ¾ bats, ¾ bats must be placed header on as well to create the corner. In English bond the stretcher course on one side of the quoin becomes the header course on the other.

READY REFERENCE

COMMON BONDS

All brick bonds are designed to stagger the joints between courses to give the wall the maximum strength.

Stretcher – also called running bond, it has bricks overlapping by half their length. Alternate courses finish with ½ bricks at each end.

Header – formed by laying bricks header on, with each course overlapping by half the width of the header face. Alternate courses finish with two ¾ bats laid next to each other at each end.

English – alternate courses of headers and stretchers. Queen closers maintain bonding at each end of the header course.

MARKING OUT FOR A CORNER

1 *Bricks can be used to hold the string lines on already laid concrete but use profile boards if a trench has to be dug. Check with a builders square that all the lines cross at right angles.*

2 *Laying the bricks dry is the best way of checking your calculations after setting out is completed, the width of your finger being a good guide to the eventual thickness of the joint.*

3 *After the first course is laid there are two things to check: the bricks must be horizontal (you can tap down any out of alignment) and their faces must be truly perpendicular.*

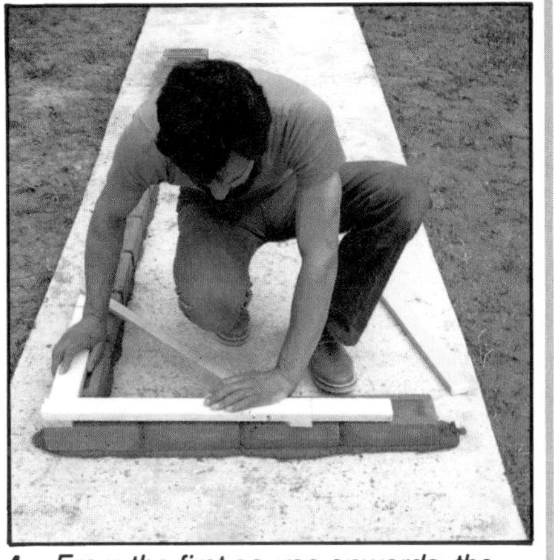

4 *From the first course onwards, the squareness of the corner must be checked so that any adjustments can be made immediately to prevent the wall leaning out (called an overhang).*

Keith Morris

A bond may also have to be altered if the bricks don't fit the actual length of the wall. When this happens you have to break the bond as close as possible to the centre of the wall. If the length differs by 56mm or less don't use a ¼ bat (this is considered bad building practice) but use ½ and ¾ bats instead, making sure you place them so that no straight joints occur.

PROFILE BOARDS

These are placed to give accurate lines when digging the trench for the foundation. Strings attached to nails in the top of the boards define the width of the trench and must cross at perfect right angles where the corner is to be built.

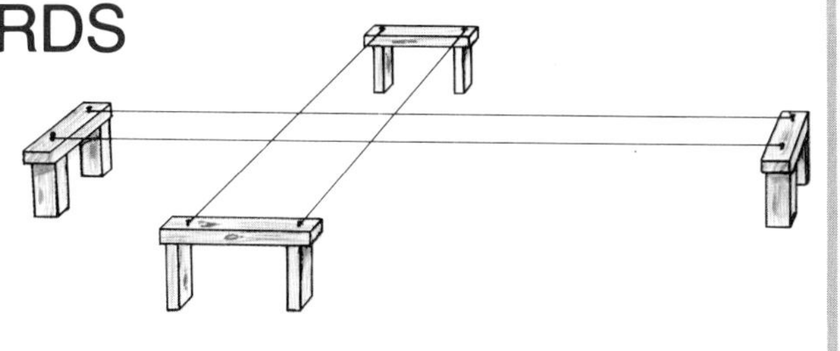

Craig Warwick

The end of the wall

If you're building a wall as a boundary, or enclosing a corner of your garden, it may have to meet existing walls at one or even both ends. In such situations you have to tie in the bricks with the other wall(s), so this may affect your choice of bond for the new wall — it's always better if the new matches the old. It also means you have to match levels, and before you lay your first new brick you have to chip out bricks from alternate courses of the existing wall to provide for 'toothing-in'. Even if you can satisfactorily match the bond pattern, the old bricks may be a different size, so to make a proper connection expect to cut bricks to odd lengths to tooth in. More about this in another section.

If your wall comes to a free-standing end you must create what's called a *stopped end*. This requires careful checking for vertical alignment, and needs to be finished off to make a clean, neat face. But it is important to make sure the end is strong enough — and to do this you actually increase the width by a half brick to create a 'pier'. In effect, instead of cutting a ½ bat to finish off each alternate course, you lay the last brick in alternate courses at right angles to the wall face. By adding a ½ bat next to it you create a squared-off end — a simple pier.

Piers for support

It's not just at the end of a wall that you may need the added strength of a pier. To give a wall extra support, particularly on a long run, you need piers at regular intervals. For instance, walls of half-brick thickness need piers that project by at least half a brick every 1.8m (6ft). To do this in a wall built in stretcher bond, you will have to alter the bonding pattern to accommodate the pier, and add cut bricks to ensure the correct overlapping is maintained. For how to do this see pictures page 95. If the wall is over 12 courses high, a more substantial pier is needed: three bricks are placed header on in the first course, and ¾ bats are used on the pier and either side of the middle stretcher in the second course.

One brick thick walls need piers at less frequent intervals — in fact every 2.8m (9ft) — but the pier has only to project by half a brick (see diagram).

Where piers occur, the foundation must be dug slightly wider at that point (about half a brick wider on both sides and beyond the end of the pier).

Method of building

Planning how you're going to lay the bricks is, of course, only the theoretical side of bricklaying. In practice, to make sure the wall stands completely perpendicular and the corners and ends are vertical it's most important to follow a certain order of work. Lay at least the first course of bricks dry so that you're sure they all fit in (see also pages

TURNING A CORNER

1 On each face of the wall check that it is perpendicular by holding the level at an angle. Tap bricks in or out.

2 As each course is laid use the gauge rod to check that the wall is rising evenly, with equal horizontal joints.

3 Check that the corner is vertical by using the spirit level straight. Hold it steady with your foot.

4 Lay the first course from the corner to the stopped end following a line and cutting bricks as needed.

5 Build up the stopped end so the courses are stepped by the correct overlap. This is called 'racking back'.

6 Raise the line to the next course between the racked corner and stopped end and fill in between.

TIP

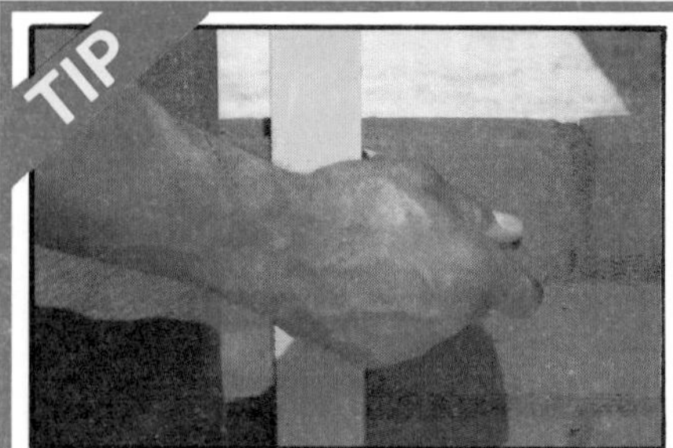

7 To make sure you get vertical joints in line, mark the position of each one on the whole brick above it.

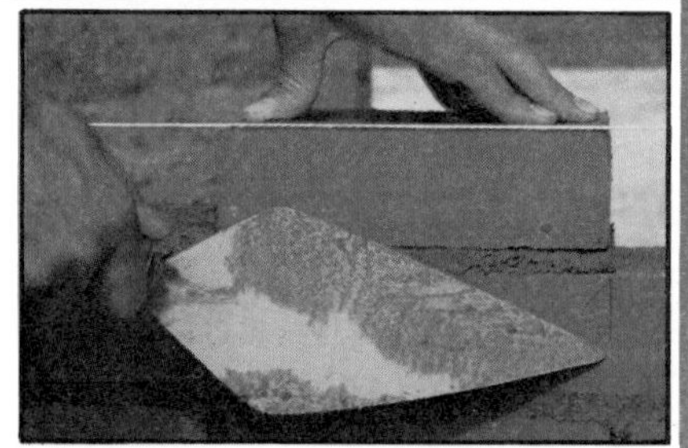

8 In the next course, align the edge of the brick with the mark. The joints will then be the right width too.

Keith Morris

READY REFERENCE

A BUILDERS SQUARE

Essential for checking that corners are 90°, this is simply three pieces of wood cut in the proportions of 3:4:5 (ie, a right-angled triangle). Nail them together with a half-lap at the right-angled corner and with the longest side nailed on top of the other two sides.

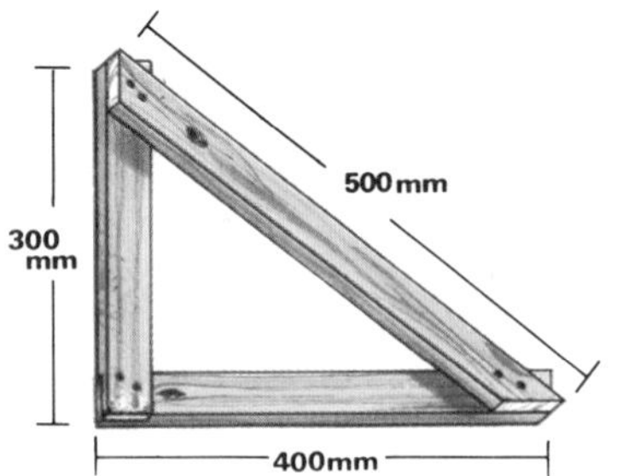

WHAT CAN GO WRONG

You can lose the horizontal because you didn't check often enough as the wall was rising. With every course

- use the spirit level
- use the gauge rod

The wall can lean out or in because you didn't check the vertical with the spirit level. Use it when

- racking back
- starting a new course

When laying to a line the last brick won't fit because the vertical joints further back along the course were not the same width. So make sure that in each course

- the joint width remains constant
- vertical joints line up in alternate courses – use the gauge rod for this.

LAYING TO A LINE

When you build up two corners at opposite ends of a wall, the process of laying bricks in between is called 'laying to a line'. To fix the line for each course use

- a bricklayers line and pins
- twine tied around two spare bricks
- triangular profile boards (good for beginners as you can see that the courses are rising evenly).

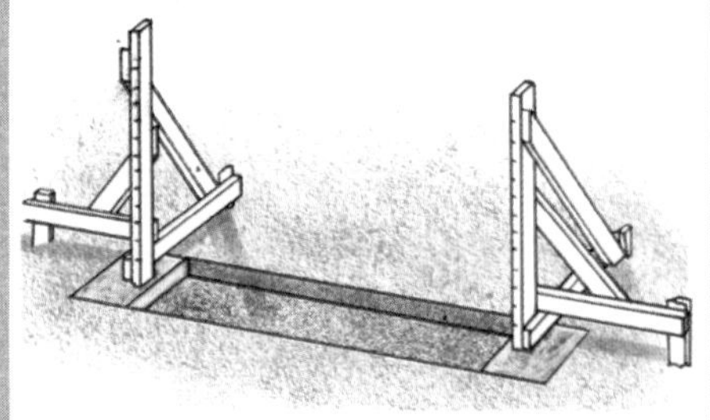

A PIER IN STRETCHER BOND

1 *In order to prevent any straight joints you have to break the bond. On the first course place two bricks header on, then place a ½ bat so it spans the joint equally.*

2 *On the second course to get the pattern right you have to lose a quarter from each of the stretchers on either side of the ½ bat. So cut and place two ¾ bats.*

3 *On the pier itself, the second course is not tied on but is merely a stretcher laid across the projecting two headers. This gives a pier the same width as the wall.*

Keith Morris

8 -13). Another big problem is that it is difficult to lay a line of bricks with each vertical joint exactly the same width — an inaccuracy of just 1mm in each joint between a line of 10 bricks will mean that the last brick at the corner or end will project over the one underneath by 10mm. The best way to avoid this happening is by 'racking back' — build up each corner or end first, stepping the bricks upwards and checking the vertical each time. When the bricks reach the required height, start filling in. Any slight inaccuracies can be accommodated by the joints in the middle part of the wall where they'll be less noticeable as long as you make sure the bricks overlap each other by as close to half a brick as possible.

Making the corner square

Marking out the corner for the foundation is the first priority—and it's vital that it is square. Using profile boards and strings—explained in Foundations, pages 30 -34—is the best way to start. Set the boards for each line of the corner about 1 metre (3ft) back from the actual building line (see diagram page 15). The strings must cross at right angles (90°) and to make sure that they do, use the 3:4:5 method (see *Ready Reference* page 16) to make yourself a builder's square. This is a large set-square made by nailing together three 75mm×38mm (3in×1½in) softwood battens cut into lengths of 450mm (18in), 600mm (2ft), and 750mm (2ft 6in) so that the sides are in the ratio 3:4:5. This is a manageable size but it can be made bigger if you prefer.

Laying out the corner

When you have the profile boards in position, dig the foundation trenches — see pages 30 - 34 — and lay the concrete. Allow it to 'cure' for at least 5 or 6 days before laying the first course of bricks. This gives it time to harden properly (although it needs about 3 weeks to reach its full strength). The next step is to mark out the actual building lines on the concrete, again using the profile boards and strings (see pages 30 -34 again).

Lay the first course of the entire wall, starting at the corner and working outwards first along one wall line and then along the other. In building a half-brick thick wall in stretcher bond it is easy to turn the corner simply by laying two bricks to make a right angle. Check the angle using the builder's square.

Once the first course is laid, check again with the spirit level to make sure that all the bricks are sitting correctly. Add a little mortar or remove a little from underneath any bricks which are out of true. At the same time, check again that all the bricks follow your building line, and tap them into position if they don't.

CORNERS AND STOPPED ENDS

Left: At a stopped end in stretcher bond, the cut side of the ½ bats on alternate courses are hidden by the mortar joints.

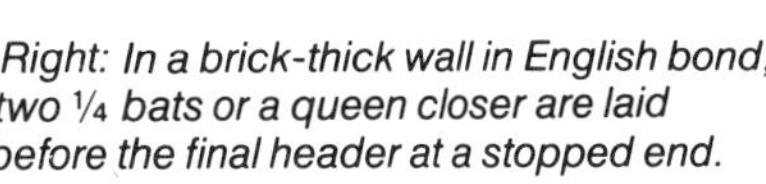

Right: In a brick-thick wall in English bond, two ¼ bats or a queen closer are laid before the final header at a stopped end.

Below: At a corner in a stretcher bond wall, the header face is seen on one side while the stretcher shows on the other.

Below: The arrangement is the same at a corner, but to maintain the bonding the course on the other face becomes stretchers laid side by side.

Craig Warwick

Once this is done you can remove the lines and start building up the ends and corners.

Putting in the piers

If you're going to need piers at any point along the wall, don't forget to plan them in from the beginning. In a half-brick thick stretcher bond wall a pier is tied in by two bricks laid header on in alternate courses. The courses in between are not tied in but consist of a single stretcher laid parallel to the wall for the pier, and a ½ bat and two ¾ bats replacing two stretchers in the face of the wall.

A pier at a stopped end in a half-brick stretcher bond wall is made using a stretcher face at right angles at the end. The course is completed with a ½ bat and on the alternate course two stretchers are used parallel to the wall.

Checking as you build

One of the most useful checking tools you can make yourself is a gauge rod (see pages 8 -13) and as you build up the corners and ends check each course with the rod to make sure the horizontal joints are consistent. If you're aiming for a wall of about 12 courses in total, build up the corners and ends to about 6 or 7 courses first before starting to fill in between them.

To step the bricks correctly, lay 3 bricks along the building line for every 5 courses you want to go up — so it's best to start by laying 4 bricks along each side of the corner and in from each end (see pictures 5 and 6 on page 16).

Filling in

Once you've built up corners and ends properly racked back (stepped with the correct overlap), the rest of the wall can be filled in course by course. Although you can lay the bricks normally, checking each time with the level and gauge, a good tip here is to string a line between bricks already laid at each end, then lay the bricks in between to this line. A bricklayer's line and pins (the pins are specially shaped to slip into a mortar joint) is ideal, but a string can be hooked around a brick at the correct height and then anchored under a loose brick to give a start line to follow.

If you over-mortar a brick and it protrudes above the level, gently tap it down with the end of the trowel handle and scrape off the excess mortar squeezed out of the joint. If a brick does not stand high enough, remove it and the mortar underneath, then replace it with fresh mortar.

Getting the last brick into the line can be quite tricky and a good tip is to scrape the mortar onto the end of bricks at each side — then squeeze the brick in.

PIERS

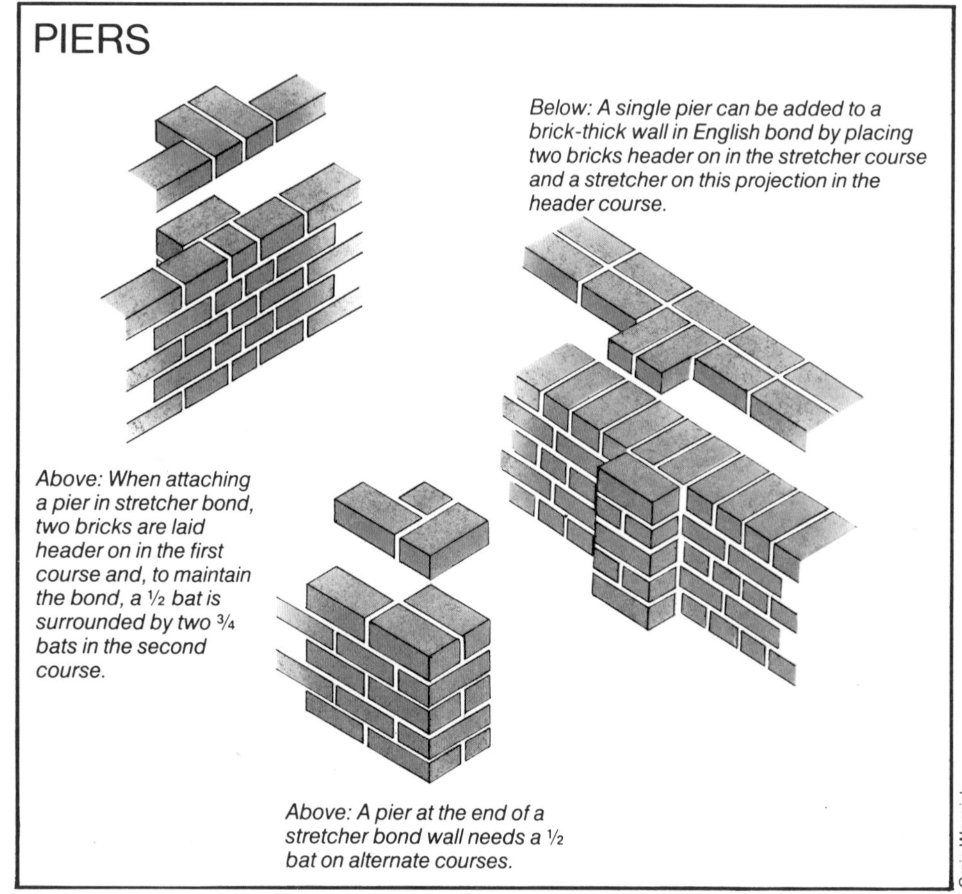

Above: When attaching a pier in stretcher bond, two bricks are laid header on in the first course and, to maintain the bond, a ½ bat is surrounded by two ¾ bats in the second course.

Below: A single pier can be added to a brick-thick wall in English bond by placing two bricks header on in the stretcher course and a stretcher on this projection in the header course.

Above: A pier at the end of a stretcher bond wall needs a ½ bat on alternate courses.

Craig Warwick

READY REFERENCE

BRICK CUTS

The most common cut bricks, used in different bonds, are:

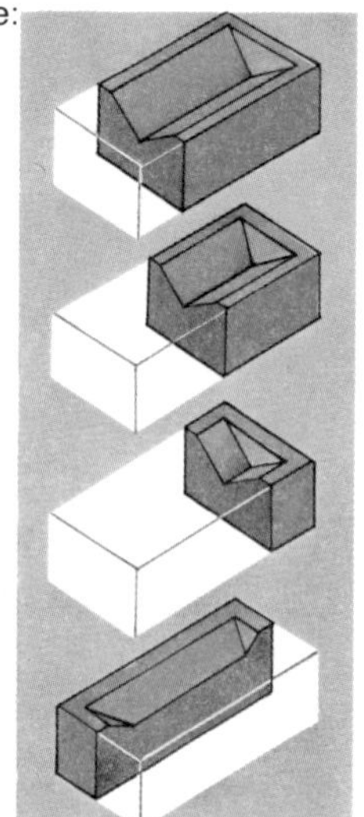

¾ bat – cut widthways, ¼ removed

½ bat – brick cut widthways, ½ removed

¼ bat – brick cut widthways, ¾ removed (the same size as a ½ queen closer)

queen closer – brick cut in two lengthways

CUTTING BRICKS

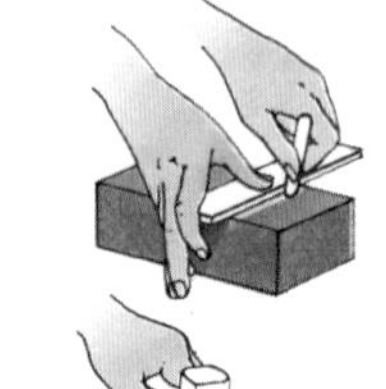

Mark cutting line on each face with chalk

Nick the line all round; with hard bricks nick each face at least twice

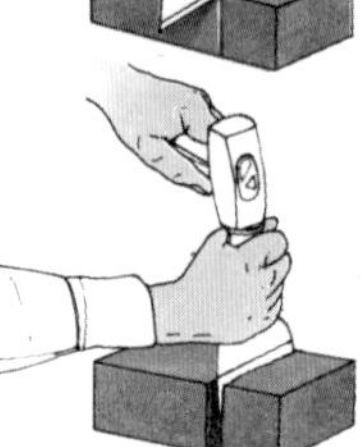

Place brick bottom up on grass, sand or newspaper. Put bolster on nicked line, give sharp blow with hammer

DON'T use the edge of the trowel to cut bricks because

- it's an expensive tool
- it's rarely accurate

DID YOU KNOW?

A pig in a wall means the wall isn't level.
An overhang means the corner leans out.
A batter is when a corner leans back.

LAYING SPECIAL BONDS

Brick bonds are crucial to the strength of a wall, and they also provide a decorative face. With careful planning, even complex patterns are easy to make.

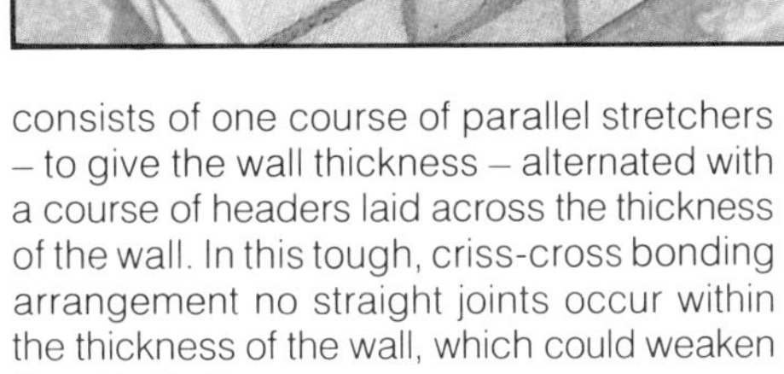

When you're building a brick wall it's vital that you lay each brick so that it overlaps and interlocks with its neighbours. You can't simply stack the bricks, one on top of the other; this would create vertical planes of weakness and, under load, the wall would soon collapse.

Instead, the bricks must be arranged so that the vertical joints (called perpends) in one course don't coincide with those in the course above and below. This arrangement, called a 'bonding pattern', ensures mainly that the weight of the wall and any load bearing on it is evenly distributed, but it can also be a decorative feature.

Types of brick bond

There are many bonding arrangements you can use, depending on the type of wall you're building, how strong it's to be, and whether its appearance is a consideration. They're basically all variations on the 'half-lap' bond, in which all bricks lap half a brick over the bricks in the course below, but there's also 'quarter-lap' bond, in which the bricks lap over a quarter of a brick length.

Stretcher bond

The most straightforward and common arrangement is called 'stretcher' or 'running' bond, which is used mainly for 112mm (4½in) thick – or half-brick thick – walls. Each course is identical; the bricks are laid end to end and each overlaps the one below by half its length, presenting its long stretcher face to the front face of the wall.

Header bond

The simplest type of bonding for a 225mm (9in) thick – or single brick – wall is 'header' bond, in which all the bricks are laid side by side with their ends (again often decorative) presented to the front face of the wall, and each course overlapping its neighbours by half its width. This type of bonding pattern, although attractive, is wasteful of bricks.

English bond

The strongest brick bond for 225mm (9in) thick or thicker walls – particularly if they're loadbearing – is called English bond. It consists of one course of parallel stretchers – to give the wall thickness – alternated with a course of headers laid across the thickness of the wall. In this tough, criss-cross bonding arrangement no straight joints occur within the thickness of the wall, which could weaken the structure.

Flemish bond

Where a highly decorative effect is needed in a single or half-brick thick wall, Flemish bond is often regarded as one of the most popular. The pattern consists of alternate headers and stretchers in each course – and the decorative effect can be increased by using contrasting coloured and textured headers, or even by slightly recessing some of the headers in the face of the wall, or allowing them to protrude fractionally.

Garden Wall bonds

Other common brick bonds you can use, mainly for their attractive appearance, but also because they're more economical, are really just modified versions of English and Flemish bonds. Basically they reduce the number of headers used, and introduce more stretchers yet still maintain the basic bonding patterns.

English Garden Wall bond, for example, originally used for boundary walls of one brick thickness, is one bond that reduces the

READY REFERENCE

COMMON BRICK BONDS

All brick bonds are designed to stagger the joints between courses to give the wall maximum strength. Some common bonds are:

- stretcher bond; bricks overlap by half their length (A)
- header bond; bricks overlap by half their width (B)
- English bond; consisting of alternate courses of headers and stretchers (C)
- Flemish bond; pairs of stretchers alternating with single headers (D).

Joints are staggered with cut bricks.

ENGLISH BOND WALL WITH END PIER

1 *English bond consists of a course of stretchers alternated with one of headers. Mark out the shape of the wall and pier and dry lay the first course.*

2 *Remove the bricks then re-lay them on a mortar bed. Lay the first course of the pier and the return wall first and check the level between them.*

3 *Make sure the end pier is laid perfectly square by checking with a builder's square. Note the brick that 'ties' the pier to the wall.*

6 *Work towards the pier using stretcher bond over the first header course then return to the corner laying the second half of the course.*

7 *Complete the second course then start the third course, which is a repeat of the first course. The return wall is laid as a row of stretchers.*

8 *Bond the pier with alternate courses of stretchers and headers spaced with queen closers. Lay the corners first and fill in with cut bricks.*

number of bricks that you'll need for headers; it usually consists of three or five courses of stretchers to one of headers.

Flemish Garden Wall bond is another decorative yet durable bond that consists of an arrangement of three, four or five stretchers to one header per course – each course being identical – and is especially attractive as the outer skin of a cavity wall. You can introduce headers of constrasting colour or texture, either inset or projecting, for a greater decorative effect.

When you're using these bonds in a half-brick thick wall you'll have to use cut bricks, called half-bats (see below and *Ready Reference*) for the headers.

Open bonding

Where load-bearing capacity isn't a vital consideration in ycur wall you can use an economical form of 'open' bond. It's especially good for screen walls for your garden or patio, giving a fairly solid appearance yet still admitting light and air.

Each course is laid as stretchers and the bricks are separated by quarter-brick spaces.

Using cut bricks

Any bonding pattern – except open bond – will need to include cut bricks in order to maintain the bond.

In a stretcher bond wall you'll have to include half bricks (called half-bats) in alternate courses at the end of the wall, or where you're tying a new wall into an original wall.

In a header bond wall you must insert two three-quarter bricks (called three-quarter-bats) laid side by side in alternate courses to maintain the symmetry of the pattern.

With the more complicated bonds the number – and variety – of cut bricks increases. English bond, for example, needs a brick cut in half lengthways (called a queen closer) in each header course, or two three-quarter-bats laid side by side in the stretcher course to maintain the bond.

You won't need to include cut bricks in an open-bond wall; you can maintain the symmetry by simply reducing the spaces between the bricks.

Turning corners

It's important to maintain the bonding arrangement throughout the wall for strength and symmetry, although when you come to turn a corner (called a quoin) you may have to vary the pattern over a small area (see pages 14 -18). The most crucial point is to ensure both sides of the corner are interlocked, otherwise you'd end up with two separate walls lacking rigidity.

With some bonds, corners are fairly straightforward to make. If you're building a

4 Lay the bricks header on between pier and corner. The bond changes to stretchers on the return wall, with a queen closer spacer at the corner.

5 Start the second course at the corner; line up the first brick with the centre of the header next to the queen closer in the first course and the end brick.

9 Continue to lay alternate courses of stretchers and headers until you reach the finished height of your wall. Check the level frequently with a spirit level.

10 Check the 'plumb', or vertical level, of the wall, paying particular attention to the corners of the pier. Rack back the return wall if you're continuing next day.

half-brick wall in stretcher bond, for example, you can simply turn a corner by placing a whole brick, header on, at the corner instead of filling in with half-bats at alternate courses. You can then continue to build the wall at the other side of the angle in exactly the same stretcher bond.

Other bonding arrangements, however, will require additional cut bricks so that you can tie-in the two leaves of the corner. In effect the pattern will change when you turn a corner, although it's re-established on the course above. In English bond, for example, the stretcher course on one side of the quoin becomes the header course on the other.

Piers for support

If you're building a particularly high or long wall you'll need to build in supporting columns called piers to give added strength to the structure. Build them at regular intervals throughout the wall ('attached' piers) and at the ends. For a half-brick thick wall you'll need a pier that projects from the wall by at least half a brick every 1.8m (6ft). In a stretcher bond wall you must alter the bonding pattern by adding cut bricks to give the necessary overlap, which ensures that the pier is correctly tied into the wall. To form a minimum-sized attached pier in this type of wall, you must lay two bricks header-on in the first course and, to maintain the bond, a half-bat flanked by two three-quarter-bats in the second course. Repeat this arrangement.

For one brick thick walls you'll need piers at only 2.8m (9ft) intervals, but the pier must still project from the wall by at least half a brick. To form an attached pier of this size in an English bond wall you should lay two bricks header on in the stretcher course and a stretcher on the projection in the header course.

READY REFERENCE

BRICK CUTS

When you're building a brick wall you'll need to cut some bricks in order to maintain the bond. The most common cut bricks are:

- ¾ bat, cut widthways, removing ¼ of the brick (A)
- ½ bat, cut widthways, removing ½ of the brick (B)
- ¼ bat, cut widthways, removing ¾ of the brick (C)
- queen closer, cut in two lengthways (the same width as a ¼ bat) (D).

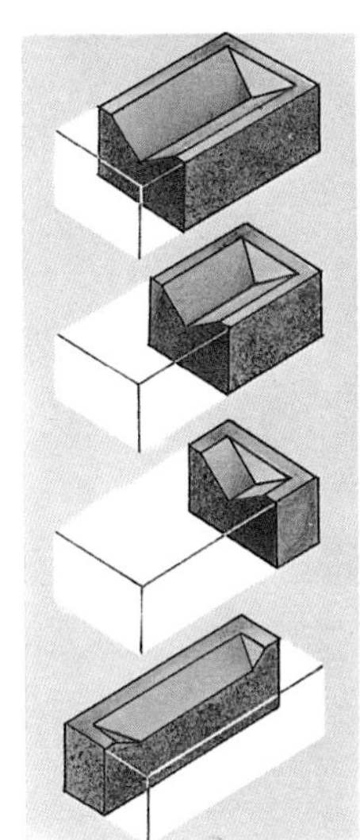

CUTTING A BRICK

To cut a brick accurately:

- mark the cutting line on each face with chalk against a rule (A)

- nick the line all round with a club hammer and cold chisel (B)

- place the brick, frog down, on a soft surface (grass, sand or newspaper) and hit the chisel sharply on the nick with the hammer (C)

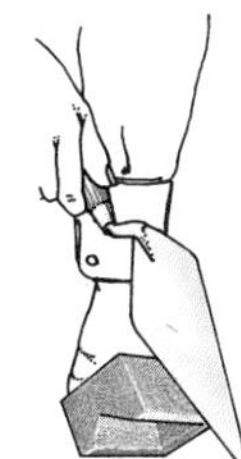

- use the side of your brick trowel to trim off irregularities in the brick (D).

FLEMISH BOND WALL WITH ATTACHED PIER

1 *Flemish bond consists of alternate headers and stretchers in each course. Mark out the wall and dry lay the bricks, incorporating an 'attached pier'.*

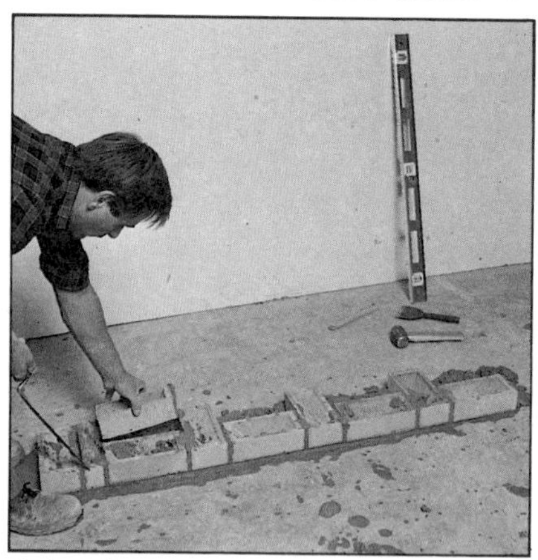

2 *Lay the facing bricks of the first course on a mortar bed, including a queen closer at a stopped end. Fill in the second half of the course.*

3 *The first course of a single brick attached pier is bonded to the wall with headers for the sides and a stretcher brick between them for the back edge.*

4 *The second course is the same as the first, but the stretchers are laid over the headers and vice versa: Lay the facing bricks first.*

5 *The second course of the pier is laid using 3/4-bats at the back corner to keep the shape of the pier, with a 1/4-bat to fill the gap that's left inside.*

6 *A stopped end – one without a pier – is formed by the inclusion of a queen closer and a brick header-on in alternate courses; then continue normally.*

7 *Build the wall as high as you'd like it, repeating the rows of headers and stretchers in alternate courses. Check the level across the face of the wall.*

8 *Check that the mortar joints are constant using a gauging rod marked off in 75mm (3in) intervals – brick height plus a 10mm (3/8in) mortar joint.*

9 *If you leave the wall overnight, rack back the end in steps so that your continuation can be bonded correctly. Finally, check the plumb of the pier.*

FINISHING BRICKS FOR WALLS

There are many ways you can finish off your wall neatly and decoratively, using a selection of special-shaped bricks made to co-ordinate with standard bricks (see pages 168 -9). Some of the bricks you can use for a straight run of walling include:

1 *Double bullnose as a stopped end*
2 *Bullnose header on flat*
3 *Plinth header (stretcher also available)*
4 *Bullnose double header on flat*
5 *Half round coping*
6 *Saddleback coping*

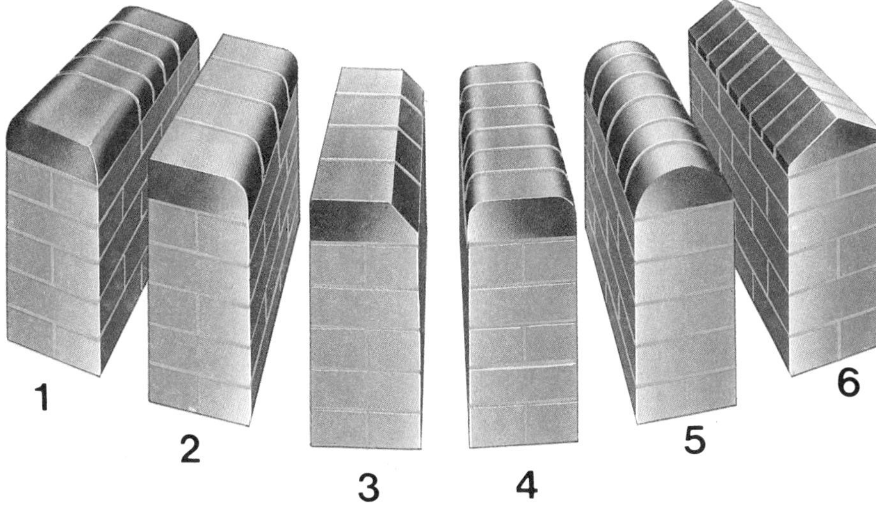

There's also a variety of special-shaped bricks for use at right-angled corners, both in left- and right-hand versions. You can use them in conjunction with the straight-run bricks above for a neatly finished effect. They include:

A *Bullnose external return on flat*
B *Bullnose external return on edge*
C *Plinth external return*
D *Cant external return*
E *Single cant (for half-brick walls)*
F *Single bullnose*

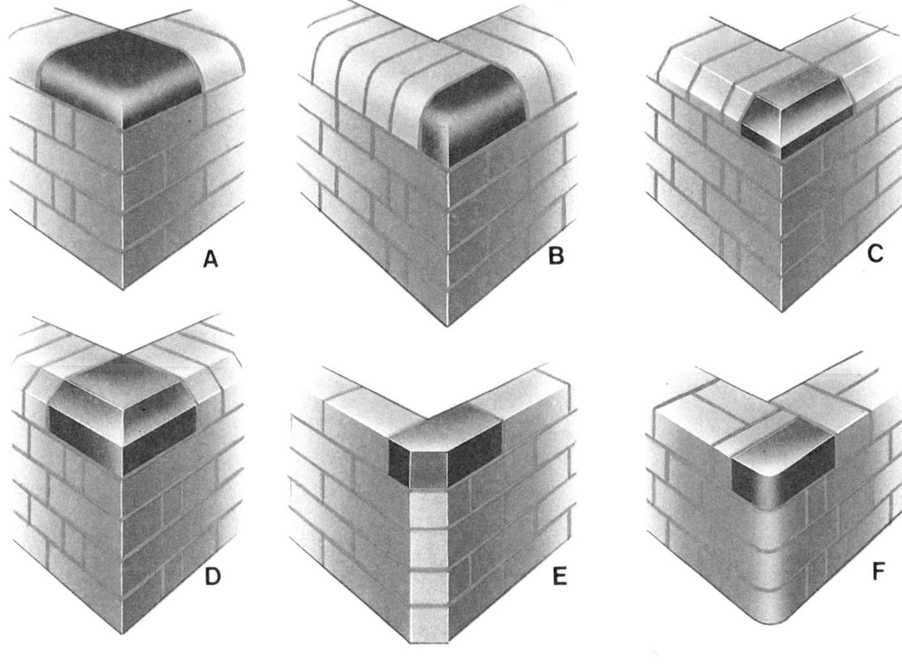

READY REFERENCE

TIP: DRY LAY BRICKS

To help you work out complex bonding arrangements accurately, and to calculate the number of whole and cut bricks you'll need, lay the first few courses, or until the pattern repeats, without mortar, maintaining finger-width joints throughout.

DECORATIVE PATTERNS

For a greater decorative effect:

- lay a random or regular pattern of coloured, textured stretchers and headers so they protrude slightly from the face of the wall (A)
- lay some cut stretchers or headers in different colours and textures so they're recessed slightly from the face of the wall (B).

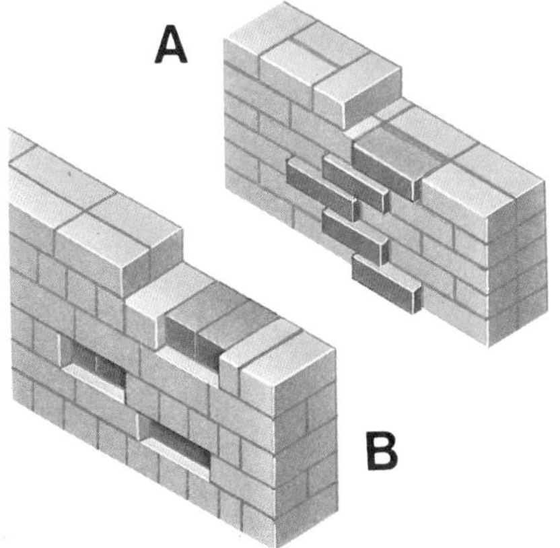

USING A LINE AND PINS

To ensure your brick courses are laid level:

- insert the flat blade of a pin into an upright mortar joint at one corner or end of your wall
- stretch the line attached to the pin to the other end of the wall
- secure the line taut with a second pin so that it's 10mm (3⁄8in) clear of the wall but in line with the top edge of the next course
- move line and pins up at each course.

CHAPTER 2

USING CONCRETE

Concrete is perhaps the do-it-yourselfer's most versatile building material. It's cheap, easily mixed from readily available ingredients and capable of forming floors, walls, paths and drives, bases for outbuildings and many other things besides. As with mortar, half the skill is in proper mixing of the correct ingredients for the job; the other half is the laying, and different techniques are used for different needs. Master these techniques and you'll have no problems, whether you're working with a small bag of dry concrete mix from the corner shop or a lorry-load of ready-mix from the local depot.

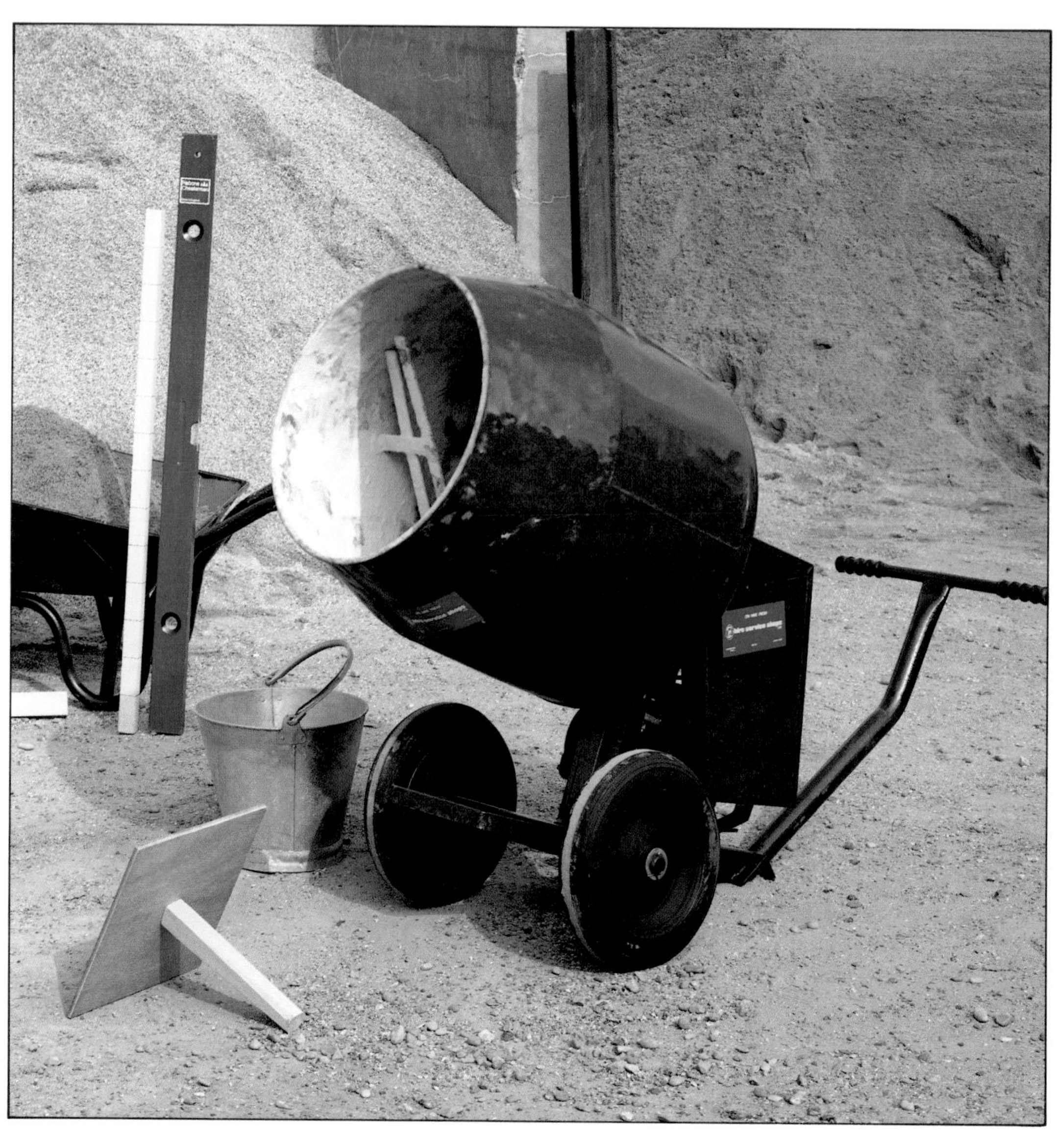

MIXING AND LAYING CONCRETE

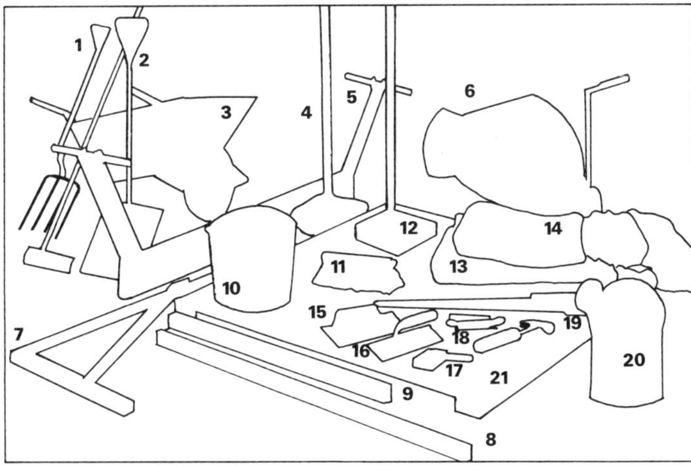

Mixing and laying concrete involves a surprising number of tools. Many of them you will have, some you can hire and others you can make. Always scrub tools well after each work session.
1 Fork For digging out foundations. You will also need a spade.
2 Shovel You will also need a rake to spread the concrete slightly proud of the formwork timbers.
3 Wheelbarrow
4 Ramming tool For compacting the subsoil or hardcore if a roller is unavailable. Fill a timer mould (about 200 x 150 x 100mm/8 x 6 x 4in) with concrete, insert a broom handle and keep it supported until the concrete has set.
5 Tamping beam. For levelling large areas. Make it from 150 x 50mm (6 x 2in) timber. Strong handles help you move the tamper more easily.
6 Hand-operated concrete mixer. The machine is pushed along a hard level surface to rotate the drum and mix the concrete.
7 Builder's square To check the corners of the form. Make one by joining three lengths of wood with sides in the proportions of 3:4:5. A good size is 450 x 600 x 750mm (18 x 24 x 30in). Use an L-shaped bracket and screws to make rigid joints.
8 Spirit level
9 Timber straight-edge For setting out fall of concrete if drainage is required. You can also use it for levelling: a 100 x 50mm (4 x 2in) straight-edge is sufficient.
10 Buckets You will need two for measuring and two shovels of equal size for mixing. Keep one bucket and shovel for measuring out and adding cement. The second set can be used for adding ballast and water and for mixing. Using equal sized buckets gives an easy guide to quantities: a 1:5 mix needs one bucket of cement and five buckets of ballast.
11 Coarse brush You can give a textured finish to concrete by sweeping the surface with a coarse brush.
12 Punner For compacting concrete. Nail together several layers of timber and attach a broom handle.
13 Polythene sheeting Use to protect concrete from the elements.
14 Straw and sacking Either can be used to protect new concrete from frost.
15 Wood float For a textured finish.
16 Steel float For finishing concrete.
17 Measuring tape
18 Pegs and string For marking out the site.
19 General purpose saw and claw hammer For making formwork.
20 Watering can To control the rate at which water is added.
21 Mixing platform Use where there is no suitable solid area for mixing concrete. Nail boards together for the base and add side pieces to keep the concrete in place. Or fix sides to a larger sheet of 18 or 25mm (¾ or 1in) plywood.

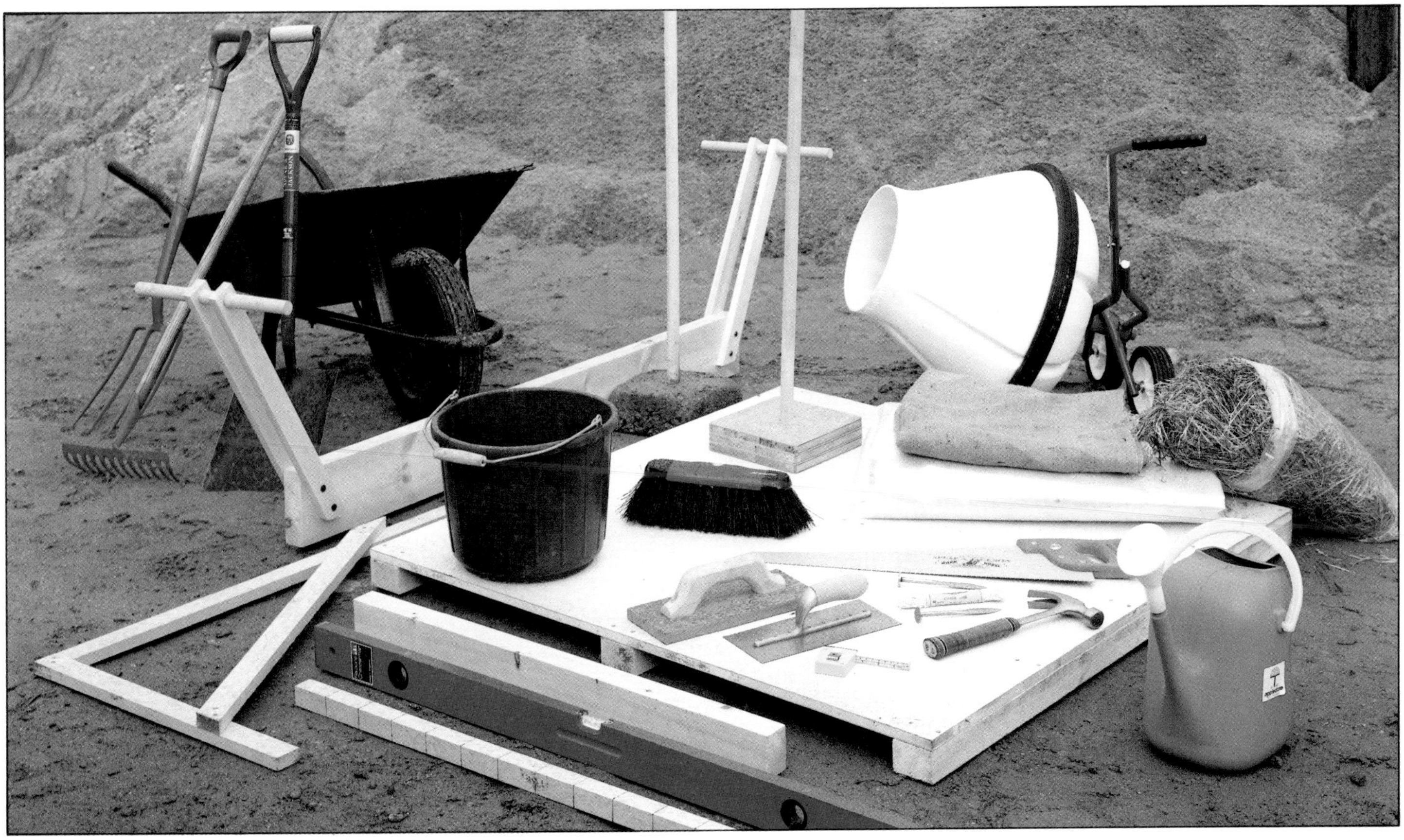

BASIC CONCRETING TECHNIQUES

One of the most versatile of all building materials, concrete is also one of the easiest to work with. The techniques for laying anything from a garden path to a patio are much the same – once you know the basic rules for mixing up the ingredients, making formwork, laying and levelling.

Concrete is made of cement, aggregate and water. Cement itself is quite a complex chemical formed by burning chalk, limestone and clay at high temperatures and then grinding the resulting clinker to a fine powder. Added to the water it becomes an adhesive and coats and binds the aggregate (clean, washed particles of sand, crushed stone or gravel — never brick — for the clays can react against the cement).

The strength or hardness of any concrete simply depends on the proportions of these ingredients. Only a small part of the water you add is used up in the chemical reaction — the rest evaporates.

Ordinary portland cement is used for most concreting work. (The name doesn't refer to the manufacturer or where it is made, it's simply that when invented in the 1820s it was thought to resemble Portland stone.)

Aggregates are graded according to the size of sieve the particles can pass through — anything from 10-20mm. Coarse aggregate has the largest stones (20mm) while fine aggregate, often described as shingle, can be 10-15mm. Sand, the third part of concrete, is also considered aggregate. (The cement holds the sand together and the combination of sand and cement holds the stones together.) In concreting the sand used is known as 'sharp' sand and graded by the sieve method. All-in aggregate is a combination of both sand and stones.

Choosing your mix

Different projects require different mixes of concrete. Three are most commonly used.

Mix A (see ESTIMATOR below and *Ready Reference* overleaf) is a general-purpose mix for surface slabs and bases where you want a minimum thickness of 75mm-100mm (3-4in) of concrete.

Mix B is a stronger mix and is used for light-duty strips and bases up to 75mm (3in) thick – garden paths and the like.

Mix C is a weaker mix useful for garden wall foundations, bedding in slabs and so on, where great strength is not needed.

The amount of water needed depends very much on how wet the sand and stones in the aggregate are. A rough guide is to use about half the amount (by volume) of cement. But add it gradually. Too much will ruin the mix and weaken the concrete.

How concrete works

New concrete hardens by chemical action and you can't stop it once it's started. The slower the set the better and it is important that after laying, exposed surfaces are covered with wet sacks, sand or polythene (and kept wet) for the first 4-6 days. Concrete also gives off heat as it sets — a useful property in very cold weather, although it would still need covering to protect it from frost.

Freshly mixed concrete will begin to set within 1-2 hours — in dry hot weather it will be faster. It takes 3-4 days to become properly hard — you can walk on it at this stage. To reach full strength, however, may take 28 days or more.

How to buy concrete

For small jobs it's best to buy the cement, sand and aggregate dry-mixed together, in

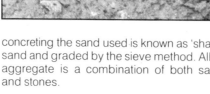

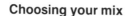

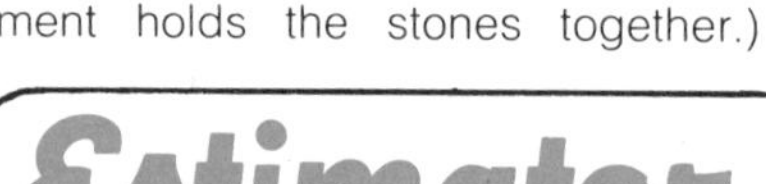

What to buy for mixing concrete

The quantities given here for sand and aggregate are rounded up to the nearest fraction of a cu metre that can be ordered. The mixes are made up by volume (see **Ready Reference**) so some sand and aggregate may be left over.

To make 1 cu metre eg 10m x 1m x 100mm	Cement 50kg bags	Sharp sand plus aggregate	OR	All-in aggregate
MIX A (1:2½:4)	6 bags	½ cu metre + ¾ cu metre		1 cu metre
MIX B (1:2:3)	8 bags	½ cu metre + ¾ cu metre		1 cu metre
MIX C (1:3:6)	4 bags	½ cu metre + ¾ cu metre		1¼ cu metre

MIXING

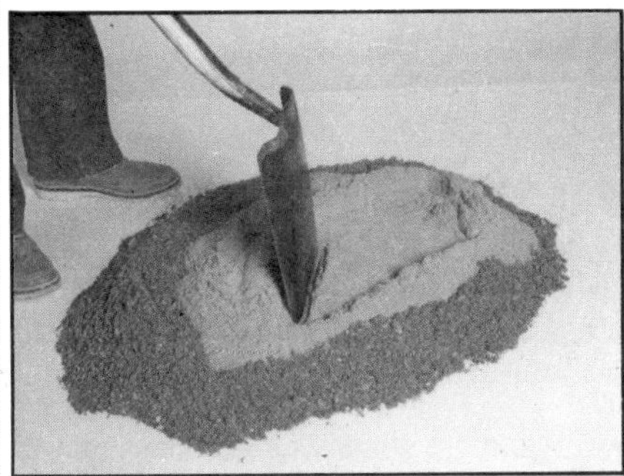

1 *Unless you're using dry ready-mix, first spread the cement over the aggregate and gradually mix by heaping into a 'volcano'.*

2 *Add about half the water to start with. Form a crater and mix in from the inside walls. Add the rest of the water gradually — not all at once.*

3 *When the concrete is about the right consistency shovel it into heaps again. Turn it into a new pile 3 times to ensure thorough mixing.*

4 *The finished concrete should look like this. When you draw the shovel back in steps, the ridges should be smooth and firm and not 'slump'.*

either 10kg, 25kg or 50kg bags. All you have to do is add water.

For larger projects, this can work out to be very expensive. Here it is better to buy the materials separately. The cement is normally sold in 50kg (just under 1 cwt) bags, though smaller (again more costly) quantities – 10kg and 25kg – are available. Both sand and aggregate are sold in 50kg bags, but it is more common to buy them loose by the cubic metre or fraction of a cubic metre. The combined or 'all-in' aggregate is available in the same way.

For really large work, however, (patios, long drives and the like) mixing the amount of concrete required by hand is extremely hard work. You could hire a powered mixer, but generally it is more convenient to buy it ready-mixed and have it delivered to the house. Check with the supplier on the minimum amount they are prepared to deliver — for quantities close to that minimum you could find it prohibitively expensive and you should consider sharing a load with a neighbour who is also carrying out building work. With ready-mix remember that you have to be prepared to lay it fast and if there is no direct access to the site and the concrete can't be tipped directly into your prepared formwork, you must have plenty of able-bodied help with heavy-duty wheelbarrows (you can hire these) standing by.

Dry-mixes have the amounts that made-up concrete will cover printed on the bag. For mix-at-home quantities, see the ESTIMATOR.

How to store materials

Under normal conditions cement will start to harden after about 30 days simply because it'll be absorbing moisture from the air. However, older cement that's still powdery inside can still be used where great strength or a high quality finish is *not* essential – but mix in a higher proportion of cement than usual, ie 1:1:2.

Cement should always be stored under cover and raised well off the ground — on a platform of wood, for example. Stack the bags closely, keep them clear of other materials and cover them to help keep the

READY REFERENCE

GETTING A LEVEL BASE

This is vital to avoid weak, thin spots in the concrete which will crack. These methods work on a reasonably flat site:

- drive 300mm (12in) pegs into the ground at 1 metre (3ft) intervals
- align their tops with batten and spirit level to match the final surface level of the concrete
- dig away soil to required depth, taking care not to disturb pegs. Use amount of peg exposed as your depth guide

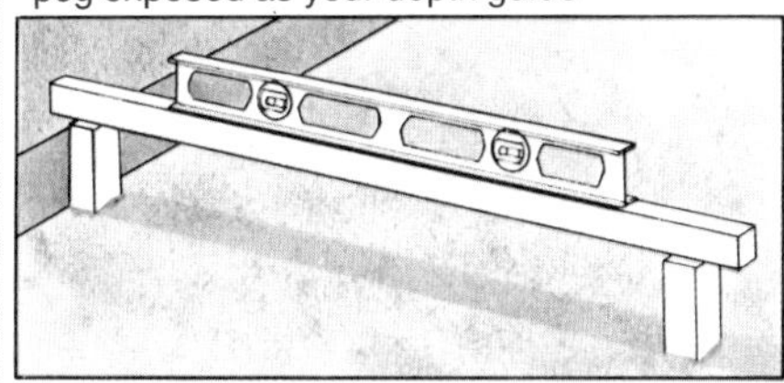

To level top of pegs over longer distances and round corners:

- tie a length of transparent hose between pegs
- fill the hose with water and drive in pegs so their tops match the water levels either end

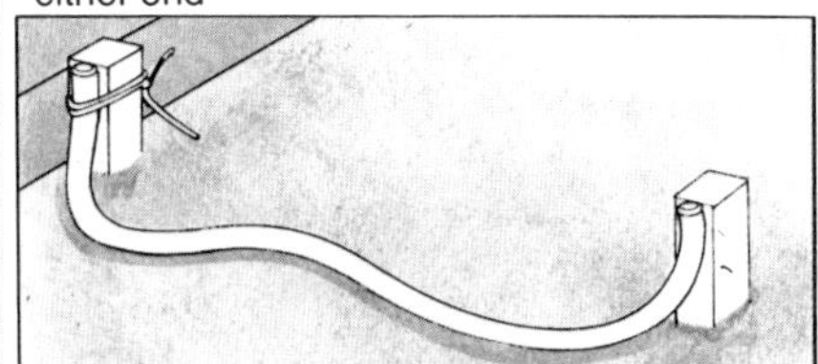

On long paths and drives:
use sighting rods made from sawn softwood – each is the same height (about 1.2 metres/4ft) and has a tee piece exactly at right angles across the top – you'll need three rods, and two helpers. Place rods on pegs and line up tops by adjusting pegs in ground.

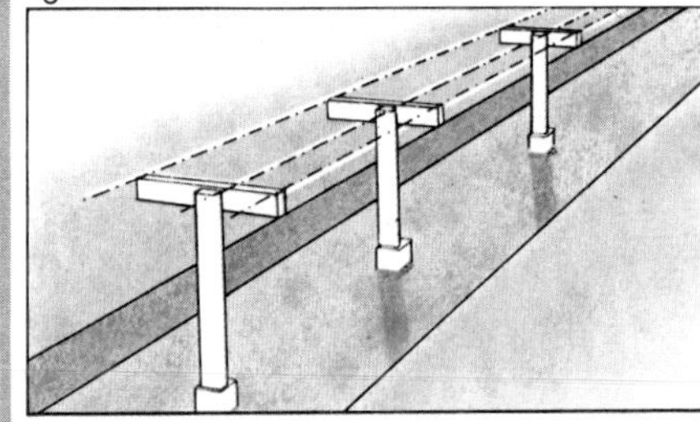

FOUNDATIONS FOR CONCRETE

- as a general rule lay rammed hardcore to the same depth as the final concrete
- on soft sub-soil excavate to twice the depth, filling soft pockets carefully with extra hardcore
- on clay, lay concrete quickly before the clay can dry out

Foundations for brickwork see pages 30 -34.

moisture out. If a bag has been part used, the remainder can be stored for a while inside a well-sealed plastic bag.

Loose sand and aggregate should be piled on a flat, dry and hard area and covered with heavy-duty plastic sheeting. It's most important to avoid the aggregate being contaminated by soil or other foreign materials. Any organic matter would decompose in the concrete leaving 'voids' which weaken it.

Site preparation

This is a major stage before you begin to erect any formwork, mix or lay any concrete. For accurate marking out, use pegs and strings to give yourself guide lines to follow. The area should be dug out and made as level as possible (see *Ready Reference).*

The big question is, how deep should you dig and how thick should you lay the concrete? To some extent this depends on how firm the soil is. For a path or patio a 75-100mm/3-4in thickness of concrete is usually enough — add a layer of hardcore of the same depth if the soil is very soft. If, however, you're building a driveway where there'll be a lot more weight on top, then 125-150mm/5-6in of concrete on top of hardcore would be advisable.

Some soils can lead to unsuccessful concreting. If your site is *clay* for example you have to concrete it as soon as possible after it has been revealed. The reason? Clay dries out quickly and then contracts. Because it will absorb water from the concrete mix, it makes an unreliable base.

Peaty and loamy soils will sink under a heavy load. Use good hardcore (see below) in the prepared area.

Made-up ground is another way of describing land that's been reclaimed. There's no knowing what was used as the in-fill, and it should always be assumed that it has minimal load-bearing capacity. Any concreting here will need good reinforcement such as hardcore, well compacted and the same thickness as the concrete you're laying on top.

Soft pockets

After you've prepared a site for laying a path or patio you could find pockets of soft soil which will cause any concrete to sink.

Large areas of soft pockets or made-up ground need something solid as a base — and this is where hardcore (broken concrete), rubble (broken brick) or a very coarse aggregate is essential. Tamp it into the ground until well consolidated — a must for areas such as drives or structural foundations taking a lot of weight.

If necessary small areas can be reinforced with a steel mesh set into the concrete. For most purposes 7mm diameter rods formed in a mesh of 150mm squares is quite adequate, and this is readily available at most builders' merchants. Rest the rods on small pieces of broken brick before you lay the concrete; make sure that the ends of the rods don't protrude from the area you're concreting, and that the mesh is completely covered.

Creating the work area

Using formwork boards to create a kind of box in which to lay your concrete has two big advantages. Firstly, it contains the concrete neatly, and secondly it gives you levels on either side to guide you in levelling the concrete itself. Although this is the most usual method of containing concrete, a brick

LAYING AND LEVELLING

1 Shovel the concrete well into the corners and only lay as much as you can finish off in one go. It's important that there are no hollows.

2 Roughly level the concrete with your shovel to a height about 6mm/¼in above the sides of the formwork. This will allow for compaction.

3 The 'tamping' board fits neatly across the formwork. First use a sawing action to level the mix, then a firm chopping action to compact it.

4 With the surface level, tap sides of formwork with hammer. This helps to compact the concrete. Fill in any hollows that result and level off again.

5 For an expansion joint use a piece of softboard the same depth as the concrete you're laying. Support it with pegs on one side.

6 Finish the concrete off on one side before you start laying on the other. Once the board is supported, hammer the pegs in deeper.

FINISHES

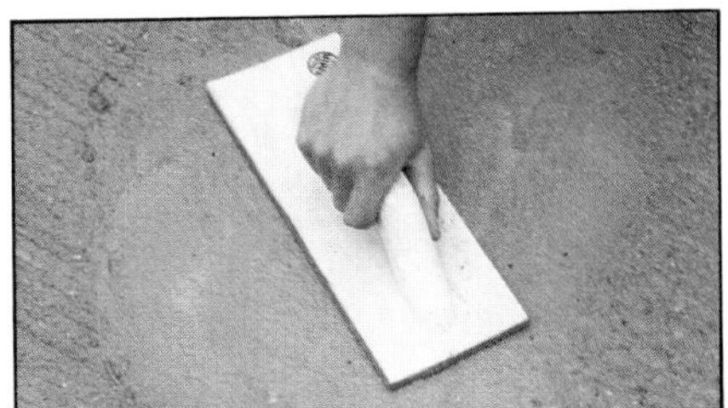

1 *Using a wooden 'float' gives you a smooth finish. Press the float down firmly as you 'scrub' the surface with circular movements.*

2 *A brushed finish leaves a much rougher surface by exposing the small stones in the aggregate. Use a stiff brush to create a pattern of straight lines.*

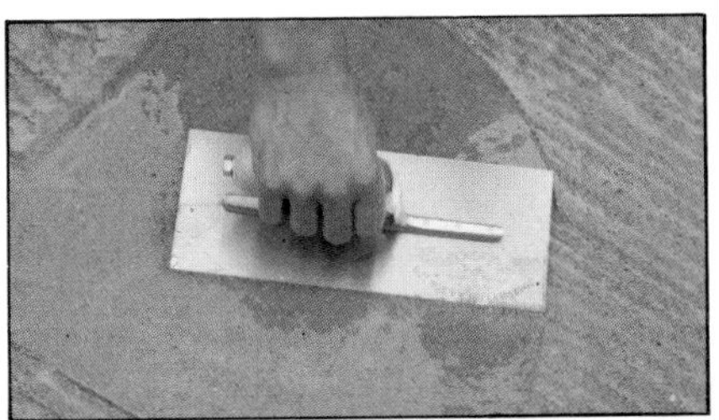

3 *For a polished finish, use a steel 'float' at a slight angle to the surface, drawing it towards you with a sweeping semi-circular action.*

surround can be used just as well — and this has the added advantage of not having to be pulled up. With bricks, however, it's more difficult to establish a completely straight and level line to follow.

For formwork use sawn (unplaned) softwood — it's called carcassing in the trade — for concrete that's to be placed below ground. It should be as wide as the depth of concrete you intend to lay and 25mm (1in) thick. Don't skimp on the thickness for it must be firm and rigid to support the weight of concrete.

Pegs are used to keep the formwork in place. They must be sturdy, not less than 50mm (2in) square and long enough to go well into the ground. Place pegs every 1 metre (3ft) against the *outside* face of the boards.

If building a raised path, formwork will give a finish to the concrete edge so you should use a timber that's planed. Unplaned timber can be used if the formwork is lined with 6mm (¼in) plywood, or if you intend finishing off the edges with more concrete after the formwork has been removed.

If you want to curve a corner in the formwork, use hardboard cut into strips as wide as the concrete is deep. This will need to be supported with pegs at more frequent intervals than softwood boards.

If you have difficulty driving the pegs into the ground (which may happen if you've put down hardcore) use lengths of angle iron instead. Alternatively drive the pegs in further away from the formwork and put timber blocks between the peg and formboard.

Expansion joints

Any large area of concrete needs expansion or movement joints to control cracking. A one-piece slab shouldn't be more than 3 metres (10ft) in any direction without a joint being included; a path should have joints at intervals of 1½ times the width of the path.

The simplest way of doing this is to incorporate a length of flexible plastic movement joint as you're laying the path. The material can be bought at most builders' merchants. Alternatively, use a piece of soft-board impregnated with bitumen – it should be the same depth as the concrete and about 12mm (½in) thick.

Drainage slopes

With a wide expanse you should have a gentle slope (1 in 60 is the general rule) so that rainwater can drain away. This is achieved by setting the forms on one side slightly deeper into the ground. To check that the slope is the same all along the form-work, set a small piece of wood (about 12mm/½in for a 1m/3ft wide path) thick on the lower side and use your spirit level to check across to the other side.

To keep the formwork on each side of a path rigid, place a length of softwood across the width at the peg points, but not so that it will make an impression on the concrete. This can be used as a guide for levelling as well.

Whether you are building a concrete path, a base for a shed or garage, a hardstanding for a car or even a large patio, the principle of formwork is the same — only the number of boxes or bays you divide the area into varies. With each stage of the job you should mix only enough concrete to fill one bay or box at a time.

As the concrete starts to dry (after 2 hours) cover the surface with plastic sheeting or damp sacking to stop it drying too quickly.

READY REFERENCE

SETTING UP FORMWORK

You need:

- planks of sawn softwood 25mm (1in) thick and wide enough to match the concrete depth
- pegs of 50 x 50mm (2 x 2in) softwood at least 300mm (12in) long
- a string line to aid setting out

Position formwork along all edges to keep concrete in place until it's set, and provide a working edge for levelling the concrete:

- hammer pegs into the ground at 1 metre (3ft) intervals round the perimeter of the area to be concreted; use foot to hold peg in position

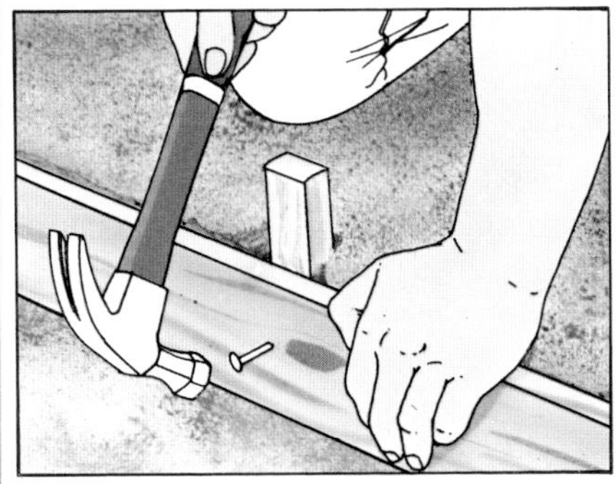

- place formwork against the pegs, aligning the boards accurately against a string line
- check levels between opposite lines of formwork with batten and spirit level, and allow for drainage slope if required
- nail the boards to the pegs

CONCRETE: WHAT TO MIX

MIX A for concrete over 75mm (3in) thick

1 bucket cement
2½ buckets sharp sand
4 buckets washed aggregate

MIX B for concrete less than 75mm (3in) thick

1 bucket cement
2 buckets sharp sand
3 buckets washed aggregate

MIX C for rough bedding concrete

1 bucket cement
3 buckets sharp sand
6 buckets washed aggregate

All mixes need about ½ bucket water; exact amounts depend on the dampness of the sand.

FOUNDATIONS for garden walls

Even if it's only for a wall to grace the garden, building a solid foundation is a must. But how deep should you dig? How wide? And what's the right thickness of concrete? Here's an easy to follow explanation of why foundations are so important, how they differ and which one to choose.

All walls need foundations to give them stability, and free-standing garden walls are no exception. The foundation is like a platform, helping to spread the weight of the bricks in the wall onto the earth base below.

Most foundations are made of concrete laid in a trench, and for a garden wall where there's no additional weight for it to carry (unlike a structural wall, for example, which may also carry part of the weight of a roof) the concrete itself doesn't need to be very thick – between 100mm (4in) and 150mm (6in) of concrete is quite enough for a wall up to a metre in height. But the thickness of the concrete is not the only thing you have to consider. How deep in the ground you place it is just as important.

For a concrete foundation to provide an effective platform which won't allow the brick wall to crack, it has to be laid on firm 'subsoil'. And you won't find this until you get below the topsoil. The depth of topsoil varies enormously from place to place, so there can be no hard and fast rules about how deep you must dig – but expect anything between 100mm and 300mm (4-12in). Once you're through to the harder subsoil, you've then got to dig out enough for the depth of concrete – at least another 100-150mm (4-6in).

In practice the other big variable is the nature of your soil. Different subsoils have different load-bearing capacities – for instance hard chalky soils can support more weight than clay (see Choosing your foundation, page 34), but sandy soils can take less. The weaker the subsoil, the wider you have to build the foundation – consult your local building inspector for advice on soil conditions in your area.

There's another important reason for digging down so deep and that is the effect the weather has on soil. In clay subsoils, for example, a prolonged dry spell will cause the clay near the surface to shrink; then, when it rains, the clay will swell. All this causes considerable movement of the ground and unless a concrete foundation has been laid deep enough it'll crack up under the stress of constant expansion and contraction. To counteract this, the foundation has to be laid *below* the point at which the weather can cause movement. Again, in different soils, this varies from 150mm (6in) to 500mm (20in) or more down, but it's advisable to consult your local building inspector to get a more precise figure for soil conditions in your area.

Foundation design

All foundations have to be designed so that they evenly transfer the weight of the wall above to the earth base below. Because of the way the wall's weight spreads out onto

David Pope

The weight of a wall spreads at an angle of 45° from its base into the foundation and then on into the subsoil. This is called the angle of dispersion.

READY REFERENCE

WHAT TO MIX

- foundations for garden walls don't need the strongest concrete mix – use the following proportions:

1 bucket cement
3 buckets sharp sand
6 buckets washed aggregate
OR
1 bucket cement
8 buckets all-in ballast

HOW MUCH TO BUY

- for small quantities buy bags of dry ready-mixed concrete (the bag will say how much it'll make up)
- for larger quantities, it's more economical to buy the ingredients separately
- although cement is available in standard size bags, sand and aggregate (or all-in ballast) have to be bought loose – and the minimum quantity you can buy is usually ¼ cubic metre.

EXAMPLE: To make 1 cu metre of concrete (enough for a foundation 20 metres long, 500mm wide and 100mm deep) you'd need to buy:
4 bags cement (50kg bags)
½ cu metre sharp sand
¾ cu metre washed aggregate
OR
4 bags cement (50kg bags)
1¼ cu metres all-in ballast

MARKING OUT

1 *Set the pegs for the profile board outside the line the foundation will follow. First hammer the pegs in, then nail a cross-piece on top.*

2 *You'll need profile boards at each end of the foundation trench so that you can string guide lines for digging out between the two.*

3 *Fix nails in the cross-pieces to establish the width of the trench. Normally this is a minimum of 300mm (12in).*

4 *Tie the line to one of the nails, then string up to the others. Loop the line round each nail and keep it taut. Don't cut the line.*

5 *The lines now mark the edges of the trench. At a later stage the building line for the wall is marked out in the same way.*

6 *To give you an accurate line to follow for digging out, sprinkle sand beneath the lines. After this remove the strings, but not the profile boards.*

Keith Morris

the foundation – called the angle of dispersion – the foundation is built so that it is wider than the wall. In fact this 'load spreading' follows an angle of 45° (see page 41) and means that the width of the foundation on each side of the wall has to be at least equal to the depth of the concrete. This is a simple rule of thumb which will help you decide how wide your foundation has to be for different wall widths. Of course, if you're building on a relatively soft subsoil, your building inspector may recommend that you build a wider foundation. Like a raft floating on water, the bigger it is the more stable it will be.

Once you've dug your trench, you'll be faced with another decision: do you just lay the minimum thickness of concrete or lay enough concrete to fill up the trench so you have fewer bricks to lay? In fact, there can be quite a difference in the amount of work involved. If your trench is 500mm (20in) deep and you only lay a 150mm (6in) depth of concrete, it means that just to get back to ground level you've got to lay some 5 courses of bricks which ultimately won't even be seen. Nevertheless, it makes no difference to the strength of the foundation – it

LAYING THE CONCRETE

1 *Once the trench is dug out, check the depth at 1 metre (3ft) intervals. The actual depth depends on the nature of the soil — see page 34.*

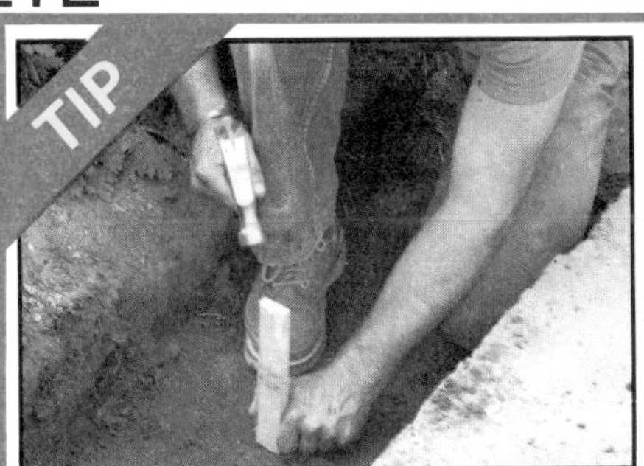

TIP

2 *Hammer in pegs at 600mm intervals down the middle of the trench. These can be adjusted to act as an accurate depth guide for laying the concrete.*

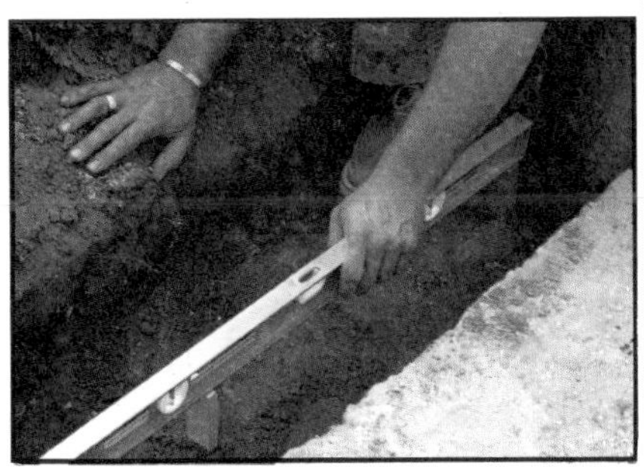

3 *Use a spirit level to check that the tops of the pegs are level. This marks the top of the foundations, and accuracy is essential. (Continued overleaf)*

simply depends on whether you prefer laying more concrete (and it'll be quite a lot more) or more bricks. Engineering bricks are recommended for any work below ground, though any special quality brick will do almost as well.

Marking out

Digging a trench foundation to a depth of about 500mm (20in) is probably the safest rule of thumb to follow if you're building a wall that's going to be more than 5 or 6 courses above ground level. (For smaller walls see Foundations for low brick walls).

And to make sure that the line of the trench is straight and the width constant, you have to mark out accurately. For this you need to set up what are called 'profile boards' at each end of the trench. All you do is string lines between the boards in the position you want for the foundation.

To make profile boards, use lengths of 50mm x 25mm (2in x 1in) timber cut a little wider than your trench. For pegs, use 50mm square (2in square) timber about 600mm (2ft) long. You'll also need nails and string.

Hammer two pegs into the ground at each end of the wall line, and nail the cross-pieces onto them. Next drive nails into the tops of the boards – to mark the outer edges of the foundation – and string lines between them, pegging the string into the ground beyond the profile boards.

These strings are then the guide lines for digging the trench, and can be transferred down to the ground using a spirit level. When the trench is dug and the foundation laid, these same profile boards can be used to create the building lines for the wall – you just add more nails and string up as before.

Constructing the trench

Remove the topsoil and dig a trench according to your marking up. To give you a guide for laying the concrete, you'll need pegs about double the depth of concrete required. These

FOUNDATIONS FOR LOW BRICK WALLS

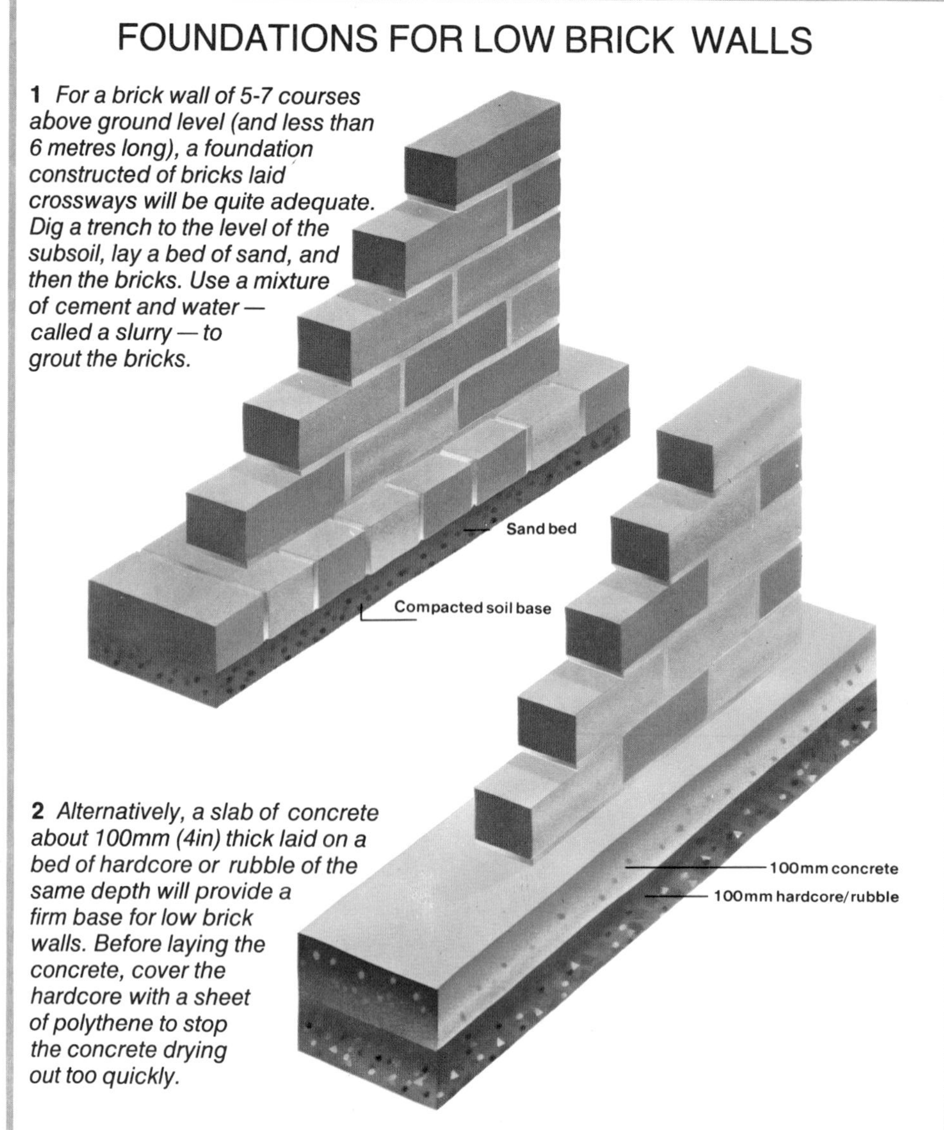

1 *For a brick wall of 5-7 courses above ground level (and less than 6 metres long), a foundation constructed of bricks laid crossways will be quite adequate. Dig a trench to the level of the subsoil, lay a bed of sand, and then the bricks. Use a mixture of cement and water — called a slurry — to grout the bricks.*

2 *Alternatively, a slab of concrete about 100mm (4in) thick laid on a bed of hardcore or rubble of the same depth will provide a firm base for low brick walls. Before laying the concrete, cover the hardcore with a sheet of polythene to stop the concrete drying out too quickly.*

4 *Fill the trench with concrete to just above the level of the pegs. Make sure it's well compacted before you roughly level it with the shovel.*

5 *Use a large piece of timber as a 'tamping' beam to finally level off so that the tops of the pegs are just visible. The surface doesn't need to be smooth.*

6 *Check the level of the foundation using a spirit level placed across a straight edge. If necessary make adjustments by adding more concrete.*

Keith Morris

should have tops cut square and should be driven into the centre of the trench at intervals of 600mm (2ft). The tops should all be precisely levelled using a builders level or a straight-edge and spirit level. Soak the trench with water and allow it to drain before the concrete is poured in. This should then be well 'rodded' (with a broom handle, for example) to ensure that the entire volume of the trench is filled and the concrete is as high as the tops of the pegs (these don't have to be removed and will eventually rot). Use a wooden float or suitable piece of timber to level the surface.. It needn't be perfectly smooth as a fairly rough surface provides a good key for the mortar. Cover the concrete with plastic sheeting or damp sacking and leave for 6 days to 'cure' – longer if the wall is more than 12 courses high. The slower the curing the stronger the concrete, so don't try and build a wall on the foundations too soon.

Concrete foundations

Concrete is ideal for foundations for no other material can take up the precise shape of the subsoil surface at the bottom of the trench and transfer the load so evenly. It should be made of 1 part cement to 3 parts sand to 6 parts aggregate (see Working with concrete, pages 26-29) with just enough water to produce a pliable consistency.

Trench foundations can be reinforced with rods or mesh – either will increase the strength of the foundation and whatever is built upon it. Both kinds of reinforcement are actually quite simple to add – you just have to make sure that the steel rods or mesh are bedded in the lower part of the concrete and not exposed at the sides or ends. In some cases, reinforcement is essential – for example, if you're laying a foundation over a drainpipe. For most garden brick walls, however, going to the trouble of reinforcing a foundation just isn't necessary – the weight of the wall doesn't justify it. What is important is that the wall doesn't crack because the ground underneath moves slightly.

Raft or slab foundations

For small walls of 7 courses or less the simplest concrete foundation is a raft or 'slab'. This is cast just below ground level (after the topsoil has been removed) in much the same way as you'd lay a concrete path. First dig out to a depth of about 200mm (8in), then add a layer of compacted broken brick or hardcore to provide drainage. Cover this with light polythene sheeting just before concreting – this will prevent the concrete drying out too quickly because of water being absorbed by the base. Then lay your concrete about 100mm (4in) deep and tamp to a level surface (see pages 31-32). Once the wall is built the concrete foundation can be hidden by soil and grass.

Brick foundations

For low walls under 7 courses high you could even avoid the expense and trouble of mixing concrete altogether, because a foundation strip of bricks laid cross-ways can be perfectly adequate. Lay the bricks on a thin layer of sand which has been well compacted and levelled in a shallow trench. This should be dug to below the level of topsoil – anything between 100mm (4in) to 300mm (12in) below ground level. Grout these together with a 'slurry' – a creamy mixture of cement and water. This is called a 'footing' course and you can lay bricks on top in the usual manner even before the slurry is hard.

Earth retaining walls

Sometimes walls built in the garden may not be free-standing but used to retain earth on one side – for example, as terraces on a sloping site or to enclose flat areas of lawn or

MARKING BUILDING LINES

1 *For a half brick thick wall (a single line of bricks) position nails on the profile board 100mm in from each side of the trench and string up between them.*

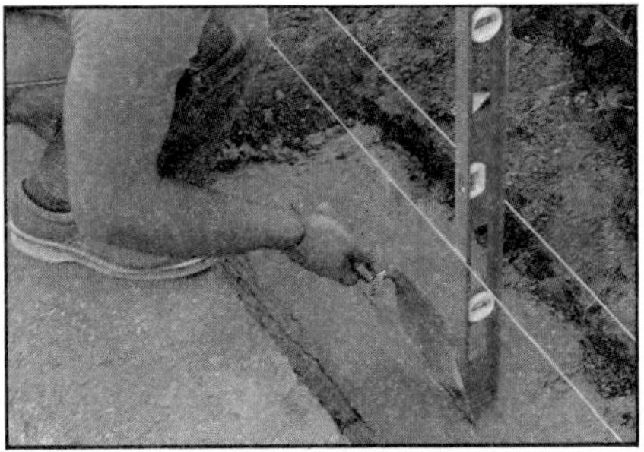

2 *These strings give you clear guidelines to follow, but it's usually worth double-checking that the building lines are positioned centrally on the concrete.*

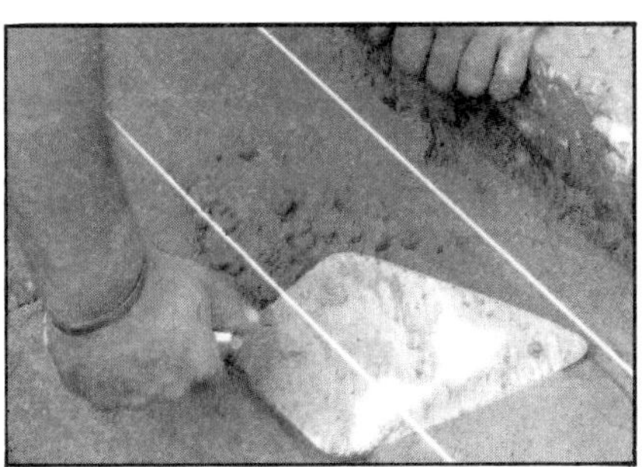

3 *Lay a thin bed of mortar directly underneath the building lines and smooth it out with the trowel. This is for marking down from the line.*

4 *A piece of scrap wood held diagonally against the spirit level gives a bit of extra support as you mark down vertically and score the mortar bed with a trowel.*

5 *Lay the bedding mortar for the first course of bricks alongside the marked line. Furrow the mortar with the tip of the trowel.*

6 *Lay the first three bricks, and then check them with the spirit level. Once the first course is complete, remove the profile lines.*

Keith Morris

CHOOSING YOUR FOUNDATION

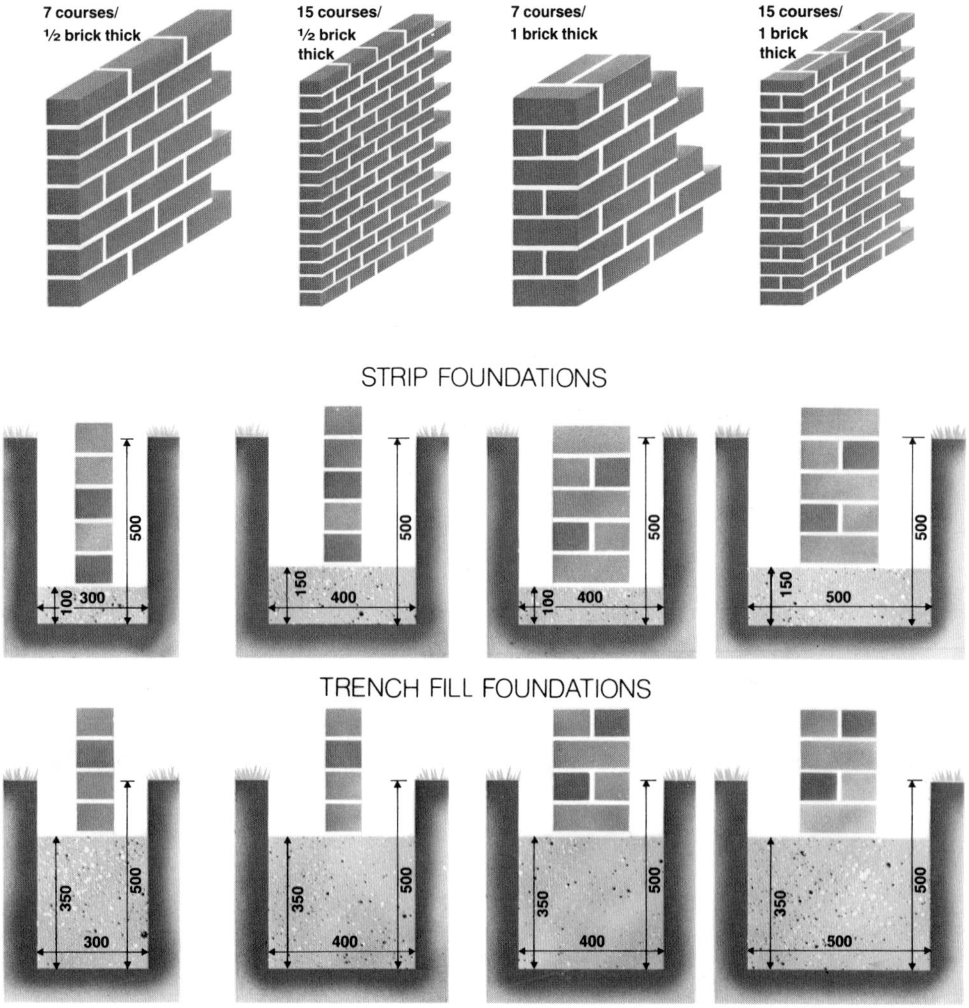

7 courses/ ½ brick thick

15 courses/ ½ brick thick

7 courses/ 1 brick thick

15 courses/ 1 brick thick

STRIP FOUNDATIONS

100 300 500 | 150 400 500 | 100 400 500 | 150 500 500

TRENCH FILL FOUNDATIONS

350 300 500 | 350 400 500 | 350 400 500 | 350 500 500

The size of foundation you lay depends on the height and thickness of the wall you intend to build, and on the load-bearing capacity of the subsoil. The chart above gives recommended dimensions for strip and trench-fill foundations for half-brick and one-brick thick walls, below 7 courses or up to 15 courses high, on a typical clay subsoil. On crumbly, loose soils the recommended widths should be doubled. See also ***Ready Reference.***

flower beds. In such cases the soil behind the wall is constantly trying to push outwards, completely changing the pattern of stress involved.

The simplest solution is to make the structure strong enough to withstand this extra pressure. With a 4 or 5 course wall this can usually be done by building the wall one brick thick (instead of ½ brick thick) and by providing 'weepholes' at regular intervals to drain excess water. These are made by removing mortar from a number of the vertical joints before it sets.

If you find that the surface of an earth-retaining wall is marked by white crusty deposits – called 'efflorescence', and caused by water carrying salts through the wall from the soil behind – dig away the earth and coat the inner surface of the wall with bituminous emulsion to create a damp barrier.

Building on a slope

It is visually unsettling and structurally undesirable to lay bricks running parallel to a slope. So, to build a wall that 'steps' down a slope, the trench foundations also have to be stepped or 'benched' into the slope. Levelling, pegging, pouring and finishing are all carried out in the same way as with a horizontal trench, but in stepped sections. You'll need form boards to frame the outer edge of each step, but otherwise the width and depth of the foundation is exactly the same as for an ordinary wall. Only the length of each step varies.

READY REFERENCE

PROBLEM SOILS

- clay subsoils are prone to shrink and swell so make the foundation trench at least 500mm deep, and lay the concrete to the width and thickness shown in the chart
- loose, poorly-compacted subsoils need foundations double the normal width to help spread the load of the wall, and should be laid as trench-fill foundations
- if the sides of your trench tend to collapse, use timber to shore them up before pouring in the concrete, or cut them wider than necessary and slope the sides.

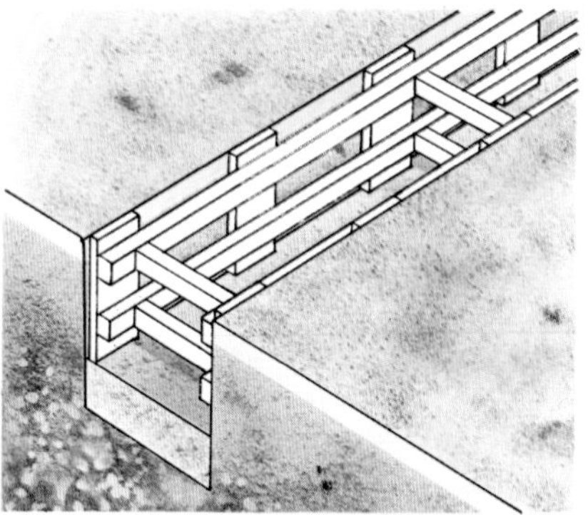

TIP: LET CONCRETE DRY SLOWLY

- first soak the trench with water and allow it to drain away
- cover concrete with damp sacking or plastic sheet
- Allow concrete to 'cure' for at least 6 days before building on it.

UNEXPECTED HAZARDS

- if the trench fills with water it must be drained before you lay the foundation concrete. Water in a small trench can be emptied with a bucket; in a large trench you may have to hire a pump
- if you discover underground pipes as you're digging the trench, dig soil away around them to the required depth. Lay the foundation as normal on either side of the pipes, then 'bridge' them with a short reinforced concrete lintel or, for narrow gaps (300mm/12in or less) lengths of paving slab laid two courses deep.

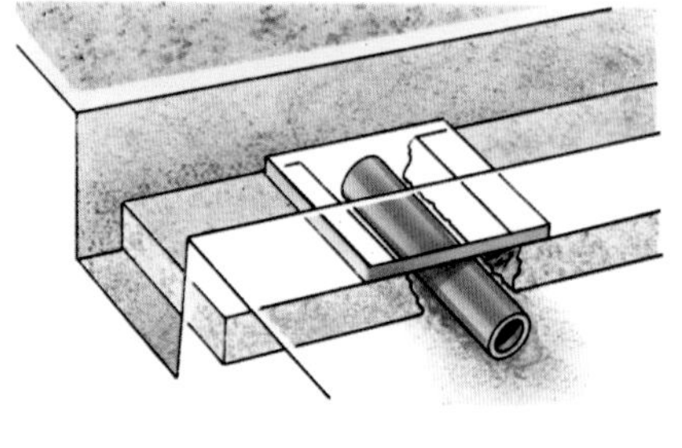

LAYING CONCRETE SLABS

Concrete is the ideal material for laying a slab for a shed, patio or driveway. Once you've mastered the techniques of mixing and casting it, you can provide a hard durable surface that will last for years.

A concrete slab can be a tough, hard-wearing base for a variety of uses in your garden, but its success is only as good as the preparation you've put into making it. Concrete consists of stone particles called 'aggregate', bonded with a Portland cement and water mix (see Working with concrete). You must mix the ingredients carefully, cast the slab on specially prepared foundations, apply a finishing texture, and allow the concrete to set properly, if your results are to be long-lasting.

Planning a concrete slab

After you've decided exactly where you want to put your square or rectangular slab, you'll have to mark out the ground accurately and prepare the foundations before laying the concrete. But if you're laying a more complex shape or a much larger concrete base, you'd be wise to make some preliminary sketches and transfer them to squared paper, to help in calculating the material required.

Before you start to lay your slab it's sensible to check with your local authority whether you're infringing any bye-laws. One of the main objections they might have is the position of your planned slab in relation to existing drains and pipe runs; as a result, you might have to re-route some of these to keep them out of harm's way.

Access to the site and the time you'll have available for laying your slab are important considerations, particularly if you're laying a large concrete drive or a garage floor, for example. With work on this scale you should use ready-mixed concrete, which is delivered in bulk ready for casting. If you go for this method, it's vital that you provide access for the lorry and space for the load to be dumped as close as possible to your site. You must have your foundations prepared so that you can cast the mix as soon as it's delivered. Any delay could mean that the mix starts to set rendering it useless.

Calculating the size of slab

Before you can mark out your slab on the ground and prepare its foundations you'll have to work out its dimensions and calculate how much concrete you're going to need. As a basic guide, the larger your slab the thicker it must be.

For an ordinary garden shed, for instance, you'll need a slab about 75mm (3in) thick, except where the ground is soft clay, when you should increase its depth to 100mm (4in). If your slab is to form the floor of a workshop, or a drive leading to your garage, a thickness of 100mm (4in) is appropriate on ordinary soils, 125mm (5in) on soft clay or other poor sub-soils.

Once you've decided on the dimensions of your slab you can estimate how much concrete you'll need and how you're going to mix it (see *Ready Reference*).

Marking out the slab

Before you start to mark out your slab on the ground, dig out and remove the top-soil, including any grass and the roots of shrubs, from the area you're going to concrete. Allow a margin of a few feet all round your proposed slab for working space.

Use strings stretched between wooden pegs driven into the ground just outside the area you're going to concrete to mark out the shape of your slab (although, for a very small slab it's possible simply to mark it out using planks positioned squarely on the ground – see photographs). Use a builder's square (see *Ready Reference*) to set the corners of your slab accurately. If you're

READY REFERENCE

SETTING UP FORMWORK

Formwork is needed to mould and retain the concrete while it sets. To set up a frame for a small square or rectangular slab:

- drive 50x50mm (2x2in) softwood pegs 300 (1ft) long into the ground just outside the slab perimeter at 1m (3ft) centres
- nail lengths of sawn softwood 25mm (1in) thick and as wide as the concrete depth to the pegs, making sure they're level and square
- butt joint the corners. Alternatively, make up the formwork to size elsewhere, and then carry it into position (see page 36).

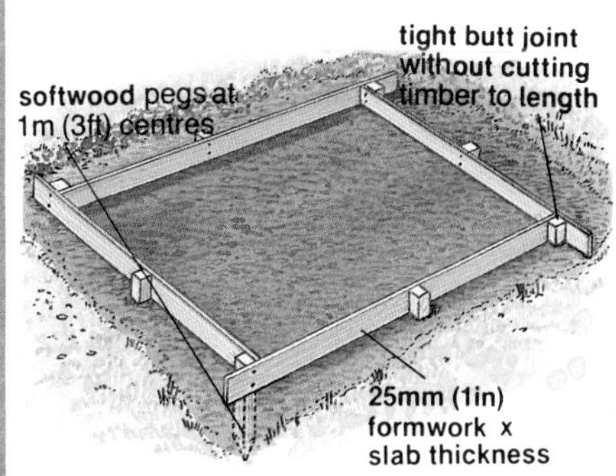

making strip foundations for a wall you can use 'profiles' to set the levels (see Working with concrete, pages 26 -29).

Once you've positioned your string lines you can dig out the sub-soil to roughly the depth of your foundation, taking it about 150mm (6in) beyond the strings to leave space for setting up 'formwork', which moulds and retains the concrete while it's hardening.

Setting the levels

While the base for a shed should be virtually level, a concrete slab patio or garage drive should be laid with a slight slope to allow rainwater to run off quickly. And of course, if the slab is near a wall you'll have to ensure that the fall drains away from it. You must allow for the slope when you're preparing the base. A gradient across the site of about 1 in 60 is about right.

You'll also have to make sure when laying a drive leading to a garage that the drive doesn't drain into the garage. If the ground is naturally sloping in that direction take it to a level below the garage floor and lay a short section of slab sloping away from the garage. Where the two slopes meet you'll have to include a channel leading to a suitable drainage point.

To establish the level of your slab over its entire area hammer 50x25mm (2x1in) softwood pegs into the ground at about 1.5m (5ft) intervals. The first peg must be one that establishes the level of the others, and its called the 'prime datum'. If your slab is to adjoin a wall you can fix the level of this peg at the second course of brickwork below the dpc, (damp-proof course) or some other fixed point of reference.

Drive in some more pegs and check across their tops from the first peg with a spirit level on a straight-edged length of timber to check their level.

You can allow for a drainage fall by placing a wedge of timber called a 'shim' under one end of the straight edge.

Once you've set the datum pegs, measure down them whatever thickness of concrete you'll need for your slab and excavate the ground, or fill in, where necessary.

Fixing the formwork

Wet concrete tends to spread out as it sets and so you'll have to fix a timber frame called 'formwork' at the perimeter of the foundations to retain it and support its edges. It must be strong enough to withstand heavy tamping, which compacts and strengthens the mix. Use straight lengths of stout timber a minimum of 25mm (1in) thick, set on edge and nailed to pegs of 32x32mm (1¼x1¼in) timber driven into the ground at the perimeter of the slab at 1m (3ft) centres. Fix the pegs outside the area that's to be con-

PREPARING THE FOUNDATIONS

1 You can mark out a small concrete slab on the ground with scaffold boards and pegs at the corners, then start to remove the topsoil or turfs.

2 Continue to dig out the topsoil or turfs until you've accurately marked out the shape of your slab. Then dig down to the depth you want the slab.

3 To fix formwork around your base butt joint four lengths at the corners. Nail battens at each corner to hold the angle; then saw off the waste.

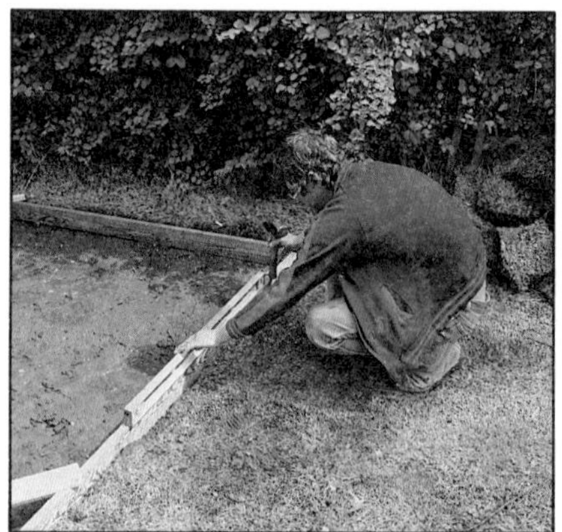

4 Position the formwork within the foundations and check across the top with a long spirit level to ensure that it's level, or sloping for drainage.

5 Tamp or roll the base of the foundations firm. If the soil is soft, add some hardcore and compact this into the surface with a sledge hammer.

6 Add as much hardcore as you need to give a firm base for the concrete. You may need to add a layer of sand to fill any voids in the surface.

creted, with their tops flush with, or slightly below, the top edge of the formwork.

You can use your string lines as a guide to positioning the formwork and a builder's square to ensure that the corners are set perfectly at right angles.

The top edge of the formwork must be set so that it's flush with the top of your finished slab; it's best to use timber that's the same thickness as your slab, otherwise you'll have to recess it into the ground. You can use your intermediate datum pegs as reference points when levelling the formwork with a builder's level, making sure you incorporate the drainage falls.

The corners of the formwork must be tightly butt-jointed (see *Ready Reference*) to prevent the wet concrete from seeping through. If you have to join two planks together end to end in order to make the required length you should again use butt joints, but back both planks at the joint with a short section of timber nailed in place and wedged with a peg at this point.

Movement joints

You can cast a slab in one piece if it's no longer than about 3m (10ft) in width or length. But if it's bigger than this, or if its length is greater than twice is width, it's usual to divide the overall slab into 'bays' that are as square as possible – or equal in size – and to include a gap called an 'expansion joint', which prevents the slab from cracking due to expansion or contraction. Fill the gap with a length of softwood 10 to 12mm (3⁄8in to 1⁄2in) thick, the depth of the slab, and cut to fit between the formwork at the sides of the slab. Treat the fillet with preservative before fitting it within the slab.

Each bay is cast separately so it's best to back up the jointing timber with a piece of formwork temporarily pegged in place for support. When you've cast and compacted the first bay, remove the formwork behind the jointing timber and cast the second bay, leaving it permanently in place.

Mixing the concrete

If your slab is too small to justify a load of ready-mixed concrete or you wish to lay it in stages over several weekends, you'll need to buy all the ingredients and mix them yourself. To decide on the volumes of cement and aggregates you'll need for your particular slab, you must first decide on the concrete mix proportions to use. *Ready Reference* gives a basic guide to proportioning, which you can relate to your own needs. Following these guidelines and using the example of the car port base given in *Ready Reference*, the volume of concrete you'd need is 1.8m^3 (63 cu ft). Materials needed are therefore going to be in the order of 1.8mx6 = 10.8 bags of cement; 1.8x0.5 = 0.9m^3 of sand; 1.8x0.8 = 1.44m^3 of coarse aggregate.

1.8x0.8 = 1.44m^3 of coarse aggregate.

Allow a 10 per cent margin for wastage to the cement to the nearest whole bag and buy 12 bags of cement. Round up quantities of aggregates to the nearest whole or half cubic metre and buy 1m^3 (36 cu ft) of sand and 1.5m^3 (53 cu ft) of coarse aggregate. Your calculations, though, should always be regarded as a guide only; exact amounts needed for a job will depend on the care you take in storage and handling and on the accuracy with which you prepare the base for the slab.

When you've an idea of the amounts of materials you'll need, you must decide on what method to use to mix them: by hand or by power mixer. Many different types of electric- petrol- and diesel-powered mixers are available for hire, and take much of the hard work out of mixing.

To get the correct consistency of concrete using a mixer, add half the coarse aggregate needed for the batch and half the water first. Then add all the sand and mix for a few minutes. Next you can add the cement and the remainder of the coarse aggregate. Finally add just enough water to achieve a workable mix. Most beginners add too much water; when it's of the right consistency the concrete should fall off the blade of your shovel cleanly without being too sloppy.

If you have to break off your work for a while, add the coarse aggregate and water you'll need for the next batch and leave the mixer running, while you are away, to keep the drum cleam.

For how to mix concrete by hand (although it's really only viable for small jobs), see pages 26 -29.

If you're mixing the concrete yourself you can store the aggregates indefinitely on a hard surface covered with a polythene sheet to keep it clean. Cement, however, must be kept dry: moisture in the air can penetrate the paper sacks and cause it to harden. Stack the sacks,under cover if possible, flat on a raised platform of planks on bricks and cover them with polythene.

Using ready-mixed concrete

Ready-mixed concrete is delivered by mixer lorry, usually in minimum loads of 3m^3 (105 cu ft). If you need this amount, or more, ready-mixed concrete is worth considering as it takes a lot of hard work out of concreting and enables you to complete fairly large projects quickly.

Your supplier will want to know the volume of concrete you'll need, at what time you want it delivered, what it'll be used for, and how you're going to use it on delivery. This information will enable him to determine an appropriate mix and give you a price. You'd be wise to seek several quotations and try to choose a depot close to your home: much of

READY REFERENCE

MAKING A BUILDER'S SQUARE

Set out square corners accurately with a builder's square, which you can make.
To do this:

- nail together three strips of 50x25mm (2x1in) softwood in the proportion of 3:4:5, so that the angle between the two shorter sides is 90°.

A convenient size for the pieces is:

- 450, 600, 750mm (18, 24, 30in) respectively
- make a half-lap joint at the 90° corner and overlap the other two corners.

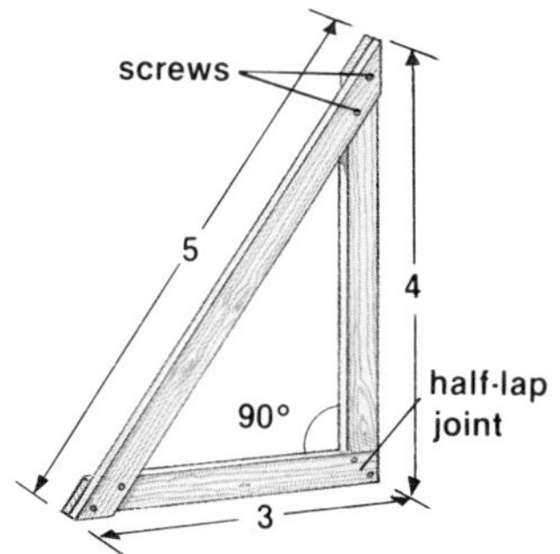

MAKING A TAMPING BEAM

To compact and level a concrete slab you'll need:

- a tamping beam made from a straight-edged length of 175x25mm (7x1in) softwood about 300mm (1ft) longer than the width of the slab
- or, for a very broad slab, a beam of 150x50mm (6x2in) timber with handles bolted on at each end so you can work standing up.

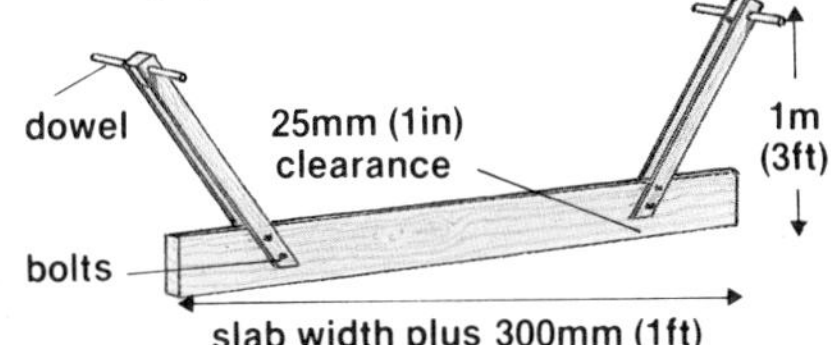

To use a tamping beam:

- lift the beam and drop it to compact the concrete as you work along the slab
- repeat this process a few times
- change to a sawing action, levelling the concrete flush with the top of the formwork.

CURING THE CONCRETE

Curing prevents the slab from drying out too quickly and becoming weak and cracking. To cure your slab:

- cover it with a large sheet of polythene
- weight down the edges – but not on the concrete – with bricks
- sprinkle sand on top to stop the sheet ballooning in the wind
- leave for 3 or 4 days before removing
- don't use the slab for 7 to 10 days.

CASTING THE CONCRETE

1 *Lay a path of scaffold boards from the concrete mix to the slab so you can take the mix by wheelbarrow without harming the ground.*

2 *Spread out the first barrowload of concrete over your foundations, using a shovel to work it into the hardcore and to avoid air bubbles.*

3 *Continue to tip barrowloads of concrete into your foundations until you've half-filled the area, just proud of the tops of the formwork.*

6 *Compact the wet concrete by lifting and dropping the tamping beam onto the concrete as you work across the slab. Repeat using a sawing action.*

7 *When you've filled the entire slab and have tamped the mix thoroughly tap the outside edge of the formwork with a hammer to settle the concrete.*

8 *You can produce a non-slip finish of fine swirls on your slab by running the back of your shovel over the wet surface.*

the cost of the concrete is in its transportation.

To receive your concrete you'd be wise to lay down a large polythene sheet to make clearing up easier afterwards. If you need to transport it any distance from the point of delivery get together as many wheelbarrows – heavy-duty ones, not light garden types – shovels and helpers as you can. It's sensible to lay a pathway of scaffold boards or planks from the pile to your site if you have to cross areas of lawn or go up or down steps.

Laying the concrete

When your formwork has been positioned you can remove the levelling pegs from within the areas, and the string lines from the perimeter, but if you're going to lay the concrete on a hardcore base you should add the ballast at this stage. Compact it well with a sledge hammer, fence post tamper or a garden roller, and leave the levelling pegs in place until you've set the foundations at the correct level.

When the base is ready, tip in the concrete from a barrow, or by the bucketful: if you're making a big slab you might even be able to get the delivery lorry to tip the mix straight into your prepared base.

Spread the concrete evenly with a garden rake to level it to just above the tops of the formwork. This allows an excess for compacting the mix. When all the concrete has been cast, compact it, using a tamping beam (see *Ready Reference*) made from a straight-edged length of 175x25mm (7x1in) softwood about 300m (1ft) longer than the width of your slab. For very large slabs use 150x50mm (6x2in) timber to make the beam, with handles fitted at each end so that you can work from a standing position rather than crouching (see *Ready Reference*).

Use the beam with a chopping action, lifting it then dropping it to compact the concrete and force out any air bubbles. Work along the slab in this way and, after a few passes, change to a sawing action, which levels any high spots and fills depressions as you move down the formwork. Continue to tamp until the concrete is even and flush with the top of the formwork.

Finishing the concrete

You can apply a variety of finishes to your concrete slab to suit its purpose. For a garage drive you can simply leave the fairly rough, non-slip, texture created by the tamping beam or you could brush the fresh concrete across its width with a stiff-bristled broom to give a more regular, but still non-slip, finish.

A smoother finish is easier to keep clean for a shed or garage floor, and you can produce this with a wooden float used in a wide, sweeping action. For a more polished effect,

4 *Draw a stout timber tamping beam across the tops of the formwork to spread the concrete roughly and to flatten any high spots.*

5 *Fill any indents left behind after you've drawn the tamping beam over the top with shovelfuls of concrete, then draw the beam across again.*

9 *Run the blade of a steel trowel along the perimeter of the concrete to prevent the edges crumbling when you remove the formwork.*

10 *After about 24 hours, when the concrete has set, remove the formwork by tapping it away from the slab with a hammer.*

you can smooth over the surface with a steel trowel after you've treated it with the wooden float, and once more when the concrete has almost set.

One of the simplest finishes to apply is to go over the surface with the back of a shovel, producing fine swirls.

Curing the concrete

After you've applied the finishing texture to your slab you should 'cure' the concrete by leaving it to set without drying out too quickly, which could cause it to crack.

Although you shouldn't attempt to lay concrete at all during frosty weather (as this can affect the strength of the slab) it's possible that a cold spell will strike when you're least expecting it. If this happens you can protect your freshly cast concrete by insulating it with a quilt of straw sandwiched between two layers of heavy gauge polythene, or you can shovel a layer of earth, sand or compost on top of your conventional curing sheet, which has the same effect.

Once you've cured the concrete properly, which normally takes about three or four days (ten in winter), you can remove the polythene. It's perfectly alright for you to walk on the slab, and you can even start to build onto it, but be very careful at the edges, which will still be weak and susceptible to chipping.

Don't put the base to full use for about ten days, when you can remove the formwork. To do this, tap it downwards with a hammer in order to release it from the slab, then knock it away from the face of the concrete edges by releasing the nails securing the butt joins at the corners.

Once you've removed the formwork you can fill in the gap it occupies with soil or lay turfs to continue your lawn.

READY REFERENCE

HOW MUCH CONCRETE?

To estimate the volume of concrete you'll need:

- first work out the area of your proposed slab by multiplying the length (A) by the width (B)
- then multiply the area by the slab thickness (C)

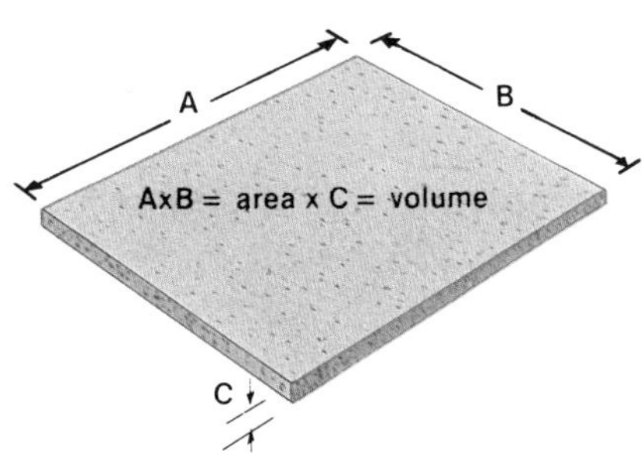

For example: a car port base 6x3m (20x10ft) by 100mm (4in) thick needs 6x3x0.1m which equals 1.8m^3 (63 cu ft) of concrete.

TIP: ROUND UP VOLUMES

For practical purposes always allow a margin for wastage, rounding up volumes to the nearest half or whole cubic metre.

CONCRETE MIX PROPORTIONS

To decide on the volumes of cement and aggregates you'll need, first decide on the mix proportions to use. For a general purpose mix suitable for a slab 75mm (3in) or over in thickness use the proportions of:

- 1 bucket of cement
- 2½ buckets of damp sand
- 4 buckets of coarse aggregate.

For 1m^3 (36 cu ft) of concrete you'll need:

- 6 bags of cement
- 0.5m^3 (17 cu ft) of damp sand
- 0.8m^3 (28 cu ft) of coarse aggregate.

If you buy combined aggregates (sand is included with the coarse aggregate) use the proportions of:

- 1 bucket of cement
- 5 buckets of combined aggregates.

For 1m^3 (36 cu ft) of concrete you'll need:

- 6 bags of cement
- 1m^3 (36 cu ft) of combined aggregates.

TIP: IF THE SAND IS DRY

Sand is nearly always delivered damp but if it's dry, reduce the amount of sand at each mixing by about half a bucket.

LAYING READY-MIXED CONCRETE

Ready-mix concrete takes a lot of the hard work out of laying a large slab for a drive or patio, or as the base for a garage. It's delivered to your house and takes only a few hours to place.

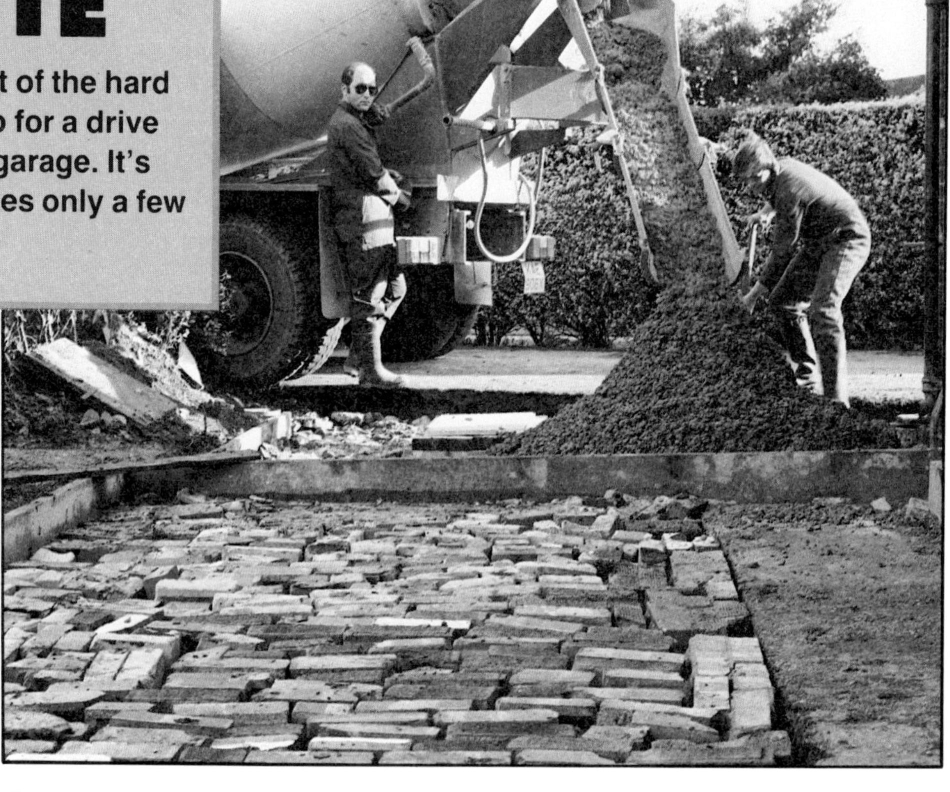

Concrete is a tough, hardwearing material that is ideal for laying as a base for a shed or garage, or as a durable surface for a patio, path or driveway. Once you've mastered the basic techniques of mixing, casting and curing the concrete (see Laying concrete slabs, pages 35-39) it's quite straightforward to make a fairly small square or rectangular slab, mixing the ingredients by hand or portable mixer. But if you're covering a much larger area the amount of concrete you'll need makes mixing your own impractical – unless you plan to lay the base in stages over several weekends.

Ready-mix concrete, which is sold in bulk and delivered to your home by mixer truck ready for casting, takes a lot of the hard work out of mixing large quantities of concrete and enables you to complete substantial projects quickly. Ready-mix has a number of other advantages over mixing your own: because the cement, sand, coarse aggregates and water are all correctly proportioned by the supplier and mixed for you, there's less likelihood that your mix will be too weak or brittle, or that you'll run out before you complete the slab. Also, you don't need to store large quantities of materials or hire mixing equipment

Casting concrete in bulk

Laying a large area of ready-mixed concrete follows the same sequence of operations as laying a small slab – you simply have to spread the mix to an even thickness over prepared foundations (see 'Foundations and formwork' opposite), apply a finishing texture and leave it to harden.

But there are some important details you should be aware of before you go ahead and order your concrete. You must provide sufficient access for the mixer truck and space for the mix to be dumped (see *Ready Reference*). If you can't arrange for the concrete to be dumped direct onto your foundations, try to get it as close as possible to your site. You must be sure that you'll be able to handle such a large amount of concrete, as well as cast it, compact it and cure it in one session, because you won't have a second chance once it has set.

Slab dimensions

The first consideration is what your slab will be used for, as this helps you decide on the correct thickness of concrete and allows your supplier to determine the correct strength of mix you'll need. Basically, the greater the load on the slab the thicker it must be. If your slab is to form the floor of a workshop or garage, or a driveway, for instance, a thickness of about 100mm (4in) is appropriate on ordinary soils. But on soft clay or other poor soils allow 125mm (5in).

On normal soils you may be able to lay the concrete direct onto the well-compacted ground, but on loose or soft soils you should include a 75mm (3in) layer of well-rammed hardcore. This can be of broken bricks or concrete, and should be topped with a 'blinding' layer of sand to fill in any voids that would be wasteful of concrete.

When you've decided on the position of your slab, and its dimensions, draw a scale plan of it on squared paper to help you estimate how much concrete you'll need.

Planning a drive

If you're altering the position of an existing vehicle access on to your property to make a new drive, you'll have to apply for permission to do so from your local authority; they'll need to make a dropped kerb and a cross-over of the pavement from the road, for which there's usually a charge.

Pay particular attention to the slope of the ground when you're planning a drive. Acute changes from one gradient to another within the drive can cause your car to hit the ground at either end or underneath.

Check also that your proposed drive allows sufficient clearance for an up-and-over garage door or that there's plenty of room for side-hinged doors to open without binding on the ground.

Draining the slab

Your slab must incorporate a slight slope to allow rainwater to run off quickly and you must allow for this when you're preparing your base.

A minimum fall of 1 in 60 to one side is adequate for a large slab, although you can form a high point along the centre of the slab with falls to both sides.

If the ground slopes naturally and your slab would drain towards a garage or house wall you should excavate it to a level below that of the floor of the building, and make a short slope away from the wall, with a gully or gutter at the lowest point to ensure run-off of rainwater to a suitable drainage point (see *Ready Reference*, page 43).

Excavating and setting out

When you've decided where you're going to lay your slab you'll have to prepare the foundations (see Laying concrete slabs). Dig out and remove the top-soil and any grass or roots within this area. Allow a margin

FOUNDATIONS AND FORMWORK

Before you take delivery of your ready-mix you must prepare the base. On firm ground simply roll the surface and cast the concrete direct; on soft ground first add about 75mm (3in) of well-rammed hardcore topped with sand to fill any voids. Set up a frame of timber formwork to mould and retain the mix while it hardens.

hardcore
formwork to support expansion joint
expansion joint of 12mm (½in) softwood
softwood pegs 50x50x300mm (2x2x12in) at 1m (3ft) centres
75mm (3in)
softwood formwork 25mm (1in) x slab thickness
corners butt-jointed
pegs 1m (3ft) apart

A *Drive stout softwood pegs at the perimeter of the foundations and nail the form boards to them from inside.*

B *When joining two lengths of formwork, butt the ends together and nail them to two pegs placed side by side.*

C *Check levels with a spirit level on a straightedge; allow for a drainage fall with a small wedge under one end.*

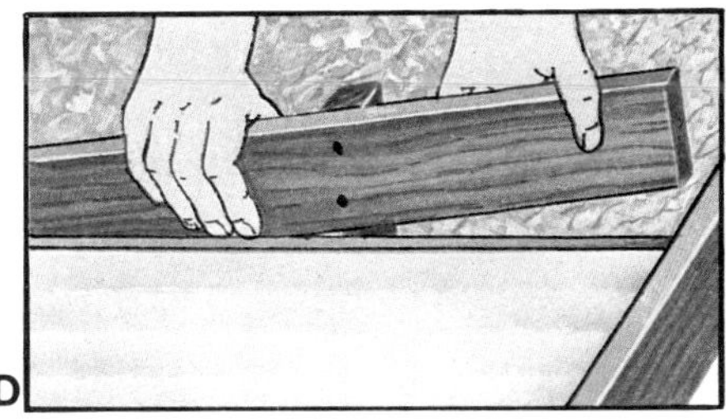

D *To form an expansion joint, cast one bay to the wood fillet, remove the form board backing it and cast the second bay.*

READY REFERENCE

ORDERING READY-MIX

To enable your ready-mix supplier to work out an appropriate concrete mix and give you a price quotation, he'll need to know:

- the volume of your proposed slab
- what you plan to use the slab for
- what time you'd like it delivered
- how you'll handle it on delivery
- what access to the site there is.

ACCESS FOR THE TRUCK

You can have your load of ready-mix dumped direct onto the foundations or as close as possible to the site, depending on access. To help you decide if you have enough room, a typical truck measures:

- about 8m (26ft) long
- 3m (10ft) tall
- 2.5m (8ft 6in) wide.

It features:

- a 2m (6ft) long chute for discharging concrete anywhere in a semi-circle behind the truck
- a 1m (3ft) extension chute for delivery to awkward or more distant sites.

Be particularly careful if you have any overhead electrical or telephone cables, which could easily be damaged as the truck manoeuvres.

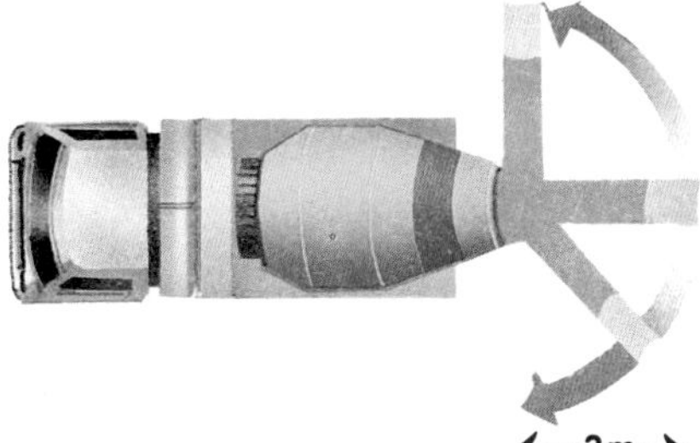

TIP: LAYING ON A SLOPE

If you're laying your slab on a slope:

- tell your supplier so that he can give you a stiff mix
- work up the slope when you're tamping, keeping a close watch on levels so that the mix doesn't bulk up below the beam
- if you're working down the slope make sure the mix doesn't slump downwards from the formwork.

CASTING READY-MIX CONCRETE

1 *Try to have your load of ready-mix concrete dumped direct onto your foundations. As soon as it's been unloaded, transfer it to barrows.*

2 *Transport the concrete to the far end of the slab, tip it onto the hardcore then spread it out. A sheet of hardboard will protect doors from splashes.*

3 *Spread out the concrete using a garden rake or a shovel. Work the mix well into the corners of the formwork and at the sides to avoid air pockets.*

6 *Fill any depressions in the concrete with shovelfuls of fresh mix, then pass over the area again with the tamping beam to level the surface.*

7 *Continue to tamp with a chopping action then work back with a sawing action to level the surface, drawing along a ridge of excess concrete.*

8 *When you've compacted the slab you can apply a finishing texture. For a polished surface, trowel over the concrete using a steel float.*

of about 600mm (2ft) around your proposed slab so you've enough room to work.

Mark out the shape of your proposed slab on the ground with strings stretched between wooden pegs, and use a builder's square to set the corners accurately.

Next, dig out the sub-soil to roughly the depth of the foundation and take it about 150mm (6in) beyond the string lines to leave space for setting up the timber formwork, which you use to mould and retain the concrete while it's hardening.

To ensure that your finished slab will be perfectly level over its entire area use 50x25mm (2x1in) wooden 'datum' pegs, taken from a 'prime datum', or fixed point that establishes the level (see pages 35 -39 and 64 -67).

Drive the pegs into the ground at 1.5m (5ft) intervals and check across their tops from the prime datum with a spirit level on a straight-edged length of timber to check their level. Remember to incorporate a slight slope for drainage; you can allow for this by placing a 25mm (1in) wedge of timber called a 'shim' under one end of the straightedge (see 'Foundations and formwork', page 41).

Measure down the pegs whatever thickness of concrete you'll need for your slab, plus about 75mm (3in) for hardcore, and excavate or fill in the ground where necessary.

If there's a drainage pipe within the area of your proposed slab it must be at least 150mm (4in) below the underside of the concrete; if it isn't you'll either have to divert the pipe run or alter the level of your slab. If you don't need to move the pipe you should, nevertheless, dig out the soil about 150mm (6in) at each side of it and fill in with pea shingle. Lay shingle rather than hardcore directly over the pipe too.

Where there's a manhole within your slab you'll have to remove the frame and cover, then reset them at the new level, perhaps even raising the manhole brickwork itself by one or more courses (see also step-by-step photographs, page 44).

The original cover may be a lightweight one that's unsuitable for supporting heavy loads, such as your car, and you may have to buy a new, stronger one.

Fixing the formwork

A timber frame called 'formwork' (see 'Foundations and formwork', page 41) is used to retain and support the edges of the wet concrete, which tends to spread out as it sets. Use straight lengths of 25mm (1in) thick sawn timber as wide as the depth of the slab, set on edge and nailed to 50x50mm (2x2in) pegs driven into the ground at the perimeter of the slab at 1m (3ft) centres.

Position the formwork using your perimeter

4 *If you're laying your slab adjoining the house you'll have to cast it in easily manageable 'bays' and compact the mix with a tamping beam as you work.*

5 *Work along the slab using the tamping beam in a chopping action to compact the concrete. This flattens high spots and reveals depressions in the surface.*

9 *The easiest finish to apply is to smooth over the surface of the fresh concrete with the back of a clean shovel, producing fine swirls.*

10 *For a more regular, yet still non-slip finish, draw a stiff-bristled broom gently across the slab to produce a finely ridged texture.*

strings as a guide and make sure their tops are flush with the finished level of your proposed slab, using a spirit level and shim to incorporate a drainage fall, as previously described. Set the corners at right angles using a builder's square (see *Ready Reference*, page 35). All lengths of timber should be tightly butt-jointed to prevent the concrete seeping out.

Movement control joints

A slab that's longer than about 3m (10ft) in length or width must include a 'movement control' joint, which prevents the slab from cracking due to expansion or contraction.

To make the joints, divide the slab into equal-sized bays and fill the gap with a length of preservative-treated softwood fillet 10 to 12mm (3/8 to 1/2in) thick, matching the slab depth, and cut to fit between the formwork. Back up the fillet with temporary formwork and pegs and cast each bay separately: once you've cast and compacted the first bay, remove the temporary formwork and cast the second bay, leaving the fillet permanently in position, and so on along the slab.

If the slab is to abut a wall or existing paving use a length of bituminous felt as a dpc and movement joint instead of timber.

Estimating and ordering ready-mix

Ready-mix is usually sold and delivered in minimum loads of 3cu m (105cu ft) and in 1/4cu m increments above that volume. To calculate how much concrete you'll need you must work out the volume of the proposed slab: multiply its length by its width (both in metres) to get the area; then multiply this figure by the slab thickness in mm, and divide the answer by 1000 to give the volume in cubic metres. For example a slab measuring 5x4m = 20sq m in area; at a

READY REFERENCE

DRAINING THE SLAB

Your slab must slope slightly to one side, or both sides, for drainage of rainwater. If the slab slopes naturally towards a garage or house wall:

- excavate the foundations to a point below the floor level
- slope the slab upwards to the building
- make a gutter or gully at the lowest point to drain water to one side.

COMPACTING THE CONCRETE

To compact and level a large concrete slab you'll need a tamping beam. Make one from a straight-edged length of 150x50mm (6x2in) timber, with handles bolted on at each end. To use the beam:

- work across the slab with a chopping action
- work back with a sawing action.

For more information about laying concrete see pages 26 -29 and 35 -39.

thickness of 150mm the volume is 20x150 = 3000 ÷ 1000 = 3cu m.

Ordering ready-mix is straightforward: you can, in fact, leave most of the calculations to the supplier, although you should make sure you'll receive a little extra in case of mis-calculations. He'll want to know the volume of concrete you'll need, at what time you want it delivered, what it'll be used for – hard standing for a car, for instance, or simply garden furniture or foot traffic – and how you're going to handle it on delivery. With this information he'll be able to determine an appropriate mix and give you a price. Obtain several quotations and try to choose a local depot because much of the cost of ready-mix is in its transportation.

You'll also need to discuss with the supplier access for the lorry and whether the load can be discharged directly into the formwork. If this isn't possible you'll have to arrange for a suitably sized space for dumping which won't cause an obstruction either to you or others. In this case you should lay a large sheet of thick-gauge polythene on the ground first to make clearing up easier afterwards.

Coping with delivery

If your ready-mix is going to be dumped some distance from the proposed site – or if the slab is very long – you'll need to gather as many helpers with heavy duty wheel-barrows as you can to transport the wet mix, (it takes 40 barrow loads to move 1 cu m). Lay down scaffold boards as runways for the barrows if you've to transport the mix up steps or across rough or soft ground. Wear old clothes or overalls, Wellington boots and gloves, when laying the concrete.

Placing and compacting concrete

Casting the concrete is a job for at least two people. But you'd be wise to enlist further aid as timing is critical.

Barrow the concrete to the furthest end of the foundations and tip it in. Work it well into the corners and around the edges with your boot or shovel to avoid air pockets. You can use your shovel or a garden rake to spread out the concrete to about 12mm (½in) above the level of the edge boards; this allows for settlement during compaction.

Try to organize your helpers as a team, using some to cast the concrete while two others follow along compacting the mix. Compaction is essential for the strength and durability of the mix; and it's done using a tamping beam (see *Ready Reference*). Position the beam so that it spans the edging formwork; and move along the slab about half the beam's thickness each time with a chop-ping action in a steady rhythm.

This tamping action will show up any high spots in the mix, which you can disperse by changing to a sawing action across the slab. Use this excess to fill in any depressions that are left by the tamping beam, and go back over the surface to level it off. Repeat the tamping process until the concrete is level with the top of the formwork.

If you're unable to stand at each side of the slab to use the tamping beam – if it's against a wall, for example – you can cast and com-pact the concrete in narrow strips.

CASTING ROUND A MANHOLE

1 *If there's a manhole within your new slab you'll have to reset the lid at the new height. Remove its temporary cover when the mix has stiffened.*

2 *Build up the level of the manhole using bricks or, if the difference isn't too great, mortar. Set the lid frame in mortar, checking that it's level.*

3 *Once you've set the frame of the manhole lid accurately you can cement it into place so that it's flush with the new surface of your slab.*

4 *Remember to choose a manhole cover strong enough to support your loads. Apply a finishing texture to the slab and leave it to set hard.*

Finishing the concrete

There are a number of finishes you can apply to the concrete slab. The final pass with the tamping beam can, for example, make a slightly ridged non-slip surface texture for a drive. Alternatively, you could brush the fresh concrete across its width with a stiff bristled broom to give a finer, more regular but still non-slip, finish.

A smoother, polished effect is easier to keep clean for a shed, workshop or garage floor and you can achieve this texture by trowelling the surface with a wooden or steel float in a wide, sweeping action.

One of the simplest finishes, however, is to go over the surface with the back of your shovel, which produces a swirling pattern.

Curing the concrete

Together with thorough compaction, curing is essential for a durable concrete slab. This means that it shouldn't be allowed to dry out too quickly, when it could crack. As soon as the surface has set enough not to be marked easily, cover the entire slab with a polythene sheet, tarpaulins, wet hessian or sacking.

Curing will take about three to four days (ten in winter, though you shouldn't really attempt concreting then unless there is no risk of frost), after which you can remove the covering. You can remove the formwork and use your slab after about ten days.

CHAPTER 3

PATHS, STEPS AND PATIOS

One of the best areas to practise your building skills is in the garden. Absolute perfection won't matter so much, it doesn't matter if you make a mess and to begin with you only have to work in two dimensions. Garden paths and patios can be laid extremely quickly using prefabricated slabs or bricks, and once you've mastered the art of laying them level you can move on to building simple freestanding or built-in steps. From there it's a relatively short step to building garden walls and creating exactly the garden you want.

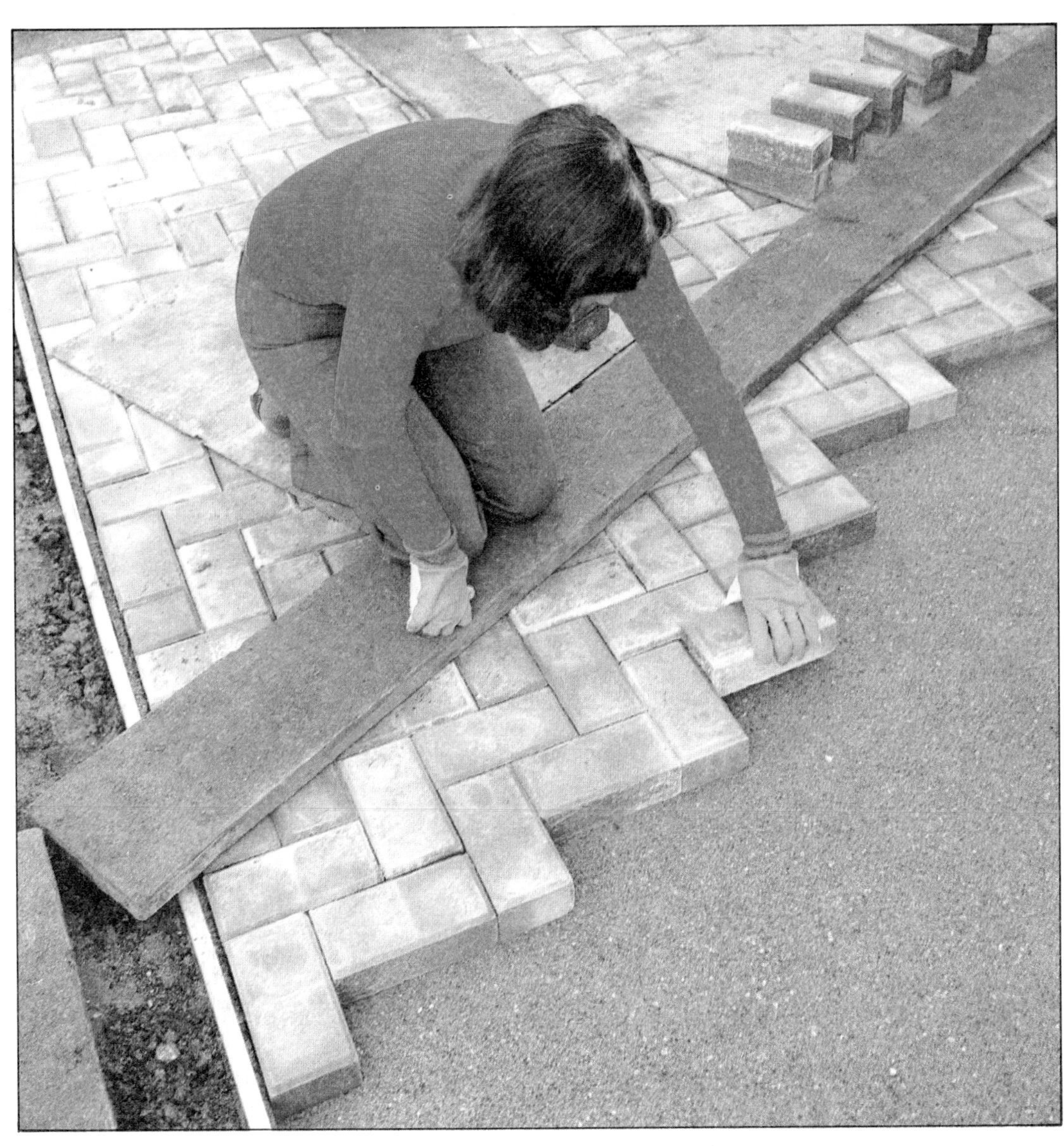

LAYING PATHS WITH PAVING SLABS

Of all the materials you can use to build a path, slabs are among the simplest to lay. The large size of the individual units means a path should not take long to complete and the range of slabs available gives you a wide choice when deciding how your path will look.

Paths are made for going places and while their function might be to prevent mud being trampled into the house or to get a wheelbarrow to the garden shed without making furrows in the lawn, how they look in relation to the garden and your house is also important.

A wide range of attractive paving materials is available for you to choose from. This section deals with the techniques for laying pressed concrete slabs. Techniques for the smaller shapes such as bricks and concrete blocks (pavers) are covered on pages 50 -53. and crazy paving on pages 61 -63.

Planning a path

Any path should have a purpose. There's little point, for example, in laying a path that skirts the garden and then seeing it ignored as short cuts are taken across the lawn.

You should also make sure your choice of material blends with the surroundings. Concrete slabs, for instance, can look out of place if you have a lot of brick walls, whereas crazy paving might complement them. If the lawn is large and you want a path straight across it, an unbroken length of slabs might look too prominent and it might be preferable to use stepping stones with areas of grass in between (remember to relate the spacing of the stones to a normal walking pace or you will defeat the purpose of the path). If the garden is dotted with trees and shrubs it might be more eye catching to curve the path around them so that it doesn't dominate the setting.

When you're designing, think of the width as well as the length — a path that's too narrow to walk on easily will remain a source of irritation. If you make it too wide it might give the garden an unbalanced look, though a wide path can look very good if flanked on both sides with an array of shrubs or flowers.

PREPARING A HARDCORE BASE

1 Set up string lines to mark the trench edges about 50mm (2in) wider than the finished path.

2 Dig out the trench so it is just a little deeper than the thickness of the slabs you are going to use for paving.

TIP

3 Using a stout timber pole, tamp the dug-out area to compact the ground and provide a firm foundation on which the path can be built.

4 Use a timber straight-edge and a builder's level to check that the trench is flat. Inset: If the ground is still soft, dig down another 75mm (3in).

5 Fill the extra depth with a layer of broken brick (rubble) or concrete (hardcore) and compact this so it is firmly bedded down.

6 Fill gaps in hardcore with sand. You can then move string lines in to mark path edges. To lay slabs on sand, add an extra 25mm (1in) thick sand bed.

LAYING ON SAND

1 *For butt joining, place edge of slab against other slabs, lower it into place and tap it into position (inset) with the shaft of a club hammer.*

2 *Check the horizontal with a builders' level and make sure the slight drainage fall is even. Over several slabs use a timber straight-edge.*

3 *Fill the gaps left between the top surfaces of the slabs by carefully brushing a mixture of soil and sand into the cracks.*

4 *To complete the job, fill the gaps at the path sides with soil and let the grass grow back, or infill with soil and then replace turf on top.*

The best way to start planning is on paper. Use graph paper to make a scale plan of the garden, marking in any fixtures such as established trees and a shed or greenhouse and obvious targets for the path such as a gate or the washing line. Draw them in ink and use pencil to plan in path shapes — they can always be rubbed out if you change your mind.

The plan will give you something to work to as well as a method of calculating the number of slabs and the amount of sand and cement you'll need. But first you'll have to decide on the pattern you want and the type of slab (home-made or bought), whether you want grass to grow between the cracks or whether you prefer the overall look that formal pointing will give.

Paving shapes

The most common concrete paving slabs are square or rectangular in shape, though you can also buy them circular or as parts of a circle (called radius slabs). These are useful for curved or meandering paths which are difficult to make with formwork. Hexagon-shaped slabs look good, too, and these can be married up with half hexagons which give a straight edge for the path's borders.

Concrete slabs can be bought in a variety of colours — anything from red, green and yellow to brown and the ordinary 'cement' grey. Some concrete slabs which are patterned to look like brick or natural stone are finished with a blend of two colours — grey over deep red and grey over buff. But the important thing to remember about any coloured slabs is that the colours won't always last. The pigments are added to the concrete during manufacture, and in time they will fade with the effect of sun and rain. In damp shady spots under trees, lichen will grow on the surface and diminish the original colours. Some slabs may also show signs of staining as a result of efflorescence — white powdery deposits brought to the surface as water dries out of the concrete. Brushing will remove the deposits temporarily.

READY REFERENCE

WHAT SIZE SLABS?

Square and oblong slabs come in a range of different sizes. The commonest are:

- 225 x 225mm (9 x 9in)
- 450 x 225mm (18 x 9in)
- 450 x 450mm (18 x 18in)
- 675 x 450mm (27 x 18in)

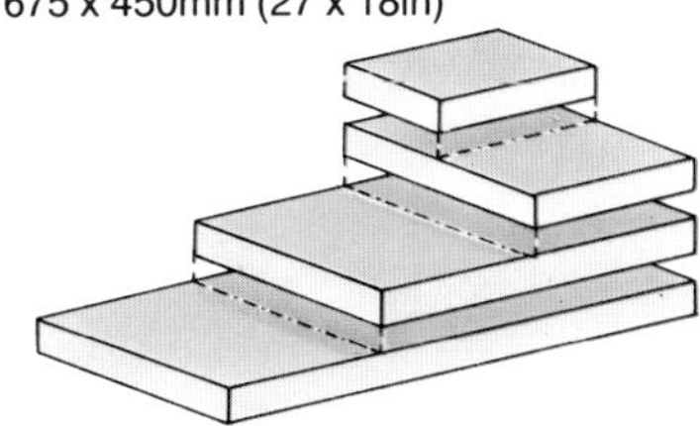

Some slabs are based on a 300mm (12in) unit, so squares are 300 x 300mm or 600 x 600mm (24 x 24in), and rectangles 600 x 300mm (24 x 12in) or 900 x 600mm (36 x 24in). Slabs over 450 x 450mm (18 x 18in) are very heavy.

Hexagonal slabs are usually 400mm (16in) or 450mm (18in) wide. Half slabs are also made, either cut side to side or point to point, and intended to be laid as shown in the sketch.

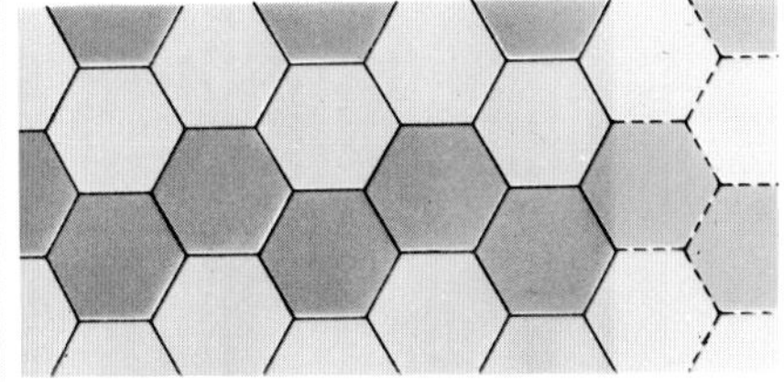

Both rectangular and hexagonal slabs are generally 38mm (1½in) thick.

LAYING METHODS

Light-duty paths – for walkers only – can be laid with slabs bedded on sand about 25mm (1in) thick. The joints should be filled with sand or soil, not pointed with mortar.

Heavy-duty paths – for wheelbarrows, rollers and heavy mowers – should have slabs bedded in stiff mortar (1 part cement to 5 parts sand) and mortared joints.

TIP: STACKING SLABS

To avoid chipping corners and edges and marking the slab faces, stack slabs on edge in pairs, face to face, against a wall, with their bottom edges on timber battens.

LAYING ON MORTAR

1 *For heavier duty use, lay slabs on pads of mortar. Place the fist-sized pads on the path bed ready to take the slab.*

2 *Lower the slab into place so there are 19mm (3/4in) wide gaps between it and neighbouring slabs. Use timber spacers to give correct joint width.*

3 *Use the shaft of a club hammer to tap the slab into position – a builders' level will tell you if the surface is even or needs adjusting.*

4 *With a timber straight-edge and a builders' level check the level across the path, making sure the drainage fall is not too abrupt.*

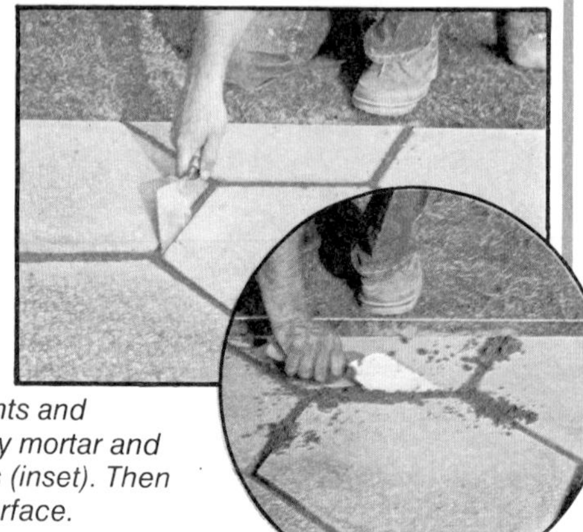
5 *Brush a dry 1:5 cement:sand mix into the joints and sprinkle them with water, or* **6** *Mix up a crumbly mortar and use a pointing trowel to press this into the joints (inset). Then draw the trowel at an angle along the mortar surface.*

Patterns of laying

You may decide on a simple chequerboard pattern using one size of slabs or a pattern with staggered joints as in stretcher bond brickwork. Alternatively you can create a more decorative path using different sized slabs. Riven surfaced slabs can be particularly effective if two sizes of slabs are used with the larger slabs set to radiate around the smaller ones, producing a square which is repeated down the length of the path. For further suggestions on laying patterns see the slab manufacturers' literature.

Cutting slabs

If the pattern you've worked out requires cut slabs (it's helpful and certainly more easy if they're half sizes), the cutting is relatively easy. After you have marked the cutting line all round the slab you place it on a bed of soft sand or even on the lawn (anything to absorb the shock) and cut a groove along the cutting line, using a bolster chisel and a club hammer. You can then split the slab by tapping the bolster with the hammer along the groove.

Cutting sections out of paving slabs is not so easy. Chipping to shape is time consuming and cast slabs are likely to fracture anyway. It's worth considering filling L-shaped gaps with two separately cut pieces, or leaving out the paving slab altogether and infilling with pebbles, stones or even cobbles set in mortar, or simply finishing off with bricks.

If you want a perfect finish for cut pieces and have a lot of cutting to do, it is worth hiring a masonry saw from a plant hire shop. Although it's possible to fit masonry cutting discs to an ordinary drill, with a lot of cutting you run the risk of burning out the motor.

Buying paving slabs

Visit local garden centres and builders merchants to see what sizes, colours and textures they have in stock. It's always worth shopping around. Your supplier should be able to give you helpful information — for example, some coloured slabs are more colour-fast than others, and he should know which ones. If local suppliers don't have what you want, remember that the cost of transporting heavy slabs over a long distance is high, so it may be better (or at least cheaper) to choose from what is available.

Prices will obviously vary depending on the type of slab — for example, hydraulically pressed slabs are more expensive than cast slabs. And when ordering, allow for a few more than the exact number required for the path; you may crack one or two during laying so it's better to have spares handy.

Preparing the base

Making a flat base is the single most important step in laying the path. And to do this you will usually have to dig out a shallow trench along the line you want the path to follow.

Digging out the topsoil, roots and any organic matter needs to be done carefully

CUTTING SLABS

1 *Mark the cutting line right round the slab with chalk; use a straight-edge so you can mark a straight line the right distance from the sides of the slab.*

and you should dig the trench to a depth just a little deeper than the slab thickness. As well as being flat, the laying surface must be firm and compacted. At this stage, an easy way of checking that the trench is flat is to use a length of straight-edged timber — a plank or a fence post, for instance — to indicate hollows or bumps which might not be obvious to the eye.

Once the trench is dug you may find that because of the type of soil, the surface is still soft. The answer is to dig out another 75mm (3in) or so and then fill the extra depth with a layer of broken brick (rubble) or broken concrete (hardcore). This layer has to be well compacted before a layer of sand or fine ash (called a 'blinding' layer) is spread on it to provide a smooth surface. If you don't want anything to grow up through the path, saturate the trench with a powerful weed-killer.

Laying paving slabs

The easiest way to control the line of a path as you begin to lay the paving slabs is to set up string lines to mark out the edges. How the slabs are bedded — whether on sand or pads of mortar — depends on the weight of traffic the path will carry and whether you intend to point the gaps between the slabs with mortar.

As you lay the slabs, use a timber straight-edge to check that each slab sits flush with its neighbours across and along the path. On level ground, you must lay them so that there is a slight slope across the width of the path — a drop of about 25mm (1in) across 1 metre (just over 3ft) will be sufficient. Check the slope by placing a 25mm thick block of wood under one end of your spirit level or batten; the bubble should then be in the 'dead level' position. On sloping ground, the slabs can be laid dead level across the path width to achieve the same effect.

Slabs can sometimes be butt jointed tightly together. However, because there is often some slight variation in the sizes of slabs, it makes sense to allow for a joint of about 9mm to 12mm (3/8in to 1/2in) to take up these minor inaccuracies. It's important that the joints be kept even — for they act as a frame for the slab shape. Spacers cut from board will give you the desired joint thickness and will also prevent the newly laid slabs closing up as you lay adjacent ones.

Finishing off

When you have positioned all the slabs you can fill the joints. If you are not pointing them, simply brush a mixture of soil and sand into the gaps. Where you want a pointed finish there are two methods you can use — and both require care or mortar stains will mar the slabs.

One method is to mix the cement and sand dry — the sand needs to be very dry — and pour or brush this into the joints. You then sprinkle the joints with a watering can fitted with a fine rose, or wait until it rains.

A better method is to mix up dry crumbly mortar and press this into the joints with a pointing trowel. Any mortar crumbs falling onto the face of the slab can easily be brushed away without staining. You can also use a piece of wood or a trowel to finish the joint so it is slightly recessed.

After pointing is completed don't walk on the path for a few days — if you tread on an edge of a slab you may loosen it and you will have to lift it (using a spade) and lay it again on fresh mortar.

To finish off the gaps at the edges you can point them where they adjoin masonry or a flower bed; where they run alongside a lawn, fill them with soil and let the grass grow back, or fill space with gravel which will drain away excess water.

2 *With the slab on a surface which will absorb the shock, use a bolster chisel and club hammer to cut a groove along the line, including the slab edges.*

3 *Place the slab face up and work the bolster back and forth along the groove, tapping it with the hammer as you go, until the slab splits in two.*

READY REFERENCE

HOW MUCH SAND?

For a 25mm (1in) thick sand bed, 1/2 cu metre (2/3 cu yd) of sand will cover about 20 sq metres (215 sq ft) – a path 20 metres long and 1 metre wide (22yd x 3ft 3in).

HOW MUCH MORTAR?

One 50kg bag of cement, mixed with 24 2-gallon buckets of damp sand will make enough mortar to lay about 30 slabs using the 'dab' method. One quarter of this mix will fill the joints between about 60 slabs measuring 450 x 450mm (18 x 18in).

GETTING THE RIGHT LEVEL

A path alongside the house should ideally be two courses of brickwork below the damp-proof course (dpc). If this is impossible, make a shallow, gravel-filled gutter about 150mm (6in) wide between the path and the wall to prevent rain from splashing the wall above the dpc.

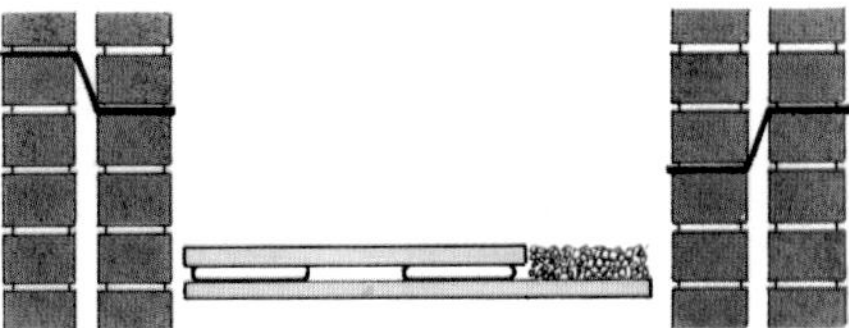

If the path runs next to a lawn, the finished level should be below that of the grass so you can mow over the edge without the blades catching the masonry.

ALLOWING FOR DRAINAGE

Horizontal paths should have a slight slope across the path width to allow rainwater to drain off the surface. A fall of 25mm (1in) per 1 metre (3ft 3in) of path width should be sufficient: use a block of wood under your spirit level or batten to check that the fall is correct.

TIP: HANDLING SLABS

Slabs are heavy. To avoid injury

- grip the edges of the slab firmly
- lift them with your knees bent and back straight, or
- hold the top corners of the slab, and 'walk' it on its bottom corners
- wear heavy shoes to protect your feet in case you drop a slab on them
- wear gloves to protect your hands when handling slabs with a very rough surface.

LAYING PATHS WITH BRICKS

Paths need to be functional but they should also contribute to the overall appearance of their surroundings. Brick and concrete pavers come in a variety of natural colours and can be laid in patterns to suit your style of garden.

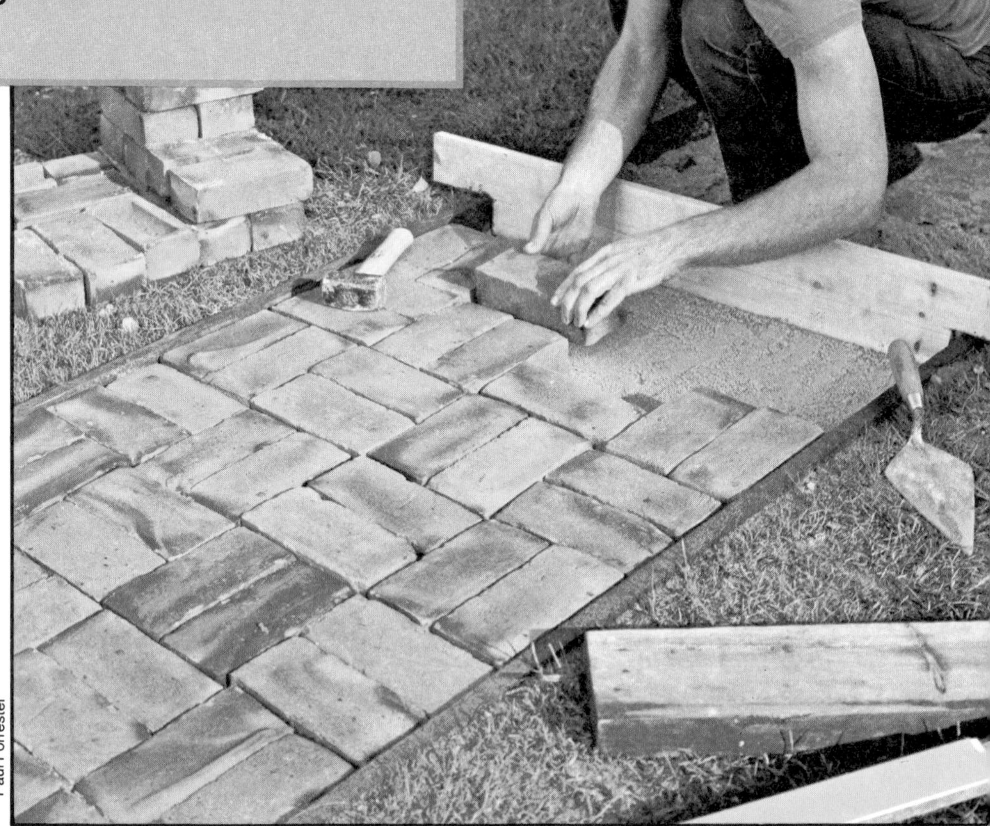

Paul Forrester

When you want to build a path with a small-scale pattern you can use bricks or concrete block paving. They can also be used to break up the larger scale pattern of a slab path. Because they are small units you can lay them to gently rolling levels and in restricted spaces where it would be awkward to lay larger slab materials.

Buying bricks and concrete pavers

These materials should be obtainable from a good builders' merchant but, with such a variety on the market, they may have to be ordered. If you live near a brickmaker who makes paving bricks it may be worthwhile enquiring direct – be sure to make it clear you want bricks of paving quality.

Clay brick pavers are produced in a variety of colours from dark red through buff to dark blue/black and a range of finishes both smooth and textured. Calcium silicate are not recommended for paving as their edges are likely to chip. Concrete bricks come in a good range of colours with smooth and textured surfaces while concrete paving blocks tend to be light in tone. Colours include buffs, greys and light reds.

Planning the path

As with paths made from other types of materials, a path of brick or concrete pavers needs to meet functional and design requirements so it will serve its purpose and be an attractive feature of the garden.

Draw up a scale plan of the garden and work out where the path will go. You can also use this to assist you in working out the quantities of materials you will need. Work out patterns that will avoid unnecessary cutting of the pavers. For dry-laid paving, interlocking is an important feature and patterns which avoid continuous straight joint lines are preferable.

When deciding on your design remember that these small units of paving can be used on their own. You may feel, for example, that the bond pattern, colour and textural patterns of clay bricks give sufficient surface interest without recourse to mixing brick types – you might, in fact, find laying rather difficult if bricks were intermingled.

Preparing the base

The foundations on which the paving is laid must be properly prepared to ensure a long-lasting path. You will have to dig out grass, soil and roots – since there will be a granular sub-base topped with a bed of sand plus a layer of pavers, you should dig down to at least 225mm (9in) below the finished level to allow for the thickness of construction.

When you are calculating the depth, remember that if the path runs next to your house the finished level should ideally be two courses of brickwork below the dpc (damp proof course) in the wall. If it's alongside a lawn, make sure its surface is below that of the grass for convenient mowing. Don't forget that on a level path you will have to allow for a drainage fall to one side of the path. Where the path adjoins a wall, the slope should be away from the wall.

To set out the levels, slopes and edge lines you can use timber pegs and string lines. To work out slopes and levels you will need a length of straight-edged timber and a builder's level.

The sub-base should consist of hard-core, which is available through sand and gravel suppliers or builders' merchants. The levels and gradients should be formed in this material so the bed of sand in which the pavers or bricks are laid can be spread evenly over the whole area.

Laying paving dry

For an even, firm path you must take care when you lay the sand base to ensure consistent compaction and perfect level. To give a regular finished surface the sand should be exactly the same throughout the work so allow it to drain before use. Cover it over during storage to minimise variation in moisture content.

You'll also have to include edge restraints for dry-laid paving at both sides of the path. These can be of creosoted or preservative-

PREPARING THE BASE

1 *Stretch strings between pegs to mark the edges of the path. Include a margin for the timber edge restraints on dry-laid paths.*

2 *Cut along the string lines using a spade or, if making a path across a lawn, a turfing iron. Use this tool to lift thin rectangles of turf.*

3 *Dig down to a level that will take a 50mm (2in) sand bed and, if the soil is spongy, a hardcore base. Compact the base using a fence post tamper.*

4 *Set creosoted boards at the sides of the trench as edge restraints. On loosely packed earth, nail the boards to stakes driven into the ground.*

5 *Check the level of the boards across the path using a builder's level. On a lawn, lower the path slightly to make mowing the grass easier.*

6 *When you have levelled the base of the trench and set the edge restraints, shovel in washed sharp sand, which forms a level bed for the bricks.*

7 *Spread out the sand over the path using a garden rake. Work only a small area at a time to avoid walking on newly-levelled base.*

TIP

8 *Make a timber spreader to level the sand bed by drawing it along the path, resting on the edge restraints and fill any hollows and level again.*

Paul Forrester

READY REFERENCE

PAVING MATERIALS

Brick pavers are made in a variety of sizes up to 225mm (9in) square and up to 65mm (2½in) thick. Some are shaped to interlock; other have chamfered edges to reduce chipping.

Clay building bricks must be paving quality (frost-resistant when saturated). Standard size is 230 x 110 x 76mm (9 x 4½ x 3in).

Concrete building bricks are the same size as standard clay bricks.

Concrete paving blocks are rectangular (200 x 100 x 65mm) or square (225 x 225 x 65mm) and edges of face side are chamfered. Special interlocking shapes are about the same size overall.

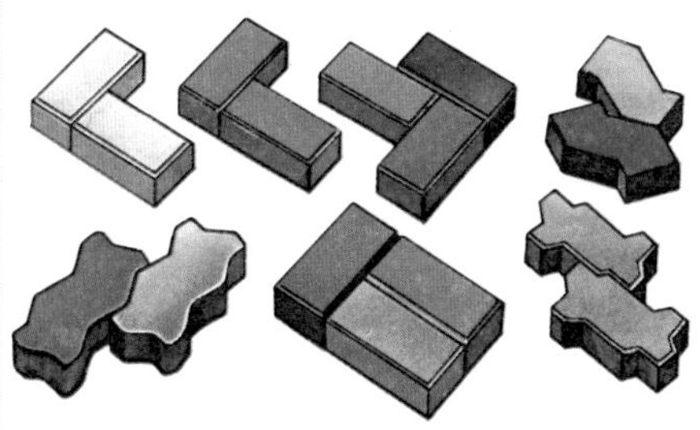

LAYING METHODS

Dry-laid (without mortar) is best for interlocking brick and concrete pavers. Needs sub-base of hardcore 100mm (4in) thick and 50mm (2in) thick bed of sharp sand.

For 1 sq metre you'll need
- 40 brick-size units laid flat
- 60 brick-size units laid on edge
- 50 small concrete pavers

Mortar bed and jointed technique can be used for standard bricks. Needs a 75mm (3in) compacted hardcore sub-base, a blinding layer of sand or fine ash under 50mm (2in) thick mortar bed (1:4 cement and sand) with 10mm (⅜in) mortar joints. To lay and point 3 sq m (105 bricks laid flat) or 2.5 sq m (120 bricks on edge) you'll need:
- 40kg bag of cement
- 17 2-gallon buckets of damp sand

TIPS: STOPPING WEED GROWTH

- under bricks place plastic sheeting on top of the hardcore
- under concrete use a long-term weedkiller

LAYING BRICKS DRY

1 Begin by laying the bricks in your chosen pattern – a basketweave design is shown here – butting up each brick to the next one.

2 Tap the bricks into place using the handle of your club hammer but don't exert too much pressure or you risk making the sand bed uneven.

3 If any of the bricks are bedded too low, remove them and pack more sand underneath, then level again.

4 Bed the bricks level using a stout timber batten held across the path, which you can tap with a club hammer, then check with a spirit level.

5 Hold your spirit level on top of the batten and check that the path is bedded evenly, incorporating a drainage fall to one side.

6 Brush sand over the surface of the path when fully laid, using a soft-bristled broom to fill the crevices between the bricks.

Paul Forrester

treated timber, or of bricks or kerb stones set in mortar.

The base should be no less than 50mm (2in) thick. When the finished surface is bedded in some of the sand will be forced up into the joints from below so the depth of the sand layer will be effectively reduced. And to complicate matters, moist sand 'bulks' in volume – the moisture acts on the sand particles to give them a fluffy texture – so the thickness of the sand may seem more than it will, in fact, be when the bricks or blocks are firmly bedded down in place. In this case you may have too little sand and will have to compensate. However, you can have too much sand – a layer which is too thick may cause surface undulation.

It's a good idea to add an extra 6mm (¼in) to 15mm (⅝in) of sand to the 50mm (2in) thickness to accommodate any unevenness. After you have levelled the first few metres of sand you can check to find out if you have added too much or too little and compensate if necessary. You should check again at frequent intervals as you continue levelling.

You will need a stout timber straight-edged board with notched ends to level the sand surface. The ends fit over the existing edge restraint and you draw the board along over the sand so that it is evenly spread over the area to be paved. This also ensures that the surface is properly compacted.

You should find laying the paving units quite simple provided you take care how you position the first few pavers. Each paver should be placed so it touches its neighbour – be careful not to dislodge a laid paver from its position – accidental and unnoticed displacement will have a multiplying effect. Similarly, don't tilt any of the laid pavers by kneeling or standing on them as the depressed edge will distort the level of the sand.

Where whole bricks or pavers do not fit at the edges, fill the spaces by cutting whole units to the required size. Gulley entries and manholes can also be dealt with by cutting bricks or pavers to fit. Alternatively, very small areas with a dimension of less than 40mm (1½in) can be filled with a 1:4 cement:sand mortar.

When all the bricks or pavers are in place you will have to bed them securely in the sand. For this you can use a stout timber straight-edge and a club hammer or, particularly useful where you are paving a large area, you can hire a mechanical plate vibrator. Once the bricks or pavers are bedded, you brush fine sand over the paving and again go over the surface with the machine to vibrate sand into the joints or re-tamp the surface with the straight-edge to ensure settlement. Once all the joints are filled you can brush the surplus sand away. The path will be ready for immediate use.

Laying bricks on mortar

This method of laying bricks is more permanent than using a sand bed, and no edge restraint is needed. Again, when you are digging out the trench, remember to allow for the drainage fall and the level of the path in relation to the dpc or lawn, and dig down deep enough to allow for the thickness of construction.

The sub-base should be a layer of hardcore which is tamped down or rolled to provide a firm foundation for the path and topped with a blinding layer of fine sand or ash. The bricks are bedded in a fairly dry, crumbly mortar mix and with spaces between them to allow for grouting. Dry mortar is brushed into the joints and then the whole path is sprayed with water. With this method, you should not use the path until a week has passed after laying the pavers. In hot, dry weather the paving should be protected against drying out prematurely by covering it with polythene sheets or damp sacking.

LAYING BRICKS ON MORTAR

1 *To bed bricks on a mortar base, trowel a fairly stiff mix in dabs onto the sand bed. No edge restraints will be needed when a mortar bed is used.*

2 *Press the bricks onto the mortar in your chosen design – here you can see a herringbone pattern – and adjust the mortar thickness so they are level.*

TIP

3 *Leave a 10mm (3/8in) gap between each brick to allow for a mortar joint, which is added afterwards. Use timber offcuts as spacers for the joints.*

4 *Check the level of the brickwork using a straight-edged length of timber and a spirit level. Adjust the thickness of the mortar if necessary.*

5 *Brush a dry mortar mix, in the same proportions as the bedding mix, into the joints until they are flush with the surface of the path.*

6 *Spray the path with clean water from a watering can fitted with a fine rose, then leave the path for about one week before using.*

CUTTING A BRICK

1 *Mark a cutting line in chalk around the bricks to be cut; place them on a sand bed and mark the line by hitting it with a bolster chisel and club hammer.*

2 *Turn the brick over and score lines on each face (inset). Return to the first line and hit it sharply with the chisel until the brick breaks cleanly.*

Paul Forrester

READY REFERENCE

PAVING PATTERNS

Here are three basic patterns you can use to lay rectangular bricks and pavers:

stretcher bond

herringbone

basketweave

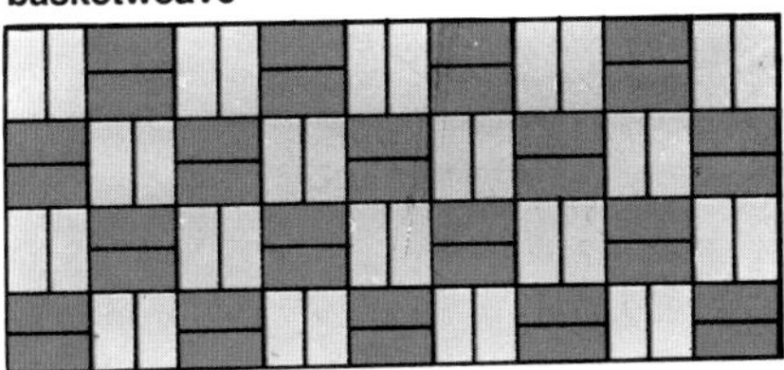

EDGE RESTRAINTS

With paths laid on a sand base, you need some form of restraint to stop the edge blocks from moving. You could choose

- pre-cast concrete kerb stones
- a band of bricks set in mortar (below)

- a sloped mortar 'haunching' ending just under the brick edges (below)

- rot-proofed timber 25mm (1in) thick fixed with 50mm (2in) square wooden stakes. These will soon be hidden by grass overgrowing the path edges

BUILDING FREE-STANDING STEPS

A garden composed of different levels will look disjointed unless there is some visual link between parts. A flight of steps not only serves the practical purpose of providing access to the various levels but also gives a co-ordinated look to a scheme.

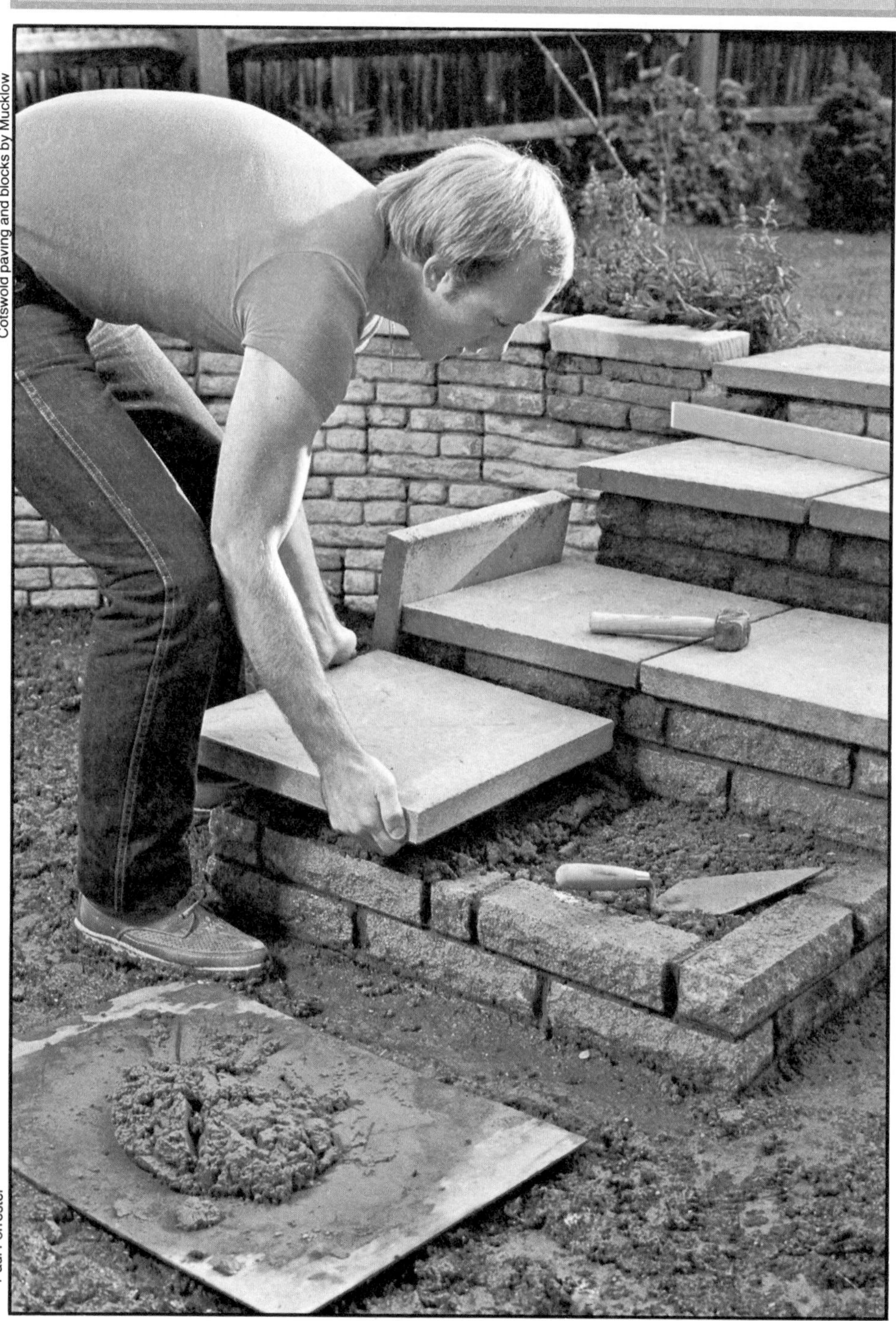

Cotswold paving and blocks by Mucklow

Paul Forrester

The main purpose of steps is to provide access from one level to another and they're an important factor in the landscaping of a garden, bringing otherwise detached areas into the overall scheme. Wide, shallow steps can, for example, double as seating for a terrace bordering a lawn, or as a display area for plants in containers, while narrow, angular steps will accentuate interesting changes of ground level.

This section deals with freestanding steps, which are designed to rise from flat ground level to a slightly higher level, such as a path to a terrace or raised lawn. Built-in garden steps describes how to construct steps which are not freestanding, but built into a bank or slope.

Types of materials

For a good visual effect, the steps should be constructed from materials that complement the style of the garden. If you've built a path, for instance, continue the run by using the same materials for the steps. Where you're building up to a wall, match the two structures, again by using the same materials.

For a formal flight, bricks can be used as risers (the vertical height of the step). Decorative walling blocks with 'riven' or split faces, or natural stone, used in the same way give a softer, more countrified look. You can top these materials with either smooth- or riven-faced slabs or even quarry tiles to form the treads (the horizontal part of the step on which you walk). Or, if you prefer, you can use other combinations of building materials for the steps; bricks and blocks, for example, make attractive treads as well as risers.

You can choose from a wide range of coloured bricks, blocks and slabs, mixing and matching them to best effect. Pre-cast or cast in situ concrete can be used as a firm base for these materials or makes a durable surface in its own right.

Planning the steps

There are no building regulations governing the construction of garden steps so you have a lot of flexibility in deciding how they will look. Sketch out possible routes – you can build them parallel to the side of the terrace so they don't extend too far into the garden. This is particularly suitable where the ground slope is steep. Decide whether the steps will be flanked by flower beds, rockeries or lawn, or linked at the sides to existing or new walls. You could build steps with double-skinned side walls and fill the gap between with soil to use as a planter.

Bear in mind the dimensions of the completed steps when choosing materials. Treads and risers should measure the same throughout the flight to ensure a constant, safe, walking rhythm. (If they are not constant the variations must be made

PREPARING THE BASE

1 Fix pairs of string lines so they are level and square over the centre of the concrete strip foundations to mark the height and width of the first riser.

2 Lay mortar along the front of the foundation and scribe a line parallel to the inside string to indicate the back of the first course of blocks.

3 Lay the first course of blocks on the mortar, checking the surface is even with a spirit level. Tap the blocks in place with the handle of a club hammer.

4 Position the first course of the side walls in the same way and check that the angle at the corners is at 90°, using a builder's square.

5 Start laying the second course of blocks at one corner, aligning them with the strings. Tap the blocks into place with the handle of your trowel.

6 When two courses of blocks are laid allow the mortar to set partially, then shovel in hardcore. Use a fence post to compact the hardcore.

Paul Forrester

visually obvious for safety reasons.) Make provision, also, for drainage of rainwater from the steps by sloping the treads slightly towards the front.

Steps should be neither too steep nor too gradual – steepness can cause strain and loss of balance while with too shallow a climb you run the danger of tripping on the steps. Steep steps, and those likely to be used by children and elderly people, require railings on one or both sides to aid balance. These can be of either tubular metal or wood and should be set at a height where your hand can rest on top comfortably – this usually works out at about 850mm (2ft 9in) high measured from the nose of the steps. Alternatively, you can build small brick, block or stone walls at each side of the flight. There should be no hand obstructions along the length of the railings or walls, or other projections on which clothing could snag. Railings should also extend beyond both ends of the flight by about 300mm (1ft); they might, in fact, continue an existing run of railings along a path.

Treads should be non-slip for safety. However, this is difficult to achieve outdoors, where they are subjected to ice, rain

READY REFERENCE

STEP ESSENTIALS

To ensure comfortable, safe walking, garden steps must be uniform in size throughout the flight, and neither too steep nor too shallow.

Risers are usually between 100mm and 175mm (4in and 7in) high. The shallower the slope the shallower the risers.

Treads should not be less than 300mm (1ft) deep – enough to take the ball of the foot on descending without the back of the leg hitting the step above. They should be about 600mm (2ft) wide for one person; 1.5m (5ft) for two people.

Nosing is the front of the tread, which projects beyond the riser by about 25mm (1in) to accentuate the line of the step in shadow.

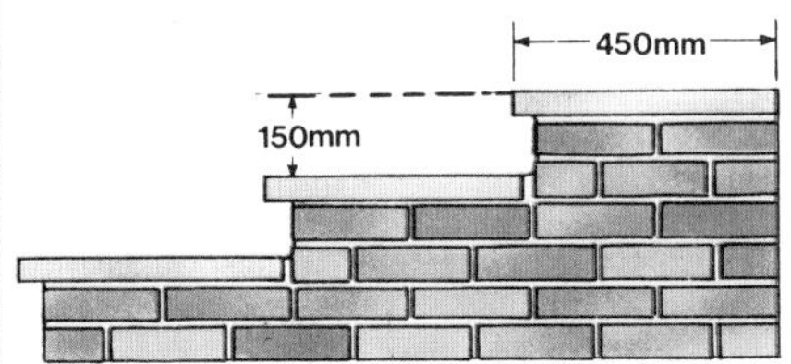

HOW MANY STEPS?

To calculate the number of steps:

- measure the vertical distance between the two levels
- divide the height of a single riser (including the tread thickness where relevant) into the vertical height to determine the number of risers.

TYPICAL TREAD/RISER COMBINATIONS

For comfortable walking combine deep treads with low risers, shallower treads with high risers.

Tread	Riser
450mm (1ft 5½in)	110mm (4½in)
430mm (1ft 5in)	125mm (5in)
400mm (1ft 4in)	140mm (5½in)
380mm (1ft 3in)	150mm (6in)
330mm (1ft 1in)	165mm (6½in)

MORTAR MIX

Bed treads and risers in a fairly stiff mortar mix: 1 part cement to 5 parts sand.

HOW MUCH MORTAR?

50kg cement mixed with 24 two gallon (9 litre) buckets of damp sand will be sufficient to lay about 7sq m of slabs – around 35 slabs 450 x 450mm (18 x 18in) – on a 25mm (1in) thick bed. Don't mix this much mortar at once or it will set before you can use it all.

CONSTRUCTING THE STEPS

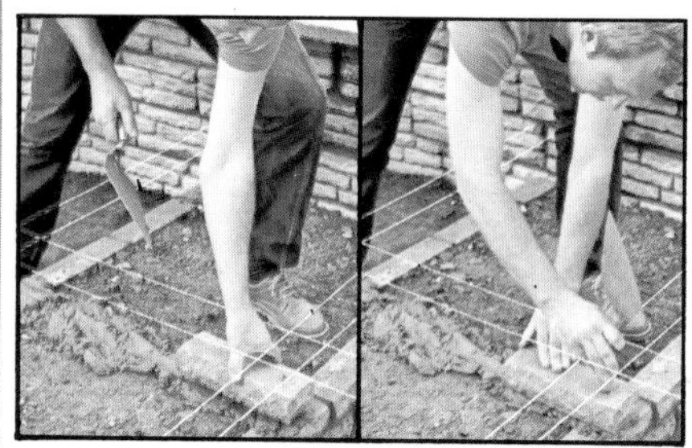

1 *Move the front strings back the depth of the tread to the second riser position, then lay the blocks on a mortar bed as described previously.*

2 *When the third and fourth courses are laid, in-fill with more hardcore and compact it thoroughly so it's flush with the top of the blocks.*

3 *After laying each course, you should check with a gauge rod at the corners that the mortar joints are of the same thickness throughout.*

4 *The slab treads can be laid on either a bed of mortar around their perimeter or on five dabs of mortar – one at each corner and one centrally.*

5 *Position the slabs squarely on the risers but projecting forward by about 25mm (1in) and sloping forward slightly to allow for drainage of rainwater.*

6 *When the slabs have been laid, fill the gap behind with a mortar fillet – but be sure not to spill wet mortar on the slabs, as this may stain them.*

Paul Forrester

and formation of moss. Proprietary liquids are available for painting on treads so they won't be slippery but a more practical solution is to use hydraulically-pressed slabs, which come with a variety of surface textures or relief designs that are both attractive and non-slip.

When you've decided on the basic appearance and route of the steps you must calculate the quantity of bricks, slabs or other material you'll need. It's best to work out your design taking into account the sizes of slabs, blocks or bricks available. Make a scale plan (bird's eye view) and a side view on graph paper to help in planning and construction.

Building methods

In order to support the steps, concrete trench foundations must be formed underneath the retaining walls. Mark out these foundations using strings stretched between pegs (see foundations, pages 30-34). Construct the flight one step at a time. When the first courses of the retaining walls have been built, a hardcore filling is used to provide a firm base for the treads. When this is rammed down, lay a blinding layer of sharp sand to fill gaps and provide a level bed for the treads. Subsequent levels are built on top in the same way and the treads laid on mortar.

When tamping down the hardcore back-filling in each level take care not to dislodge any bricks or blocks. Concentrate on the areas that will support the next risers and the back edges of the treads because these parts are subject to the most pressure.

This method of construction is suitable only for freestanding units up to five steps high. If you want larger steps, you'll have to make substantial foundations in the form of a cast concrete slab about 75mm (3in) thick, covering the entire area of the steps, and build intermediate supporting walls the width of the flight under each riser – a hardcore back-filling alone is simply not firm enough to support the extra weight. These walls can be laid in a honeycomb fashion, with gaps between the bricks in each course; this allows for drainage through the structure and uses less bricks than a solid wall. Each section formed by the intermediate walls should contain a rammed-down hardcore back-filling topped with a blinding layer of sand.

Alternatively, you can cast a solid concrete flight in timber formwork and lay the surface materials on top, bedded in mortar, but this will involve large quantities of concrete.

Larger flights built up to a wall should be joined to the wall or there's the risk of them parting company after time. 'Tooth-in' alternate courses to the brickwork or blockwork of the terrace by removing a brick from the terrace wall and slotting in the last whole brick from the side walls of the steps. On smaller flights you can tie in the steps by bedding a large 6in (150mm) nail in a mortar joint between the two structures for added rigidity.

Brick bonds

The type of brick bond you use in the construction of the retaining walls depends upon the size of the materials you use for the risers. You should, though, try to keep the perpends consistent throughout the flight for strength and for best visual effect.

It's wise to dry-lay bricks or blocks first to make sure you get the best – and strongest – bond; perpends that are very close together mean a weaker construction and you should try to avoid this if possible (see Bricklaying, pages 8-13). When choosing materials, you should take note of typical, 'safe' tread/riser combinations (see *Ready Reference*).

ARRANGING THE BONDING

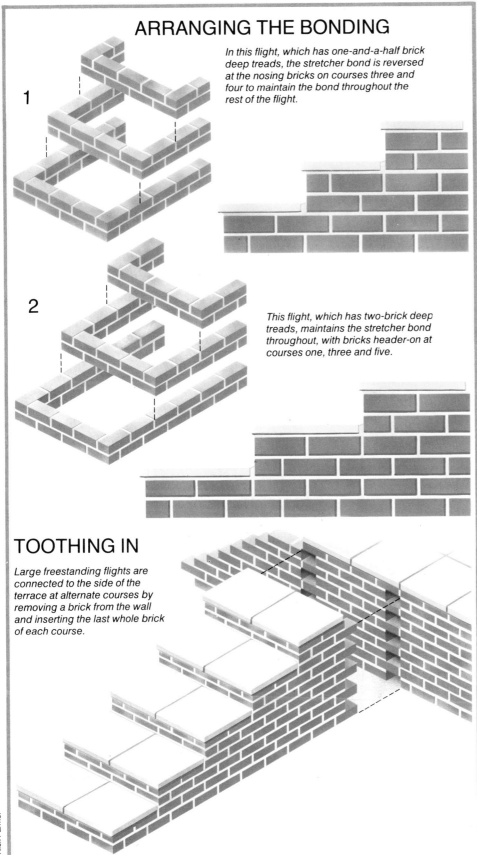

1

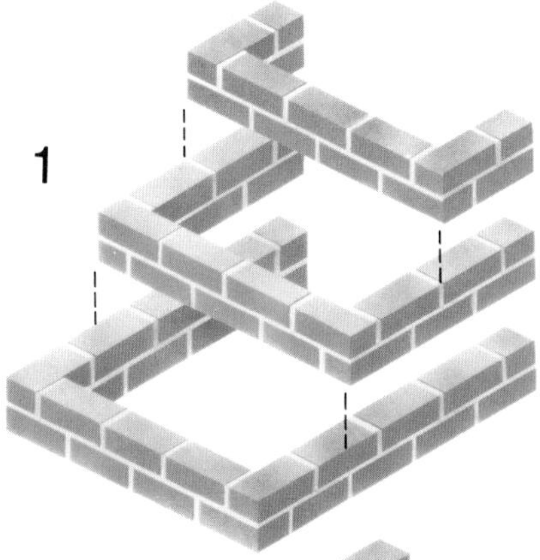

In this flight, which has one-and-a-half brick deep treads, the stretcher bond is reversed at the nosing bricks on courses three and four to maintain the bond throughout the rest of the flight.

2

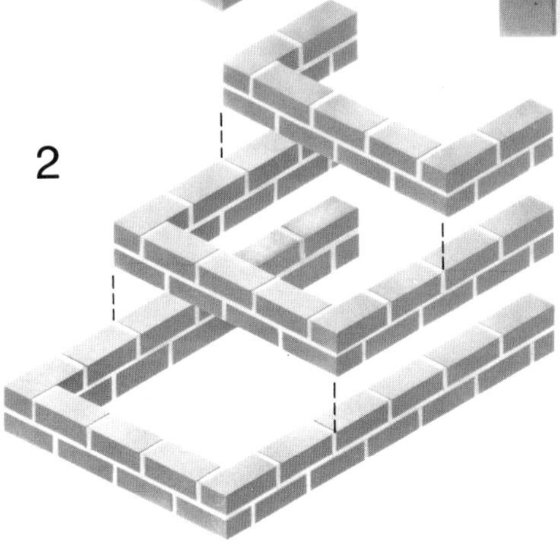

This flight, which has two-brick deep treads, maintains the stretcher bond throughout, with bricks header-on at courses one, three and five.

TOOTHING IN

Large freestanding flights are connected to the side of the terrace at alternate courses by removing a brick from the wall and inserting the last whole brick of each course.

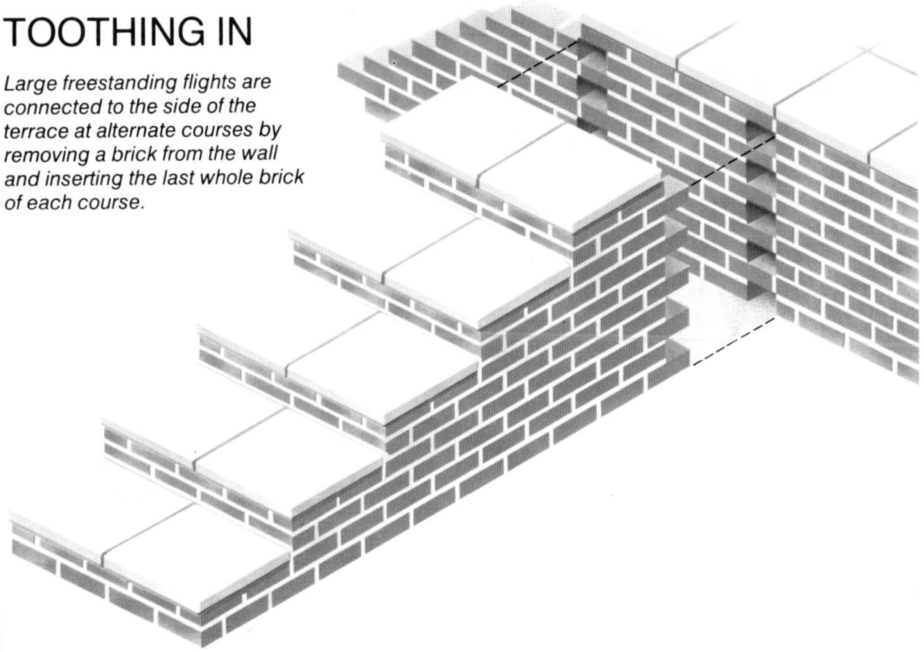

Nick Farmer

READY REFERENCE

MASONRY STEPS

A simple freestanding flight consisting of five treads requires:

- a concrete-filled trench about 100mm (4in) deep and twice as wide as each retaining wall
- broken brick or concrete back-filling

A flight larger than five treads requires:

- substantial foundations in the form of a cast concrete slab about 75mm (3in) thick, covering the entire area of the steps
- intermediate supporting walls the width of the flight under each riser
- rammed-down hardcore back-filling topped with a blinding layer of sand
- or a concrete base flight cast in timber formwork, on which the surface materials can be laid, bedded in mortar.

DRAINAGE

Rainwater must not be allowed to collect on the steps. Provide drainage by:

- allowing for a fall of about 12mm (½in) to the front of each tread or, where the steps abut a wall:
- make concrete gullies about 50mm (2in) deep by 75mm (3in) wide at each side of the flight to drain from the top.

If the flight drains towards a house wall:

- make a channel at the foot of the steps, parallel to the wall, to divert water to a suitable drainage point.

CREATING BUILT-IN STEPS

Steps can be used to great effect in the garden. They not only enable you to get from one level of the garden to another with ease, but draw together otherwise separated features of the landscape.

On pages 54 -57 you'll find details of the techniques for freestanding unit steps, which lead from flat ground level to a slightly higher level and have their own support. It is also possible to incorporate steps into an existing slope or bank by using the shape and structure of the slope as the base.

The rough shape of the flight is dug into the bank and the steps are then bedded in mortar on a hardcore base. With some soils, well-compacted earth alone can make a firm enough base for the treads and risers. On a soft crumbly soil you may find it necessary to build low retaining walls at the sides of the flight before you lay the treads, to prevent the soil from spilling onto the treads.

Planning the site

When planning you should consider the site as a whole or the steps might end up looking out of place. You have a considerable amount of freedom in the design of your steps as construction rules are more relaxed outdoors than they are for buildings, but you should adhere to the design principle that the new element should fit into its setting. Because you're using the lie of the land as your foundations it's as well to plan the flight so that it traces existing gradients or skirts flower beds, trees or other features.

Make a sketch of the garden, plotting possible locations for the steps and transfer this information to a more detailed plan on graph paper, including a cross-section of the ground slope. With this plan you can calculate quantities of materials – always try to design the steps with particular sized bricks, slabs and blocks in mind to avoid having to cut them or alter the slope dimensions unduly.

Match materials that have been used elsewhere in the garden – as boundary walls or raised flower beds for instance – for a feeling of continuity. Also remember the basic rules of step design: although you should avoid creating steep steps, which can cause strain, shallow flights are also not recommended because you can easily trip on them.

Bricks, blocks and natural stone come in sizes that are suitable for building risers in one or more courses. Riven- or smooth-faced concrete slabs or quarry tiles make convenient, easy-to-use and non-slip treads for these materials. You can also use the smaller-scale materials such as bricks and blocks for the treads as well as the risers, although they'll need a much firmer base than slabs.

The colour of the steps is also important and most materials are available in a range of reds, greens, browns and greys.

Whatever combination of materials you choose, keep the dimensions of the steps constant throughout the flight – a jumble of sizes not only looks untidy but also upsets a comfortable walking rhythm and so can be dangerous.

Where your flight is larger than 10 steps it's wise to build in a landing. This will visually 'foreshorten' the flight, provide a broad resting place and more practically, will serve to 'catch' anyone accidentally falling from the flight above. You should also include a landing when changing the direction of a flight at an acute angle.

Include railings to aid balance on steep or twisting flights, or those likely to be used by children and elderly people. They can continue the run of existing fences or walls along a path for a sense of unity.

You should also allow for a slight fall towards the front of each tread so that rainwater will drain quickly away. Don't slope the

READY REFERENCE

STEP ESSENTIALS

Steps should conform to the following dimensions to ensure comfortable, safe walking:

- **risers** between 100mm and 175mm (4in and 7in)
- **treads** not less than 300mm (1ft) deep; a minimum of 600mm (2ft) wide for one person, 1.5m (5ft) wide for two
- **nosing** projecting beyond risers by 25mm (1in)
- **flights** not more than 10 steps without a landing.

MORTAR MIX

Bed treads and risers in a fairly stiff mortar mix: 1 part cement to 5 parts sand.

HOW MUCH MORTAR?

One 50kg bag of cement mixed with 15¾ 3-gallon (14-litre) bucketfuls of damp sand will be sufficient to lay about 7sq m of slabs on a 25mm (1in) thick mortar bed. Don't mix all the mortar at once or it will set before you can use it all – mix about a quarter at a time.

AVERAGE DIMENSIONS

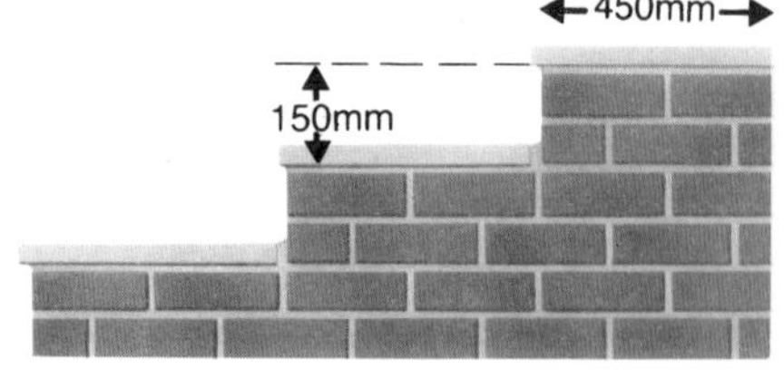

PREPARING THE SLOPE

1 *To measure the vertical height of the slope stretch string between a peg at the top and a cane at the base; check that it's level with a spirit level.*

2 *Set string lines from the top to the bottom of the slope to indicate the sides of the flight, ensuring they are parallel and that the flight is straight.*

3 *Mark the nosing for each tread with string lines stretched across the slope; check that they are level and that the angle with the side strings is at 90°.*

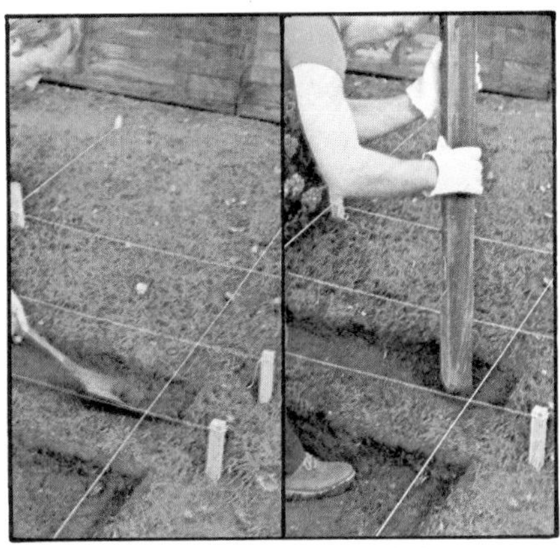

4 *Dig out the rough shape of the steps, taking care not to dislodge the nosing markers and compact the earth using a fence post as a tamper.*

5 *Continue to dig out the steps, working up the slope. Use each cut-out as a standing base for excavating the next, but be careful not to crumble the edges.*

6 *Dig below and behind the nosing strings to allow for the depth of the slab treads and the thickness of the block risers, thus defining the step shape.*

treads to one side as this can give the flight a lopsided look.

Planning awkward slopes

Your plans for building in steps will seldom run true, as your slope probably won't be regular in shape. If the riser height doesn't divide equally into the vertical height of the slope your steps will have inconsistent dimensions – not only unattractive but also likely to upset constant walking pace. One solution is to remodel the slope. Use earth from another part of the garden placed at the top to increase the slope's height; remove earth from the top to decrease its height. Any extra earth must be compacted before it can be 'stepped'.

You can often use any undulations to your advantage: because you're using the firmed ground as your foundations you're able to build much longer and twisting flights. So base your plan on the shape of the ground rather than vice versa.

Marking out the flight

Before you can mark out the flight on the slope you must calculate the number of steps you'll need by measuring the vertical height of the bank.

To mark out the steps stretch strings between pegs from the top to the bottom of the slope to indicate the width of the flight. You can then set other strings across the slope to indicate the top edges of each step's nosing. You should check the level of these nosing strings using a spirit level.

Constructing the flight

Starting at the base of the slope, excavate the rough shape of the first step, digging behind and below the nosing marker to a depth that will allow for a hardcore filling and the thickness of the tread and riser.

Compact the earth, then use the cut-out as a standing base for excavating the next step. Work in this way up the slope. When the whole flight has been excavated in this way, and compacted, lay a hardcore base (if necessary) followed by the treads and risers. Use the back of each tread as a base for the next riser and back-fill with hardcore, which should be well compacted with a fence post tamper, taking care not to dislodge the newly-laid riser.

The first riser should ideally be laid on a cast concrete footing to support the weight of of the flight, preventing it from 'slipping', although on small flights where the soil is firm this might not be necessary. The footing should be about 100mm (4in) deep and twice the thickness of the riser.

An alternative way to build steps into a slope is to cast a concrete slab flight in timber formwork and either face it with bricks, blocks, slabs or tiles, or leave it bare.

LAYING THE STEPS

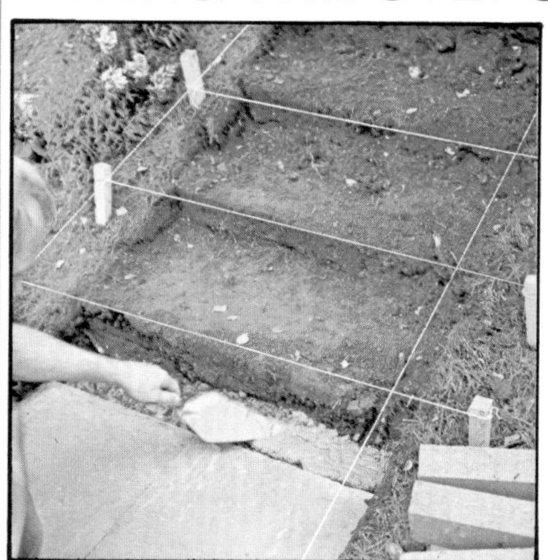

1 *Lay any slabs at ground level where a path run continues then add hardcore under the first riser position. Compact and lay mortar on top.*

2 *Lay the blocks for the first riser on the bed of mortar, making sure they're level and square. Tap down into place with the handle of your club hammer.*

3 *Fill the gap between the blocks and the soil with hardcore, then compact to the top level of the riser. Take care that you don't dislodge the blocks.*

4 *Shovel hardcore onto the first tread position and compact well using a fence post. Check the level of the foundation, incorporating a fall to the front.*

5 *Lay the first two slabs on mortar and check that they are level. Set them forward by about 25mm (1in); the nosing marker should align with their top edge.*

6 *Bed the next course of blocks for the second riser in mortar on the back of the first tread; try not to splash mortar on the slabs as this will stain.*

7 *Back-fill with hardcore and tamp down thoroughly, then lay the next two slabs; a bed of mortar at the perimeter of each slab makes a firm and level bed.*

8 *Continue in this way to the top of the flight, bedding each riser on the tread below it. Set the last tread level with the ground at the top of the slope.*

9 *Brush a dry mortar mix into the gaps between the slabs, point all the joints and brush off any debris. Leave the steps for 7 days before using.*

LAYING CRAZY PAVING

Garden paving doesn't have to be all squares and rectangles. With crazy paving you can have a more informal look in any shape you fancy, and you can use it on wall surfaces too.

Crazy paving is a versatile material that can be used as a resilient and attractive surface for patios, driveways, garden paths or steps, or as a decorative feature in an otherwise plain paving scheme.

It's simply broken paving slabs and you can often buy it quite cheaply, by the tonne, from your local as demolition material. Slabs bought in this way will usually be a heavy duty variety used for pavements and have a rather dull grey colour and a relatively plain finish, but you can add interest in the way you lay the pieces. Local building contractors can often supply broken slabs of various textures in greens, pinks, reds and buff tones in sufficient quantities for use in paving. Natural stone can also be bought to make up crazy paving.

Planning crazy paving

Before you can begin to lay your paving, sketch out some ideas for its overall shape, size and, in the case of paths, its route through the garden. Transfer your final design to graph paper so that you can use this to estimate the total area to be covered and place an order for the correct quantity of paving. You can also use your plan as a blueprint for ordering and laying the slabs.

Because of its irregular profile crazy paving, unlike conventional square or rectangular slabs, can be used to form curves, such as a winding path, a decorative surround to a pond, or an unusual-shaped patio. Although you have a lot of freedom in your creative design you mustn't allow the paving to appear out of place with its surroundings. If your garden is strictly formal, for instance, avoid complex curves or too 'busy' a surface texture — the mix of angles could clash. Small areas of random paving can, on the other hand, give a plain scheme a visual 'lift'.

Although crazy paving has an overall random design, it must be placed with some precision to avoid an unbalanced look. The best way to plan out an area of paving when you've decided upon a basic site is to dry-lay it when the base is complete.

Separate the pieces that have one or more straight sides for use as edging and corners and lay these first, choosing only the largest pieces — small ones tend to break away.

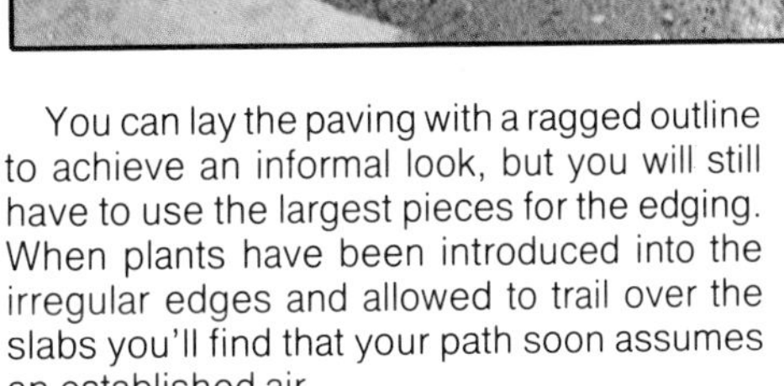

You can lay the paving with a ragged outline to achieve an informal look, but you will still have to use the largest pieces for the edging. When plants have been introduced into the irregular edges and allowed to trail over the slabs you'll find that your path soon assumes an established air.

It's not important to make a regular joint width between the pieces — in fact, the paving will probably look much more natural if the joints vary. Fit the smaller inner pieces together like a jig-saw puzzle, mixing colours to best effect. When dry-laying the slabs avoid a continuous joint line across the path as this can be jarring to the eye and weakens the structure.

Laying the paving

The methods of laying crazy paving are similar to those using regular paving slabs (see pages 46 -49 and 64 -67 – with a firm base being the first requirement.

To prepare the base remove the topsoil and compact the area using a roller or tamper. If the ground is soft or crumbly add a layer of hardcore and compact this into the surface. A blinding layer of sand added to the top accommodates any unevenness in the hardcore base and acts as a firm bed for the slabs.

Lay the slabs on generous mortar dabs under each corner or, with smaller pieces, on an overall mortar bed.

Work your way across the dry-laid surface,

READY REFERENCE

PATH LAYING

When laying a path in crazy paving:

- place the larger straight-sided slabs first as the path edging, if these are to be straight
- place the largest irregularly-shaped slabs down the centre of the path, fill in between the edging and centre slabs with small broken fragments.

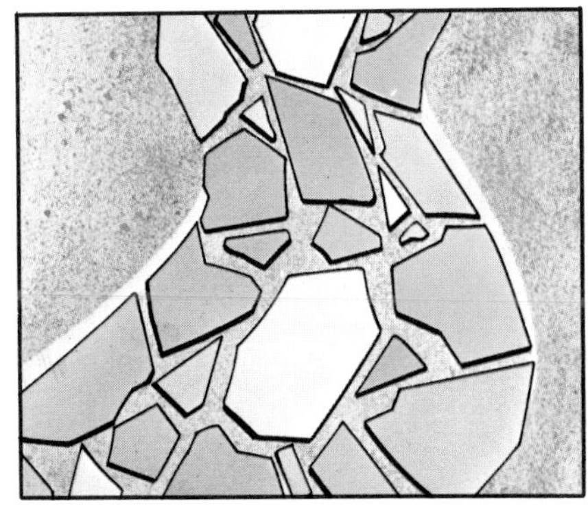

SLAB THICKNESS

Remember when laying crazy paving that the slabs might be of different thicknesses and you'll have to accommodate these variations in the thickness of the mortar bed.

PREPARING THE BASE

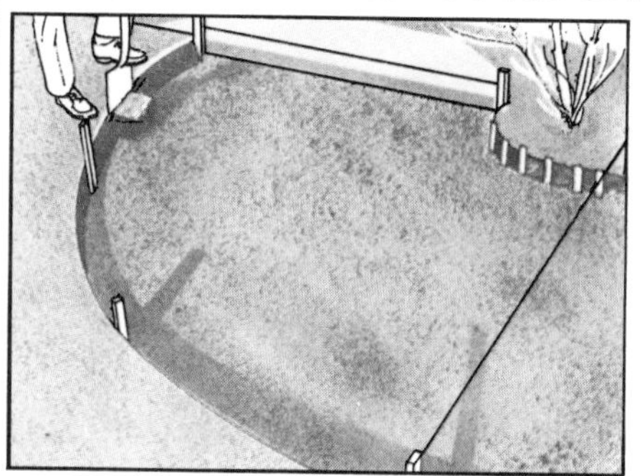

1 *Mark out the area to be paved using pegs and string, and remove the topsoil (above). Then tamp the subsoil down all over (below), adding hardcore if needed.*

2 *Having thoroughly compacted the surface and filled any hollows, check that it is flat in both directions using a long timber straight edge. Allow for a slight drainage fall in one direction across the paved area by scraping away soil from the base so that your spirit level bubble is slightly off centre.*

3 *Spread a 50mm (2in) thick layer of sand across the area (above). Then use your long timber straight edge to tamp it down to a level, even bed (below).*

bedding each stone in turn and checking the level frequently using a builder's level — don't forget to incorporate a slight drainage fall to one side of a path or to the front of a patio or step treads. Tap the paving in place using the handle of your club hammer: as you do this some mortar will be squeezed up into the joints, which can be anything up to 25mm (1in) wide. You needn't scrape out this mortar from the joints: it actually makes for a stronger bond.

After you've laid the slabs point between them with mortar (see *Ready Reference*). Be careful not to smear any on the faces of the slabs otherwise they will stain. Alternatively you can leave the joints mortar-free and brush in soil later in which to plant low growing plants or herbs.

If you have used crazy paving for patio or path surfaces, you may also want to give your garden a unified look by cladding steps and low walls.

On steps, start at the bottom of the flight and clad the lowest riser first. Build up the cladding from ground level, buttering mortar onto the back of each piece of paving and pressing it firmly into place against the riser. Finish cladding the riser with pieces that fit flush with the existing step surface, and then lay paving on the surface of the tread so that those at the front overlay the riser by about 25mm (1in). Clad all the risers and treads, then point between the pieces. Clad low walls in the same way as step risers.

DRY-LAYING THE SLABS

1 *When you've marked out the shape of your paved area and dug out and prepared the foundations, start to dry-lay the slabs in one corner.*

2 *Separate the straight-edged slabs from the irregular-shaped ones and dry-lay the largest pieces at the perimeter of your marked-out area.*

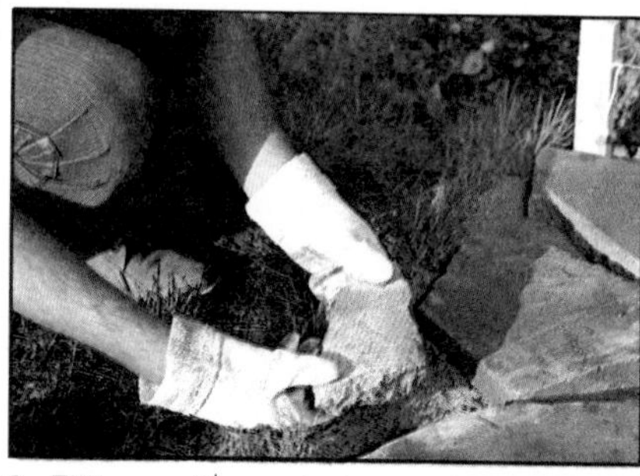

3 *Fill in small gaps between the perimeter slabs with straight-edged fragments, but avoid a run of small pieces, as it makes a weaker edge.*

4 *When you've placed all of the perimeter slabs you'll be able to see what the overall effect will be: swap them about for the best-looking plan.*

LAYING THE PAVING

1 *When you're satisfied with the positions of the slabs you can start to bed them on a fairly stiff mortar mix; lay the large ones on five dabs of mortar.*

2 *You can lay smaller pieces on a continuous mortar bed. Trowel ridges in the mortar: this aids levelling and provides a stronger bond.*

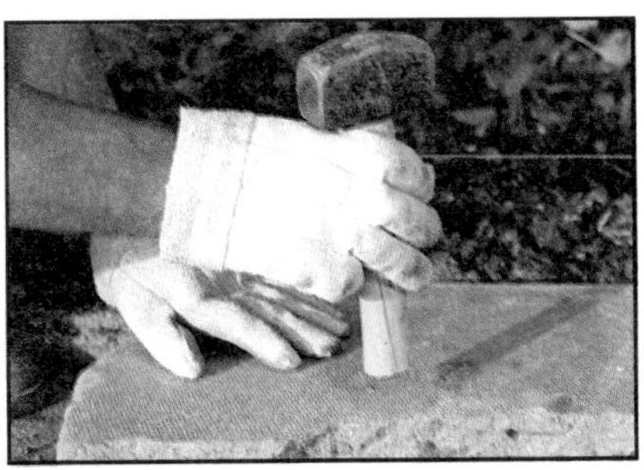

3 *Bed each slab level with the ones next to it by tapping it gently with the handle of your club hammer. Allow mortar to squeeze out between the joints.*

4 *Mortar in the perimeter slabs first then start to in-fill with smaller, irregular-shaped pieces, using different colours for a more varied pattern.*

5 *Check at intervals across the tops of the slabs with a builder's level to ensure they're bedded evenly. Tap them in place with the handle of a club hammer.*

6 *Mortar joints can be up to 25mm (1in) wide. Make the joint flush with the slabs but don't spill mortar on the slab faces, or it will stain them.*

7 *If you want to make a feature of the joints fill them with soil rather than mortar and add some low-growing plants to give a natural, established look.*

8 *After you've pointed the joints, or filled them with soil, brush over the surface of the paved area with a stiff-bristled broom to remove any debris.*

READY REFERENCE

POINTING PAVING

There are three ways to treat the joints between crazy paving:

- Flush pointing — fill the joints with mortar flush with the tops of the slabs.
- Bevelled pointing — form bevels in the mortar about 9mm (3/8in) deep at each side of the joint to outline the shape of each slab.
- Soil joints — fill the joints with soil and plant low-growing plants or herbs to blend in the paving with the rest of the garden.

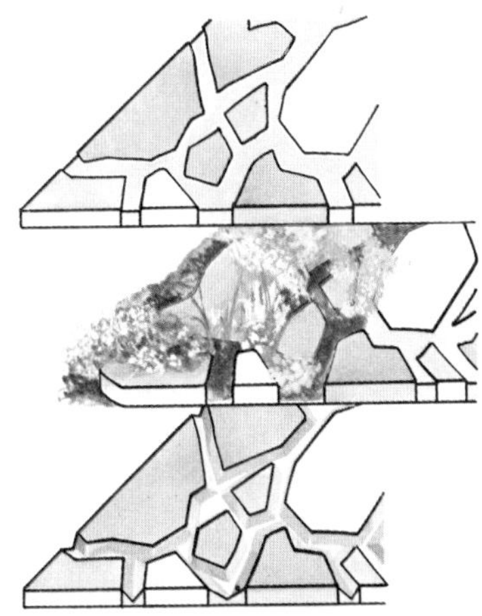

TIP: BREAKING LARGE SLABS

A delivery of crazy paving might include some pieces too large to lay. Break them by dropping them on any hard surface — easier than using a bolster chisel and club hammer. Use these tools for trimming smaller pieces.

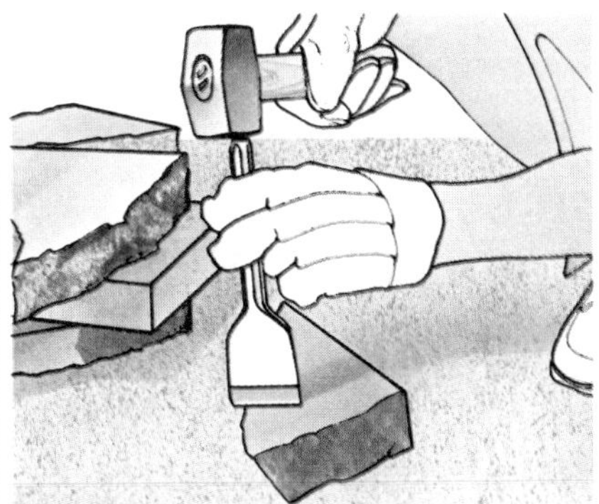

COLOURED POINTING

Make a feature of the joints in crazy paving by colouring the mortar. Additives are available for adding to the mix but you should follow the maker's instructions precisely — too great a proportion of colouring can upset the strength of the mortar.

For more information on laying paving slabs see pages 46 -49 and 64 -67.

BUILDING A PAVED PATIO

Building a patio close to the house is one way of transforming a dull and featureless garden into a durable paved area that's geared especially for outdoor living.

A patio makes a versatile summertime extension to the house, providing space for dining, entertaining, or merely for relaxing and soaking-up the sun. You must plan your patio to take advantage of the best aspect and construct it from materials that are both in keeping with the house and garden and durable enough to withstand harsh weather conditions.

Siting the patio

Patios are usually sited as close to the house as possible – ideally adjoining it – or at least nearby for easy access. The best aspect is south-facing, but whichever way your garden faces, you should examine the proposed site at different times of the day during the summer to see how shadows fall. Neighbouring buildings, or your own house, might obscure the sun and this will severely limit the use of the patio. Unfortunately there's nothing you can do about this, but if the obstruction is just a tree or tall hedge you might be able to prune it.

Some shadows can be used to your advantage: although you might want to lie sun-bathing at certain times of the day, you'll appreciate the shade while you eat. If there's no natural shade, you could attach an awning to the house wall, which can be folded away when not needed. A pergola or trellis on which you can train climbing plants will also provide shade where you need it. Or you might prefer simply to allow enough space for a table with an umbrella or a swing seat with a canopy.

What size patio?

In theory a patio may be as large or small as you wish, but in practice you'll be limited by available space. Try to relate the dimensions of the patio to the needs – and size – of your household. Measure your garden furniture and allow enough space around it so that you won't be cramped: the patio must measure at least 2.4m (8ft) from front to back to enable you to position furniture and allow free passage. In general a patio measuring about 3.7m sq (12ft sq) is big enough to take a four-seater table or four loungers.

To work out the size and position of the patio make some preliminary sketches of the garden with the proposed patio in various locations. When you've decided upon a suitable scheme transfer your ideas to a scale plan on graph paper. Cut out a paper template of the patio and use it in conjunction with the plan to help you decide upon the best position.

What type of surface?

There is a wide range of paving materials available in various shapes, sizes and colours and you should choose those which blend with materials used around the house exterior and garden for a sense of unity. The main requisites are that the surface is reasonably smooth, level and free-draining. Whatever your choice of paving, avoid too great a mix – two types are usually sufficient to add interest without making the surface look cluttered. You can, however, include confined areas of small-scale materials such as cobblestones and granite setts to add a textural change to an otherwise flat scheme composed of larger slabs.

Cobblestones – oval pebbles – can be laid in three ways: on a continuous mortar bed over hardcore foundations; on a bed of dry mortar 'watered in' by watering can; or loosely piled on top of each other on compacted earth. Granite setts are durable square-shaped blocks with an uneven

PREPARING THE BASE

1 Set a long prime datum peg in a hole 300mm (1ft) deep at one side of the patio. Its top should be 150mm (6in) below the house dpc.

2 Set a second peg at the other side, level with the first. For a very wide patio use intermediate datum pegs set 1.5m (5ft) apart.

3 Use a string line to set timber pegs accurately in line with each other so that they outline the proposed perimeter of the patio.

4 Check with a builder's square to make sure that the corners of the patio are perfectly square. Adjust the peg positions if they are not.

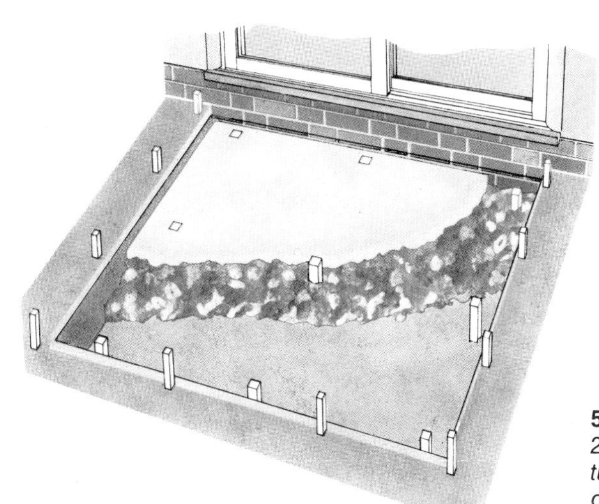

5 Dig out the site to a depth of about 230mm (9in), saving the top-soil and turf. Then compact the earth with a garden roller.

6 Drive in pegs 1.5m (5ft) apart over the entire area. Check that they are level with the datum peg and each other using a spirit level.

7 Fill the hole with hardcore, then compact it thoroughly to a depth of about 125mm (5in) using a tamper. Don't disturb the pegs.

8 Rake out and roll a 50mm (2in) layer of sand over the hardcore. The peg tops should be level with the sand surface.

LAYING THE SLABS

1 *Start to lay the slabs at the corner marked by the prime datum peg. You can lay them dry on the sand bed, without using any mortar.*

2 *As you progress across the patio you should check frequently with a builder's level that the slabs are bedded evenly on the sand.*

3 *Lift up any slabs that are unevenly bedded and trowel in some more sand until the bed is filled out and the slabs are flush with their neighbours.*

4 *Alternatively you can lay the slabs on dabs of fairly stiff mortar, one placed under each corner of the slab and one under the centre.*

5 *Position the slab carefully on top of the mortar dabs. Space out the slabs using offcuts of timber 9mm (3/8in) thick for pointing later.*

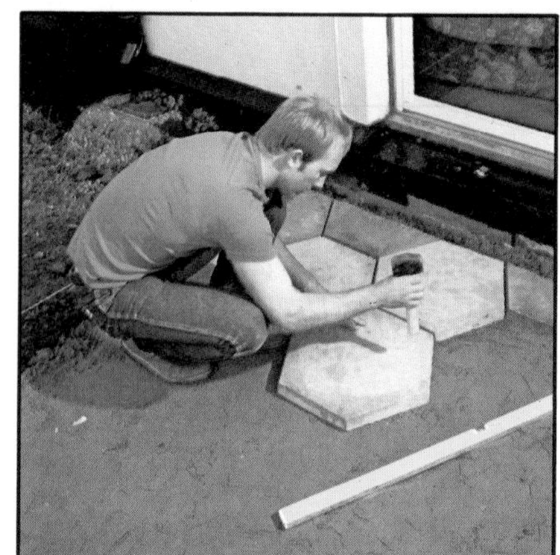

6 *Bed down the slabs using the handle of your club hammer. If any of the slabs are too low, remove them and add more mortar to the dabs.*

7 *Stretch the string lines across the patio every second course to help you align the slabs accurately.*

8 *When all the slabs are laid, brush a dry mortar mix between the joins and remove any excess from their faces. This will form a bond when watered in.*

9 *Water in the dry mortar mix with clean water from a watering can fitted with a fine rose. Avoid over-watering or you'll wash away the mortar.*

surface texture, and can be laid on sand or mortar.

Other small-scale paving materials that can be used in large or small areas include concrete blocks, brick pavers and paving-quality bricks. They're available in various colours and the blocks also come in a range of interlocking shapes. Lay these materials in patterns for best effect and use coloured mortar joints as a contrast.

You can also use special frost-proof ceramic tiles for a patio surface but they are very expensive and need a perfectly flat base if they are to be laid correctly. Consequently, they're really only suitable for very small patios.

Probably the simplest of surfaces is one made of a solid slab of cast concrete. Although the concrete can be coloured with pigments, many people find its surface appearance unattractive.

Concrete paving slabs are probably the best materials for a simple rectangular patio. Made with reconstituted stone, they're available in a range of reds, greens, yellows and buff tones with smooth, riven or patterned faces. Square, rectangular, hexagonal and half-hexagonal shapes are also made and they're easy to lay on a sand bed. Broken concrete slabs, known as crazy paving (see pages 61 -63 for details), can also be used as a patio surface, laid on a mortar bed.

Link the patio to the rest of the garden by building walls, paths and steps (see pages 46 -60 inclusive) in matching or complementary materials.

Marking out patterns

Whatever paving materials you choose you have enormous flexibility in the design of your patio. There's no reason why, for instance, it should be square or rectangular – most of the materials previously described can be laid in curves or can be cut to fit other shapes and angles.

Sketch out some patterns and dry-lay the paving in both width and length to test how the designs work in practice. Adjust the pattern or the dimensions of the patio to minimise the number of cut pieces you use. This will ensure the surface looks 'balanced'.

Concrete slabs can be laid in various grid and stretcher bond patterns but, for a more informal effect, whole and half slabs can be used together in a random fashion. Crazy paving should be laid with larger, straight-edged pieces at the borders and smaller fragments inside. You can lay bricks in herringbone or basket-weave designs.

Setting the levels

Draw your plan on graph paper, then use it to transfer the shape of the patio onto the site. Use strings stretched between pegs to mark out the perimeter of the patio. You must also drive pegs in to represent the surface level of the patio, so to ensure they're accurately placed you have to drive a 'prime datum' peg into the ground against the house wall (if the patio is to abut the house). The peg should indicate one corner of the patio and should be set in a hole about 300mm (1ft) deep with its top 150mm (6in) below the level of the house damp-proof course. If the soil is spongy you may have to dig deeper in order to obtain a firm enough surface on which to lay the foundations.

Set a second peg in the ground against the wall to mark the other side of the patio, and check that the level corresponds to that of the prime datum peg by holding a long timber straight-edge between the two and checking with a spirit level.

All other marking-out strings and pegs should be taken from the base line formed between these two datum pegs. Indicate squares and rectangles by stretching strings from the two pegs and checking the angle with a builder's square. Plot out curves by measuring from the base line at intervals and driving in pegs at the perimeter, or use lengths of string or a long hosepipe to mark the curves. Circles and half-circles can be marked out by taking a string from a peg placed as the centre of the circle: you place the first slabs or other paving along the string, then move it around the radius and set the next row.

If size permits, you could incorporate planting areas in your patio by leaving sections un-paved, or simply place tubs of plants on the perimeter.

Laying the paving

When you've marked out the shape of your patio, remove the topsoil (which you should save for use elsewhere in the garden) from within your guidelines and set intermediate datum pegs at 1.5m (5ft) intervals over the entire area of the excavation. These pegs have to be sunk to the level of the prime datum peg, using a spirit level on a batten between the pegs. Now is the time to set the drainage fall away from the house.

When you've set the levels, fill the hole with hardcore, which you must compact thoroughly by rolling and tamping. Then a layer of sand rolled out flat over the hardcore brings the level of the foundation up to that of the peg tops, and provides a flat base for the paving.

On a site that slopes away from the house you'll have to build a low retaining wall of bricks or blocks; the ground behind it can be filled with hardcore and then paved. However, where the ground slopes towards the house you must excavate the patio site in the bank (forming a drainage fall away from the house), and build a retaining wall.

READY REFERENCE

PREPARING THE FOUNDATIONS

Patio paving requires foundations of compacted hardcore covered with a blinding layer of sand. The excavation should be:

- about 150 to 200mm (6 to 8in) deep to allow for 75 to 125mm (3 to 5in) of hardcore plus the sand and paving
- about 230mm (9in) deep if the patio is to be built up to a house wall, so the paving can be set 150mm (6in) below the dpc.

ALLOWING FOR DRAINAGE

To ensure run-off of rainwater the patio surface must slope by about 25mm in 3m (1in in 10ft) towards a suitable drainage point, which might be an existing drain or soakaway.

On ground sloping away from the house: construct perimeter walls for the patio from brick, block or stone, to form a 'stage'. Infill with hardcore and a blinding layer of sand and gravel, then pave. The walls must be set on 100mm (4in) deep concrete footings.

On ground sloping towards the house: excavate the site for the patio in the bank, forming a fall away from the house. Build a retaining wall to hold back the earth.

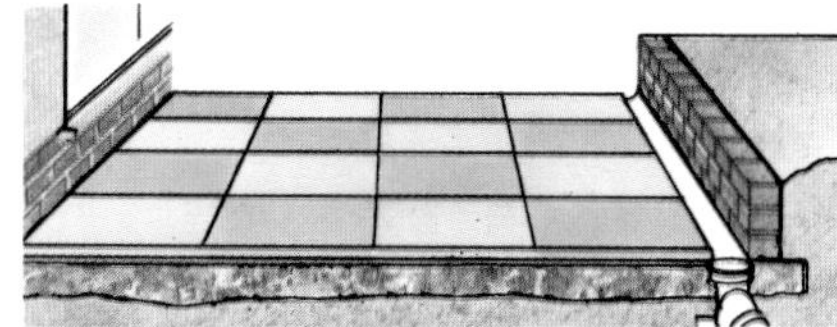

INSPECTION CHAMBERS

The patio must not interfere with access to drainage inspection chambers. If drain covers are within the patio area:

- build it up to the level of the new surface
- cover it with loose-laid slabs for access.

TIP: PAVING ROUND TREES

If you're paving around a tree or shrub to make a garden feature, keep the paving at least 300mm (1ft) from the trunk to allow rainwater to reach its roots.

For more information on laying paving slabs see pages 40 -49.

LAYING BLOCK PAVING

Concrete block pavers can be used to make a durable and decorative surface for a drive, path or patio. They're quick and easy to lay in an interlocking pattern on a sand bed – and no mortar is needed.

A path or patio must be durable enough to withstand fairly heavy traffic – from people, wheelbarrows, other garden equipment and furniture; and a drive has the additional weight of a car to contend with. So the surface must be tough and long-lasting.

But traditional surfacing materials – commonly cast concrete, asphalt and paving slabs – tend to give a plain, slab-like appearance, which often detracts from other features. 'Flexible' paving, however, is a method of making a hard surface that's both attractive and easy to lay.

Using concrete pavers

This type of surface uses concrete paving blocks, small-scale units that are very tough and can be laid in numerous patterns. Each block interlocks with its neighbours to form a solid, firm and decorative surface capable of supporting fairly heavy loads.

Concrete paving blocks are made in a variety of shapes (see 'Types of block') some with a rough texture for a more natural effect, others with smooth faces for a formal setting. They're also made in a choice of reds, blues, greys, charcoal and buff tones, and some have a mottled effect that resembles old brick. The simplest types of blocks are rectangular – usually about 200 x 100mm (8 x 4in) and about 65mm (2½in) thick. Some have a bevel or 'mock joint' on their top edge so that the shape of each block – and the bonding pattern – is accentuated when laid in a large area.

Other types – approximately the same size overall as the rectangular ones – incorporate zig-zag edges, some sharp, some gently rounded, which mate together when laid. There are basically three bonds in which you can lay concrete blocks: herringbone, stretcher and parquet (see 'Forming patterns with blocks').

By using the irregular-shaped blocks you can also create a rippling effect across the area of paving.

Herringbone is the strongest bond and ideal for drives, where the wearing and load from cars is considerable; stretcher and parquet bonds aren't as tough and so they're better for areas that will receive lighter traffic.

It's also possible to create variations on standard bonds, by laying blocks in pairs, for example, or mixing two bonds in an area of paving.

Laying the blocks is straightforward: they're simply positioned on a layer of sand between permanent edge restraints (see 'Site preparation') and are bedded down by vibration, using a special machine which you can hire. Suitable edge restraints are: existing paving, kerbstones set in concrete, a house or garden wall, pegged boards or a row of bricks. In most types of flexible paving no mortar is needed, either to bed the blocks or to point the joints between them. For this reason it's quite simple to remove the blocks whenever you like and reposition them on a new sand bed in a new bonding pattern.

Preparing the base

The base for your paving must be firm and level. Firstly, clear all weeds, and loose materials from the area. Dig out any soft spots and fill the holes with firmer soil – or even broken rubble – then compact the surface thoroughly using a garden roller.

If you're making a drive or other area that's to be used for vehicular traffic you should excavate the site and lay a firm base of at least 100mm (4in) of clean, fine hardcore, which you should also compact thoroughly.

Areas that are only to be used for foot traffic, such as paths and patios, will probably only need a base of compacted soil, unless the surface is clay or soft, peat soil; here, the base should be the same as for drives.

The base must be at least 115mm (4½in)

TYPES OF BLOCK

Block paving may be fired clay (A) or concrete (B, C and D – like small paving slabs). The clay ones are usually rectangular, but the concrete types come in a range of interlocking shapes as well. The surface texture is fairly coarse, and most have chamfered top edges to give the effect of a mock joint when laid.

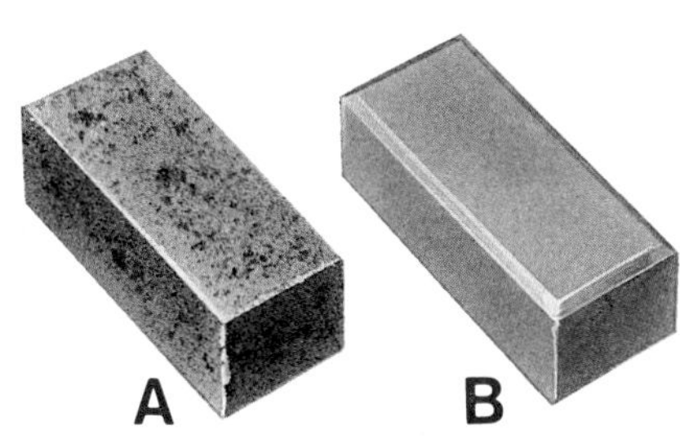

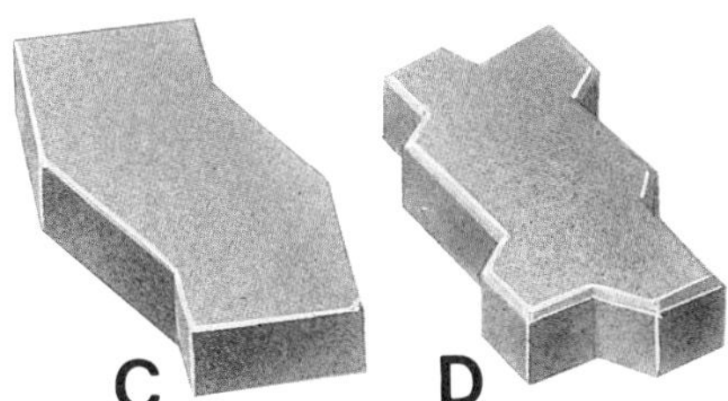

Forming patterns with blocks

1 *Interlocking blocks such as type D above look highly attractive when laid in a parquet bond of alternating pairs. They can also be laid in a herringbone pattern.*

2 *So-called 'fishtail' blocks (type C above) look best when laid in stretcher bond. The effect is of undulating lines in one direction, aligning joints in the other.*

3 *The simplest way of laying rectangular blocks is in stretcher bond, with successive rows having their joints staggered by half the length of a block.*

4 *An alternative to stretcher bond that looks very attractive with the rectangular block format is a simple herringbone pattern. Half-blocks form the edging.*

READY REFERENCE

CONCRETE BLOCK FORMAT

Concrete block pavers are available in a variety of shapes, colours and textures (see 'Types of block'). They measure:

- 200 x 100mm (8 x 4in) overall
- 50, 60 or 65mm (2, 2¼ or 2½in) thick.

FOUNDATIONS FOR BLOCK PAVING

Concrete block paving must have firm, level foundations. These should consist of:

- a well-compacted sub-base of soil for light use
- 110mm (4½in) of well-compacted hardcore for heavy traffic
- plus 50mm (2in) of sharp sand levelled and smoothed over the sub-base
- suitable edge restraints.

EDGE RESTRAINTS

To retain the edges of your paved area:

- fix creosoted timber formwork at the perimeter
- bed kerb stones or bricks in concrete
- or lay the paving against an existing wall or fence.

DRAINAGE FALLS

Your paved area must slope away from house walls. To allow for this:

- fix timber 'datum' pegs into the sub-base at 1.5mm (5ft) intervals so their tops are level with the sand bed
- check across the pegs with a spirit level on a straightedge
- incorporate a fall of about 1 in 50 with a 25mm (1in) thick wedge of timber (shim) under one end of the level.

TIP: LAYING BY HAND

For small, decorative, non-loadbearing areas of paving it's not really worth hiring a plate vibrator. Instead:

- lay the blocks as normal on sand
- bed each in place using an offcut of wood just larger than the block and a wooden mallet.

TIP: CUTTING BLOCKS

To cut paving blocks:

- chop along the cutting line with a club hammer and bolster chisel, or
- use a hydraulic stone splitter or guillotine.

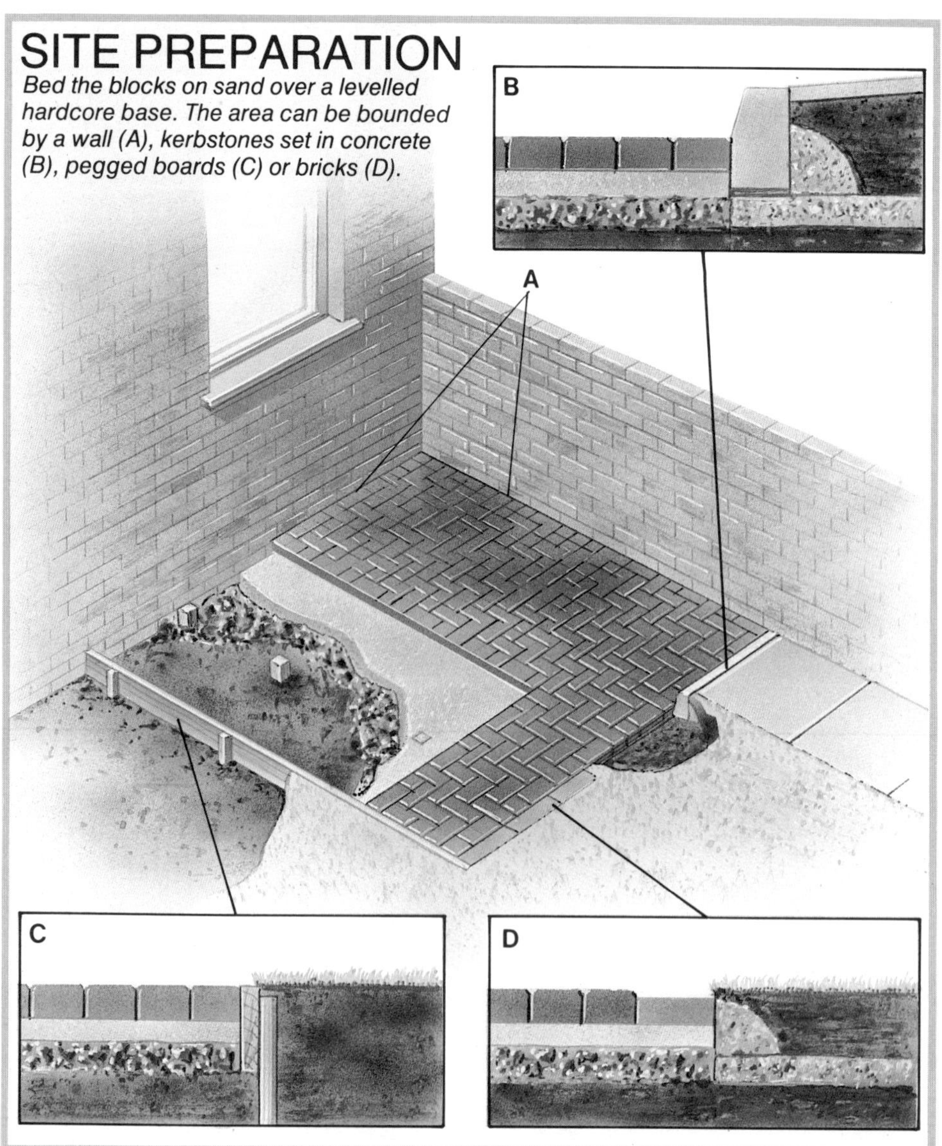

SITE PREPARATION

Bed the blocks on sand over a levelled hardcore base. The area can be bounded by a wall (A), kerbstones set in concrete (B), pegged boards (C) or bricks (D).

below the level of the completed paved surface. The top surface, in turn, must be at least 150mm (6in) below the house damp-proof course (dpc) to prevent moisture rising in the walls or rain splashing up above the dpc.

Setting levels

Unless there's a natural slope to the site you'll have to excavate the base so that the finished surface will slope to one side – or at least away from the house walls – with a fall of about 1 in 40, to ensure efficient drainage of rainwater.

To set the levels over the area of the base drive 300mm (1ft) long timber pegs into the ground at about 1.5m (5ft) intervals and set them at the correct level, taken from a 'prime datum', or fixed point of reference: two bricks below the dpc is adequate, for example. Use a long timber straightedge with a spirit level on the top edge to check that the pegs are set at the correct depth. Place a small wedge or 'shim' of timber under one end of the spirit level to incorporate an adequate drainage fall (see 'Site preparation').

Edge restraints

The edges of your concrete block paving are best retained, by the walls of your house, garden walls or existing paving: simply setting them against earth won't prevent eventual spreading of the blocks. If your dpc is lower than 150mm (6in), however, you must leave a 75mm (3in) wide channel between the wall and the paved surface to stop rainwater from splashing up and soaking the wall above the dpc. In this case you'd need to fit an edge restraint.

If there isn't a natural or existing edge restraint you'll have to install permanent formwork. Precast concrete, concrete path edging stones or a row of bricks on edge can be used, bedded in a strip of concrete, or you could use creosoted lengths of 38mm (1½in) thick softwood screwed or nailed to stout 50 x 50mm (2 x 2in) pegs driven into the ground outside the area you're going to pave. The depth of these battens should equal the thickness of the blocks plus the sand bed.

The sand bed

When you've levelled and compacted the sub-base and have set up the formwork you can lay the bedding sand directly on top. For this you'll need sharp (concreting) sand. As a rough guide to amounts, 1 tonne of sand is ample for 10 sq metres (110sq ft) of paving.

Although the final thickness of the sand bed should be 50mm (2in) you'll have to add more to allow for compaction of the paving.

You'll find it more convenient to work if you divide the load of sand into separate piles and position them at intervals along the site, away from the point at which you want to start laying the blocks.

Spread out the sand evenly over the sub-base, between the edge restraints, using a garden rake, then 'screed' or smooth the surface to the correct level with a straight-edged length of timber that spans the width of the area you're paving. The top of the sand bed should be levelled to about 50mm (2in) below the finished paving (and therefore the level of the edge restraint) when you're using 65mm (2½in) thick blocks, and about 45mm (1¾in) below when using 60mm (2¼in) thick blocks. To allow for this depth you can cut notches in the straightedge to form 'arms' that rest on the formwork; the body of this 'spreader' should be the thickness of the blocks plus about 15mm (½in) for compaction of sand (see 'Site preparation' for more details).

Where you're using an existing wall or fence as an edge restraint you'll have to set a temporary screeding batten on the sub-base, which you can remove after levelling. Fill the groove left by the batten with more sand and level the surface.

Screed the sand in areas only about 2m (6ft) ahead of the blocklaying for convenience and avoid walking on the sand during or after you've levelled the surface.

Laying the blocks

Start to lay the blocks in your chosen pattern against the edge restraint nearest to your pile of blocks. Bed each block up to its neighbours, without any gaps. The blocks can be fairly rough, so you'd be wise to wear thick gloves.

As your area of paving enlarges you should work from a plank laid across the blocks as a kneeling board, to spread the load. Lay plank runs also for transporting barrowloads of blocks from the main pile to the laying edge

LAYING THE BLOCKS

1 *Unless the blocks abut a wall, you'll need some sort of edge restraint. Here preservative-treated boards are nailed to stout timber pegs.*

2 *Start placing the blocks by hand in the pattern you want (here, a herringbone pattern). Leave gaps at the edges to be filled later with cut blocks.*

3 *As laying proceeds, kneel on a board to spread your weight. This avoids pressure on individual blocks, which could bed them too deep in the sand.*

4 *If you find that occasional blocks are obviously sitting proud of their neighbours, lift them and scrape away some of the sand beneath before replacing them.*

5 *Then tap the offending block back into place using the heel of a trowel or club hammer. Similarly, if a block is sitting too low, lift it and add more sand.*

TIP

6 *To avoid having to traipse to and fro for more blocks, fill a barrow with blocks and park it nearby. Use boards to spread the barrow's weight.*

7 *With stretcher and herringbone patterns, you will have to cut blocks to fill the edge gaps. Hold a block over the gap and score the cutting line on its surface.*

8 *You can cut paving blocks by hand with a brick bolster and club hammer, but if you have many to cut a hired hydraulic splitter will make light work of the job.*

9 *Finish off the laying sequence by placing the half blocks in position round the edges of the area as you cut them. Set them level with their neighbours.*

to avoid disturbing the blocks you've already laid but haven't compacted.

Cutting the blocks

Although you should use whole blocks wherever possible for maximum strength, you'll certainly need to cut some to size and shape where the paved area contains obstacles such as drains or inspection covers, and where it meets the edge restraints.

It's possible to mark the individual blocks to size and cut them using a club hammer and bolster chisel but you can hire a hydraulic stone splitter or guillotine, which will make the job much easier. Keep the guillotine close to the edge of the paving for convenience. If you must stand it on the paving, rest it on a board so you don't mark or upset any of the blocks. If you use a hammer and chisel be sure to wear gloves and goggles to protect your eyes from flying fragments.

Compacting the blocks

When you've laid a large enough area of blocks bed them firmly into the sand using a plate vibrator fitted with a rubber sole plate. This will settle the blocks into place and force sand up into the joints, without damaging them.

The plate vibrator should have a plate area of 0.2 to 0.3sq metres (2 to 3sq ft), a frequency of 75 to 100 Hz and a centrifugal force of 7 to 20kN, to ensure the blocks will be bedded correctly. Most machines of this type will fit easily into the boot of a car.

Make two or three passes over the paving with the plate vibrator in order to bed the blocks to the correct level, but avoid lingering in one place or you might sink them too low. Also, don't take the machine closer than 1m (3ft) to the unrestrained edge you're laying or you're likely to form a dip in the surface.

Finishing the paving

Finally, simply brush sand onto the paved surface and make a few more passes with the plate vibrator to force the sand down between the blocks.

If you're laying a very small, mainly decorative, area of blocks that won't be used for vehicles – a narrow border around a flower bed, for instance – it's acceptable to lay them without using the vibrator, although the job is more laborious and the results won't be as durable

Instead lay a thinner sand bed, moistened with water from a watering can fitted with a fine rose, and level and compact this with a straight-edged tamping board. Lay the blocks as previously described but bed each as level as possible using a wooden mallet with an offcut of timber just larger than the block (see *Ready Reference*). Water more sand into the joint, again using your watering can, to complete the area of paving.

COMPLETING THE JOB

TIP

1 Vibrate the blocks into the sand bed with a hired plate vibrator. You can avoid marking the blocks by running the machine over some old carpet.

2 After the first passes with the plate vibrator, scatter sand over the surface and brush it well into the gaps between the individual blocks.

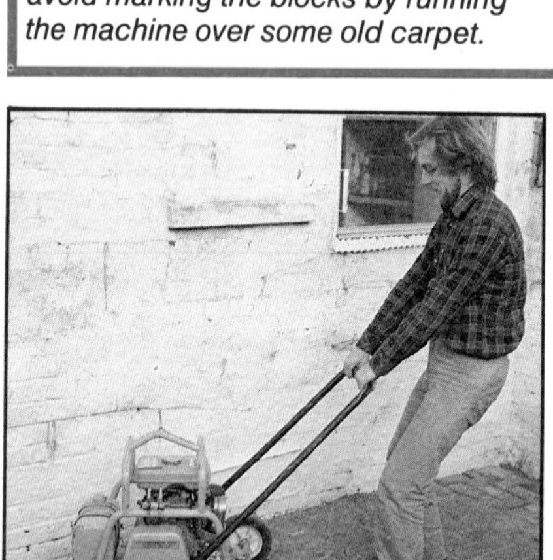

3 Run the plate vibrator over the surface again, making two or three passes over each area, to compact the sand between the blocks thoroughly.

4 Finish off the paving by brushing off excess sand with a soft-bristled broom, taking care not to brush out the joints. The paving is ready for immediate use.

Immediately you've laid the blocks and vibrated them into place, your paved area is ready for use. Inspect the surface after a few months to make sure there aren't any areas that have subsided fractionally. If there are uneven areas you may simply be able to lever out the relevant blocks and re-bed them on more sand.

Paving irregular areas

Because of the small size of the blocks, you can use them to good effect to pave irregularly-shaped areas. Where necessary the blocks can be cut at an angle to form neat edges, and circles can be formed by laying the blocks with wedge-shaped joints instead of parallel-sided ones – rather like forming a brick arch. To achieve perfect curves, lay the blocks using a string line attached to a peg at the centre of the curve, so that by holding the string taut you can align them accurately.

Coping with steps

If you have used blocks for a path or patio, you may want to link these areas with steps paved in the same way. Because of the small size of the blocks it is vital to bed them on mortar rather than on sand, particularly at the edges of the treads. You can still use sand to fill the joints between the blocks and so maintain the overall sense of unity.

Building with blocks

Similarly, you may want to build dwarf walls at the edges of your paved area, and while you could use any garden walling blocks for this there is nothing to stop you using two or three courses of paving blocks instead. The rectangular ones are laid just like bricks with pointed mortar joints between, but the interlocking and fishtail types can also be built up into walls if they are overlapped by half a block and aligned carefully.

CHAPTER 4

WALLS AND WALLING

Using the basic techniques of bricklaying, it's a short step to tackling simple garden walls – freestanding, or to retain earth and create split-level effects in sloping gardens. Perhaps the simplest type to build is the screen block wall, using square blocks available in a range of attractive pierced designs. If you want earth-retaining walls, you must build more sturdily and remember to allow water to drain from behind the wall (unless you want to create your own dam). Decorative arches can top off your handiwork, and there is a range of decorative finishes you can apply as well.

BUILDING A SCREEN BLOCK WALL

If you want privacy in your garden but don't like the idea of looking at a blank brick wall or a solid timber fence, then a wall made from pierced screen blocks could be the answer.

A garden wall made from pierced screen blocks offers a measure of privacy, yet doesn't shut out light and welcome cooling breezes. The blocks are quick and easy to lay, and give you a wall that's equally attractive from both sides. This makes them extremely good for marking out areas within the garden, without forming a heavy-looking solid barrier.

Screen blocks can also be used for boundary walls, though obviously they cannot provide total privacy. And while they're not intended to be load-bearing, they are able to support, say, the corrugated plastic roof of a carport, or the beams of a pergola or awning of light-weight construction.

Choosing the blocks

It's best to write off for manufacturers' brochures, or to visit a local builder's merchant, to find out exactly what's available in your area. You'll find several geometric patterns available (see *Ready Reference*): some are self-contained within each block, others need a set of blocks to complete the design. Solid patterned blocks are also sold, and can be set at intervals throughout the wall for added interest.

Most blocks are made from white concrete (to enhance their light appearance) and measure 300mm (1ft) square by 100mm (4in) thick. Because of the nature of their design, there are no half blocks, nor can blocks be cut.

To strengthen the wall, you have to install supporting columns called 'piers' at regular intervals along its length; while these can be built using bricks it's more common to construct them from special 'pilaster blocks'. These are basically hollow 200mm (8in) cubes, which incorporate slots to take the screen blocks. Four types of pilaster are available: those forming intermediate piers have a slot on two opposite sides; end pilasters have a single slot; corner pilasters have a slot on two adjacent sides; finally, there are pilasters with a slot on three sides, which are used at T-junctions. In addition, you'll find half-pilaster blocks to leave the tops of the piers level with the tops of the screen blocks, and to complete the wall there

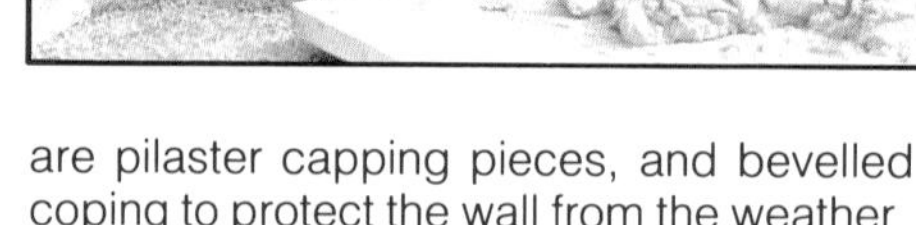

are pilaster capping pieces, and bevelled coping to protect the wall from the weather.

Designing the wall

Before you buy any materials, plan your wall carefully, making a scale drawing on squared paper to enable you to calculate what you need.

Start by deciding what happens at the base of the wall. The simplest option is to lay a concrete strip foundation (see Chapter 2 pages 30-34) finished off at ground level. However, here the foundation will be left on show, so you may prefer to stop the foundation below ground level. But this also presents a problem: part of the first course of blocks will be buried in the ground, which makes the wall look sunken. It's therefore best to build a low wall in bricks or solid concrete blocks, and to top this with a flat coping to form a sort of plinth on which the screen wall proper can be built.

The next step is to work out the arrangement of the blocks. Screen walls are always laid with what's called a 'stack bond': that is without staggering the vertical joins, as in conventional brick and blockwork. This makes the wall quick and easy to build, but does mean that its dimensions are restricted. The wall must be a whole number of blocks high, and the length must be an exact number of blocks plus pilasters (remember, there are no half blocks and blocks cannot be cut). What's more, stack bond gives the wall little intrinsic strength, and so it's vital that you incorporate piers at the intervals recommended by the manufacturer of the blocks.

Unfortunately, even then the wall may not be strong enough for some situations. High walls, walls subject to strong winds, boundary walls (especially those next to a public highway) and the like, require additional reinforcement, and here, once again, it's best to follow the manufacturer's recommendations. In general, though, if the wall is inside the garden, has piers at no more than 3m (9ft) intervals, and is over 1.8m (6ft) high, the pilasters must be reinforced using 50x50mm (2x2in) angle iron of the sort used as fence posts. All boundary walls should have the pilasters reinforced in this way, and if the wall is over 1.8m (6ft) high, then 60mm (2½in) wide welded steel mesh made for the purpose and available from a builder's merchant should be bedded into the horizontal mortar joins to stop the vertical joins 'zipping' open. This mesh should also be used to tie the blocks into piers made of brick.

Storage and handling

Working from your scale drawing, you can order the necessary materials. First, though, there are a few points to bear in mind about handling and storage.

BUILDING A PIER

1 *When you've prepared your wall's foundations, stack your blocks nearby, mix some mortar and trowel on a bed where the first pier is to go.*

2 *Position your first pilaster block squarely on its mortar bed and use your trowel to scoop off any excess mortar that's squeezed out.*

3 *When you're satisfied that the first block is bedded evenly, trowel a mortar bed around its top rim, on which to lay the second block.*

4 *Lay the second block on top of the first and bed it down. Scrape off excess mortar but don't smear any on the face of the blocks or it will stain.*

5 *Continue to build up the blocks until you reach the height you want for the pier. If it's higher than about seven blocks, include reinforcement.*

6 *Check at intervals that the pier is built level and square by holding your spirit level against and on top of the blocks; adjust if necessary.*

READY REFERENCE

SCREEN BLOCK FORMAT

Screen walling blocks are made from white concrete and can be used for garden and boundary walls or semi-load-bearing walls. To make a wall you'll need:

- standard blocks (1) measuring 300mm (12in) sq x 100mm (4in) thick, with a variety of cut-out geometric patterns and some solid patterned blocks
- hollow 200mm (8in) cubes called 'pilasters', with slots in one side (2) or in two opposite sides (3), to make end and intermediate piers
- pilasters with slots in two adjacent sides (4) for corners
- pilasters with slots in three sides (5) for T-junctions.

To finish off the wall you'll need:

- capping pieces (6) to complete the pier tops
- coping stones (7) to protect the top of the wall from the weather.

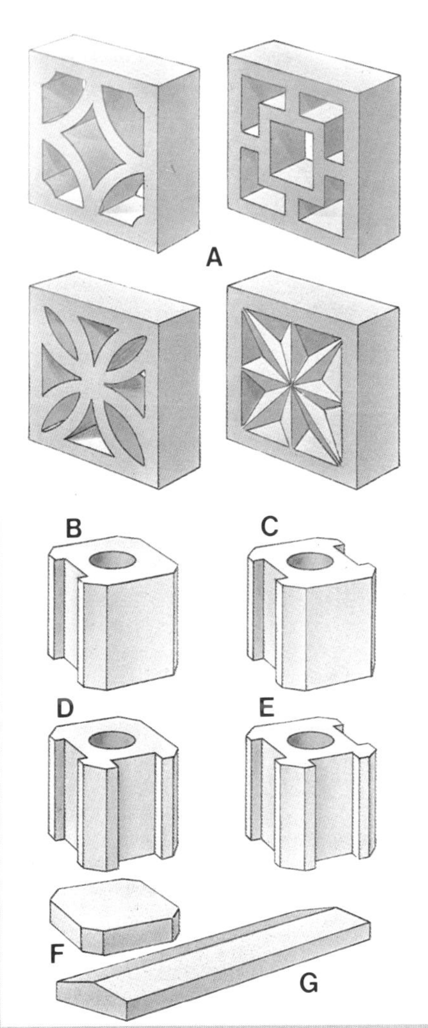

BUILDING THE WALL

1 Once you've built the first pier you can build your wall out from it. Lay a bed of mortar about three blocks wide on the foundations.

2 Butter one edge of your first block with mortar and trowel furrows in the surface to aid positioning of the block and adhesion of the mortar.

3 Position the buttered edge of the block in the slot in the side of the lowest pilaster block; then trowel mortar onto the opposite side.

6 Continue to build up the wall, checking constantly with your level that it isn't bulging or crooked; then stack up the second pier.

7 Fill the cavity down the centre of the piers with mortar for extra strength, then bed capping stones in mortar on top.

8 Finish off the top of the wall with coping stones bedded in mortar. These have bevelled tops so that rainwater is thrown clear.

The edges and corners of the blocks are particularly vulnerable to damage, so handle them with care. Any chipping will result in unsightly joins in the finished wall. As for storing the blocks until they're needed, stack them on edge with alternate rows at right angles to each other to stop them toppling over, then cover them with a tarpaulin or plastic sheeting to protect them from the rain so they're not soaking wet when you come to lay them.

Finally, remember that concrete is rough on the hands, and that the edges of the blocks can be sharp. It's therefore best to wear thick gloves when handling them.

Laying the foundations

The first step in building your wall is to lay the foundations (see Chapter 2 pages 30 -34). These should be twice the width of the wall; that is 200mm (8in) with wider sections for the pilasters. Their depth depends on the height of the wall and on the condition of the ground, but in general a 125mm (5in) layer of well-rammed hardcore topped with 125mm (5in) of concrete is sufficient. Reduce this to 200mm (8in) overall – 100mm (4in) of concrete – if the wall is less than 900mm (3ft) high.

The concrete used is a 1:2½:4 mix of Portland cement, sharp sand, and coarse aggregate. Alternatively, if you prefer to use mixed aggregate, you'll need five parts of this to one part of cement.

Where you're going to reinforce the pilasters, rather than set the reinforcing bars into the foundations, set in what are called 'starter bars'. These are short lengths of 10mm (⅜in) diameter mild steel rod, bent at right angles to hold them fast in the concrete. About 300mm (1ft) of rod should protrude above the surface of the foundations, and it's important that these stubs are vertical.

Once you've completed the foundations, leave the concrete to harden for three or four days before proceeding to build the wall.

Mortar and pointing

Screen blocks are best laid using a mortar made from one part masonry cement (this contains a plasticiser to make the mix more workable) and five parts builder's sand. Alternatively, you can use a 1:1:6 mix of Portland cement, lime, and builder's sand. For small jobs, however, it's often more convenient to use a dry-mixed bricklaying mortar.

Using any of these mortars, you can obtain a neat finish by scraping off the excess as each block is laid, then, having built a reasonable amount of wall, run a piece of plastic tubing along the joins to give them a half-round profile.

For a more attractive finish to the mortar joins it's better to rake them out after the

4 *Butt the second block up to the first and scrape off any excess mortar that's squeezed out of the vertical joint or out of the mortar bed.*

5 *Continue to lay more of the first course until you reach a point where you need another pier. Check that the wall is level across its top.*

9 *If you're building a second wall from one of your piers but won't finish it in one day, 'rack back' the corner to leave a stronger bond.*

10 *When your screen wall is complete, point the mortar joints between each block with a length of plastic tube to give a neat, half-round profile.*

mortar has stiffened then point the joins with a coloured mortar.

Building up the piers

Begin your wall construction by building up at least one pair of pilasters to a minimum of three courses high. If you're including any reinforcement, tie lengths of angle iron firmly to the starter bars using stout wire, having cut it to finish about 50mm (2in) below the tops of the pilasters.

You now just thread the pilaster blocks over the reinforcing bars, and bed them in mortar, ensuring that they're accurately centred, as well as vertical and level. Make sure, too, that the spacing between pilasters is correct. Check this by dry-laying the first course of screen blocks.

With each pilaster block in place, fill the pier with a slightly runny concrete mix of one part cement to three parts sharp sand.

Laying the blocks

You can now start to lay the screen blocks, each course being laid working inwards from the pilasters. It's just like ordinary blocklaying but without the bonding. With the two end blocks in position, stretch a string line between them and use this as a guide to laying the rest of the course.

Continue in this way building up one course after another and extending the pilasters as necessary. Check frequently with a spirit level that the blocks are correctly aligned, laid to a horizontal and standing truly vertical. As the wall rises, check, too, that it's not 'bellying out' by laying a long, straight-edged batten diagonally across its face.

Lastly, once the wall has reached its final height, top the pilasters with the special capping pieces, and lay coping stones on top of the screen blocks, bedding both in mortar in the normal way.

READY REFERENCE

FOUNDATIONS FOR SCREEN WALLS

For walls over 900mm (3ft) high, foundations should be 200mm (8in) wide – and wider where there are piers – and should consist of 125mm (5in) of well-rammed hardcore topped with 125mm (5in) of concrete.

For lower walls, foundations should consist of 100mm (4in) of well-rammed hardcore topped with 100mm (4in) of concrete.

REINFORCING THE WALL

Walls over 1.8m (6ft) high subject to strong winds, and boundary walls, especially if they're close to a public highway, need extra reinforcement. This should consist of:

- starter bars consisting of lengths of 10mm (3/8in) diameter mild steel rod bent at right angles and set in the foundations so that about 300mm (1ft) of metal protrudes
- lengths of 50mm (2in) sq mild steel angle iron tied to the starter bars with stout wire (A)
- the hollow pilasters slot onto the bars and lengths of 50mm (2in) wide steel mesh are bedded into the horizontal mortar joins to stop the vertical joins opening (B).

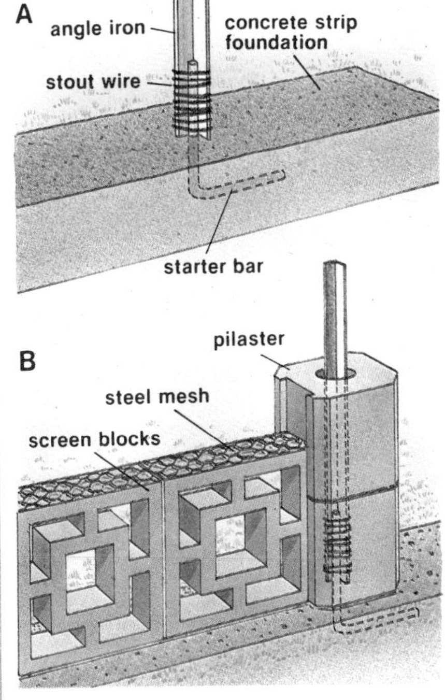

CONCRETE MIXES

For the foundations, use a concrete mix consisting of 1 part Portland cement to 2½ parts sharp sand and 4 parts coarse aggregate, or 5 parts mixed aggregate to 1 part of cement.

BUILDING RETAINING WALLS

Regardless of whether your garden is flat or sloping, earth-retaining walls are an ideal way of remodelling it to create interesting features such as a raised lawn, a sunken patio or terraced flower beds.

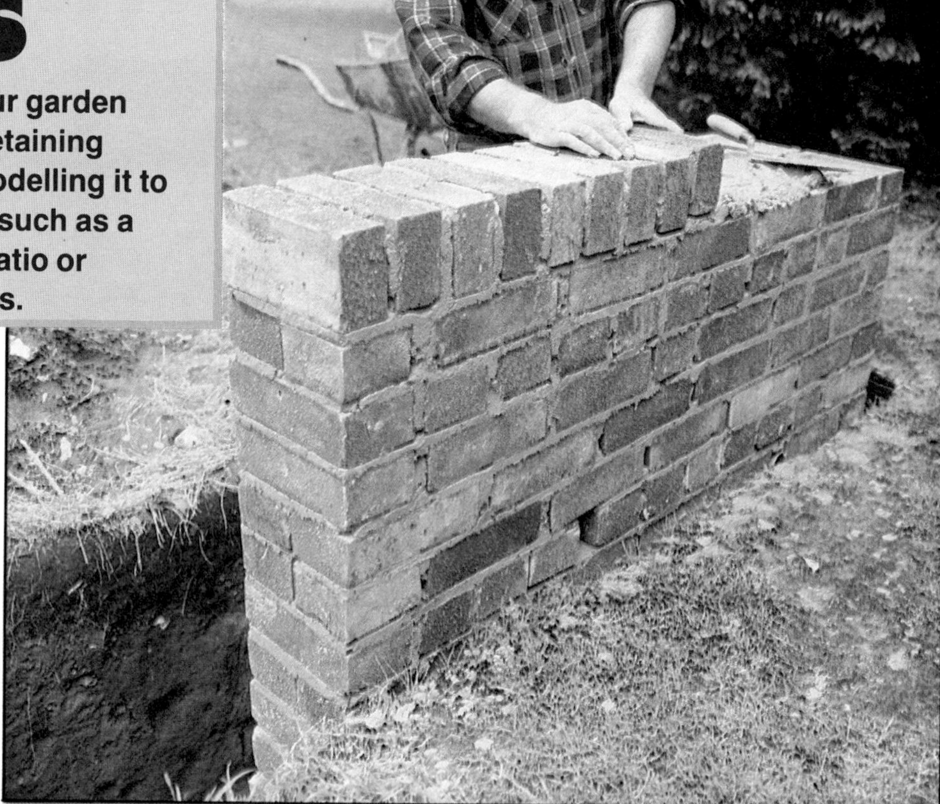

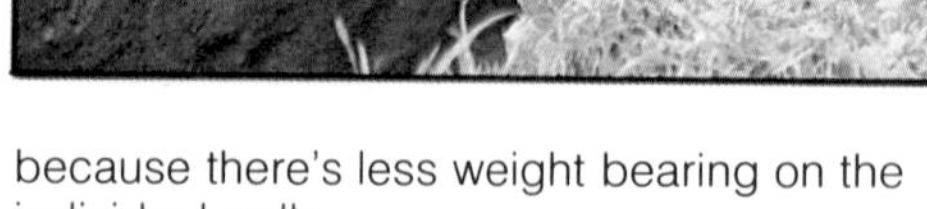

A sloping garden, although it may be an attractive, natural-looking feature of your property, can be hard work to keep in good order. You'll probably find it tiresome to work on, especially if you have to carry heavy tools and equipment such as the lawnmower to the top.

You can, however, landscape the shape of the bank into a series of flat terraces, connected by steps which not only offer easier access when gardening but also give you greater flexibility in planning your planting arrangements.

But you needn't only alter the shape of a sloping plot; it's also possible to re-style a flat, featureless site by building a raised flower-bed, lawn or patio. By digging out areas of your garden you can even create sunken features.

However you design your new scheme you'll have to incorporate in it a solid, load-bearing barrier called a retaining wall, which must be strong enough to support your remodelled earth and prevent the soil from spilling onto the lower level.

Planning a retaining wall

The scale of your wall largely depends on the amount of earth it's to retain and the steepness of the bank. But if it's to be over about 1.2m (4ft) high, you should consult your local authority. They may demand that you include some form of safety measures for the structure or that you adhere to certain building standards, especially if it's part of a boundary wall, where it could affect public access.

Start planning by sketching out your landscaping ideas, and try to keep your terraced or sunken features in scale with the rest of the garden. If you simply want to border a shallow sunken patio or path, for instance, or create a raised lawn or flower-bed in a level garden, you could build a fairly low wall, 300 or 600mm (1 or 2ft) high, and lay flat coping stones on top to use as informal seating, or as a display for garden ornaments or plants in containers.

A gradual sloping site with two or more terraces will allow you more scope for varied planing than a single, high 'platform' would, and also forms a much stronger structure because there's less weight bearing on the individual walls.

Where you're building a flight of steps in a bank (see Chapter 3 page 58) to connect your terraces, you may need to include 'stepped' retaining walls at each side to stop the earth from spilling onto the treads.

When you're remodelling your ground don't forget to set aside any topsoil and re-use it for any new planting beds. The remaining areas that you excavate for foundations must be well consolidated or compacted, then levelled, on a layer or hardcore, so that they're firm enough for normal traffic without danger of subsiding.

Laying the foundations

The prime requirement for your wall, whatever building material you use, is to build it on adequate foundations set below the frost line (see *Ready Reference*).

In effect, your foundations should be a cast concrete strip or 'raft' foundation (see Chapter 2 pages 30 -34) the length of the wall and about twice its width. For example, for a typical 225mm (9in) thick brick wall 1.2m (4ft) high, built on clay soil, you'll have to lay your concrete 500mm (20in) wide and 150mm (6in) thick. Set the entire foundation in a trench about 500mm (20in) below soil level. In very loose soil you'll have to increase the width of the strip, or build a 'key', which projects down at the toe, or outer edge of the

READY REFERENCE

PARTS OF THE WALL

A retaining wall must have enough mass and sufficiently solid foundations to resist the lateral, or sideways, pressure of piled earth and the water that builds up within it. A typical retaining wall up to 1.2m (4ft) high should consist of:

- cast concrete strip foundations (A) the length of the wall, 500mm (20in) wide, 150mm (6in) thick, and set in a trench 500mm (20in) below soil level
- rigid construction using bricks (B), blocks or stone
- adequate drainage from the back of the wall (C).

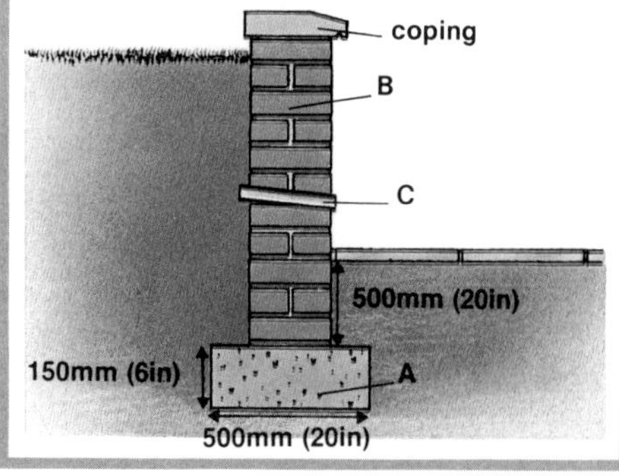

BUILDING UP TO GROUND LEVEL

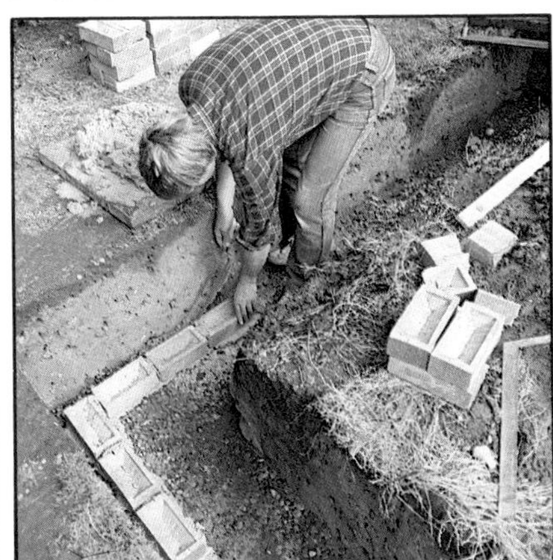

1 *Start laying the first bricks at a corner if one is planned, at one end of the wall otherwise. Bed down the bricks in the outer 'skin' first.*

2 *Having laid several bricks on each side of the corner, lay the first course of the inner 'skin' alongside them. Note how the bond is arranged.*

3 *With the first course complete at the corner, check with a builders' square that there is a perfect right angle inside and outside the corner.*

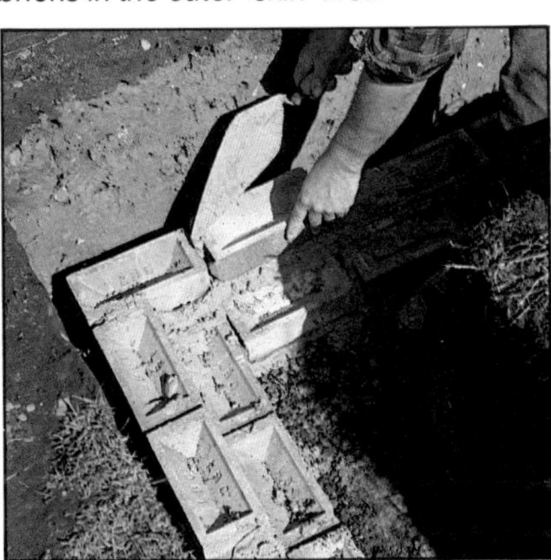

4 *Return to the corner, and start to lay the second course of brickwork in the outer skin, scraping off excess mortar with the side of the trowel.*

5 *After laying two or three courses of stretchers, lay the next course as headers – a bonding pattern known as English Garden Wall bond.*

6 *As you add each course to the wall, check that it is level, that the faces of the wall are truly vertical and that the corner is a true right angle.*

7 *When the footings reach ground level, form drainage holes in the wall by bedding short lengths of plastic waste pipe in a generous mortar bed.*

8 *Check that the pipe slopes down towards the outer face of the wall at a slight angle so that it will drain water away efficiently.*

9 *With the piece of pipe in place, you will have to cut the bricks in the inner and outer skin of the wall to maintain the bonding pattern.*

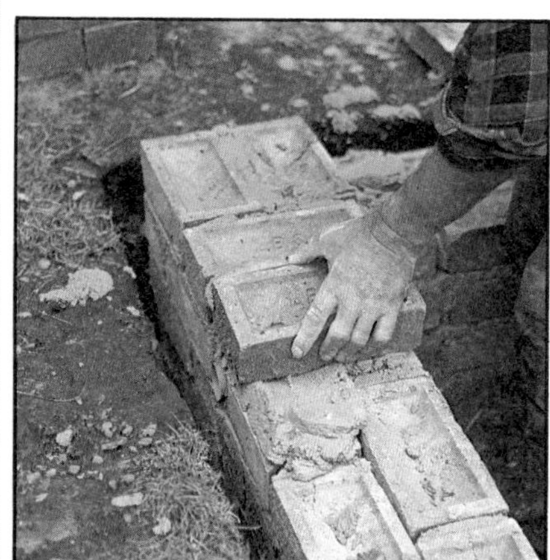

10 *Continue building up the wall above ground level, alternating three to five courses of stretchers with a course of headers.*

11 *As an alternative to building in short lengths of pipe to provide drainage, you can form weep holes – simple gaps left between the bricks.*

12 *Where you have left weep holes in the ground-level course, lay the next course as stretchers to bridge them and maintain the wall's strength.*

foundation, to help prevent the wall sliding forwards under pressure from the retained earth (see *Ready Reference*).

Choosing walling materials

Your earth-retaining wall must have enough mass, as well as sufficiently solid foundations, to resist the lateral, or sideways, pressure of the retained soil and the rainwater that collects in it (see *Ready Reference*). So long as you provide this strength, you can build your wall from most common building materials – bricks, concrete or stone blocks, cast concrete, and even timber. Which you choose depends on the visual effect you want to achieve and on the conditions you're building in.

Bricks must be dense and durable to withstand the damp conditions to which your wall will be exposed, and can give a neat, formal appearance in a garden that has rigidly defined areas, such as lawn, patio and rockery. Choose 'special quality' or engineering bricks (see pages 168 -9) which are quite impervious and ideal in wet surroundings. Ordinary quality or common bricks are far too porous and susceptible to frost damage, although if your wall's going to be fairly small and in a sheltered situation you can use the more attractive second-hand ordinaries in conjunction with a water-proofing treatment (see below).

The strength of a brick wall is in its bonding and the mortar mix used. A brick retaining wall, therefore, must be built a minimum of 225mm (9in) – or one whole brick – thick in a tough bond such as Flemish, English, or English Garden Wall bond (see pages 8 -13) for strength.

Concrete blocks, which are much larger than bricks and much lighter to handle enable you to build a high wall relatively quickly. They're available either solid or with hollow cavities to take reinforcement (see *Ready Reference*, page 82). Make sure you choose dense quality blocks that are suitable for use underground.

If you don't like the plain, functional look of concrete blocks you can clad the completed wall with a cement render, or coat it with a textured masonry paint. A rendered finish, though, is likely to crack eventually in damp conditions. Alternatively you can just use the blocks underground and continue the wall above ground with bricks. Concrete blocks should be laid in stretcher bond to give the strongest structure.

Decorative concrete walling blocks are suitable for low retaining walls; they're available both in brick size and in the larger 215x440mm (9x18in) size, and usually have a split-stone or riven face for a more natural, softer look. They also come in a range of reds, greens, and buff tones for a more attractive finish. You should only use this

READY REFERENCE

DRAINING THE BANK

To prevent a build-up of rainwater behind the retaining wall:

- leave vertical joints free of mortar every 1m (3ft) just above lower ground level to act as weep holes, or
- bed 75mm (3in) diameter drainage pipes in the wall every 1m (3ft) just above lower ground level
- additionally, bury lengths of pipe in pockets of gravel behind the wall to drain water to the sides of the wall
- include a trench 200mm (8in) wide, filled with layers of compacted bricks, pebbles or gravel for rapid drainage.

TIP: WATERPROOF THE WALL

Although it's impossible to waterproof an earth-retaining wall totally, you can reduce the risk of serious damp penetration by:

- painting the back face of the wall with two coats of bituminous paint, or
- tacking a 250-gauge thick polythene sheet to the back face of the wall.

PREVENTING LANDSLIDES

In loose soils your wall may tend to be pushed forward by the weight of the retained earth. To prevent this:

- form a 'key' at the toe of your strip foundation
- set your wall and foundation at an angle or 'batter' – no more than 1 in 5 – into the bank
- build the bank side of the wall in steps, becoming narrower at the top.

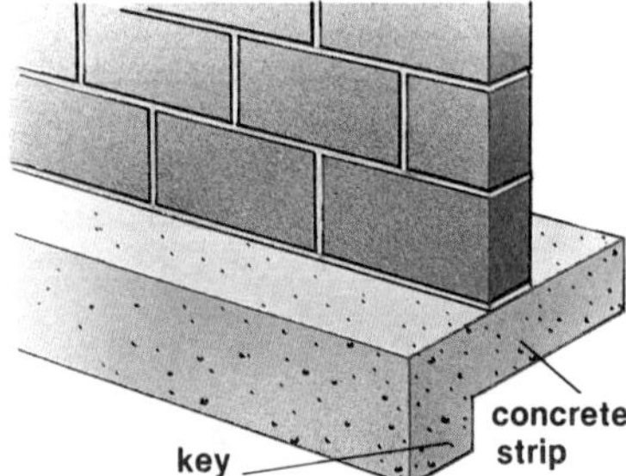

INSTALL MOVEMENT JOINTS

A long retaining wall will need a break in the bond, called a movement joint, which allows for seasonal expansion or contraction and prevents the masonry from cracking. Joints should be:

- the height of the wall
- filled with strips of expanded polystyrene
- pointed with a weak mortar mix to conceal the polystyrene filling.

In a brick wall:

- leave joints every 3.6m (12ft)

In a block wall:

- leave joints every 1.8m (6ft).

COMPLETING THE WALL

1 *Instead of English Garden Wall bond, you can use Flemish bond; each course has alternate stretchers and headers laid as shown.*

2 *When your wall has reached the height you want, finish it off with a soldier course – a course of bricks laid on edge to form a coping.*

3 *Since the top of the wall is the most exposed, ensure that a complete layer of mortar is 'chopped' down between each of the soldiers.*

4 *Complete the soldier course by pointing neatly between the bricks. Rounded joints, formed with a piece of metal or hosepipe, look neatest.*

5 *Finish the job by brushing down all the wall surfaces with a soft brush to remove any excess mortar that could stain the brickwork.*

6 *Leave the wall to stand for a few days, then start to back-fill behind it, tamping the soil down gently but firmly as the level rises.*

type of block above ground. Use the same bonding patterns as bricks for a stronger structure.

A cast concrete retaining wall is tough and durable. but it has a drab, slab-like appearance and calls for the construction of sturdy timber formwork to mould the mix while it hardens. If you think you'll need such a robust structure you can make it look a little more attractive by adding a pigment to the mix (see pages 170 -171).

You can even build a dry stone retaining wall for an unobtrusive cottage-style garden wall, although it's not suitable for holding high banks or heavy soil weights. The irregular soil-filled gaps between each stone make ideal places for introducing creeping plants to mellow the overall look of the wall. In this type of wall, each stone must be tilted downwards into the bank, forming a slanted or 'battered' wall. This will increase the strength of the structure, and will also give it a much less formal appearance.

Naturally rot-resistant hardwoods or preservative-treated timber can be used to make a wooden earth retaining wall, or you can use it as cladding for a concrete or blockwork wall. Railway sleepers, for instance, which you can often buy from specialist suppliers, can be used to make sturdy retaining walls if you pile them on top of each other, or stick them in the ground vertically, and support them with steel rods or stout fence posts set in concrete foundations. You could even stack concrete or wooden fence posts in this fashion, or in a dove-tailed design, to leave small soil-filled gaps between each post for planting.

Reinforcing the wall

Brick walls and walls made of small-scale block materials are susceptible to bulging outwards under pressure. You can reinforce them by setting hooked metal rods in their mortar joints, which project through the back of the wall into the bank where they're 'tied' to blocks of cast concrete called 'deadmen', which act as stabilisers. Timber retaining walls can be reinforced with a similar arrangement of sturdy timber braces set in the bank.

In a hollow concrete block wall you can lay a wider strip foundation on the downhill or outer side of the wall and set in it L-shaped steel rods on which you can slot the blocks for extra reinforcement (see *Ready Reference*), then fill in the block cavities with concrete.

A tall wall over about 1.2m (4ft) high must incorporate supporting columns called 'piers' (see *Ready Reference*) at each end, and also at intermediate positions along its length if it's very long. You may, though, just want to include piers in a smaller wall purely for visual effect, where the wall breaks at each side of a flight of steps, or where the wall must support a heavy gate.

In a brickwork wall you should bond the piers into the structure for strength but in solid block walls you can simply tie a stack-bonded pier to the wall by setting galvanised expanded metal mesh in the horizontal mortar joints.

A wall of hollow concrete blocks can be similarly tied to a matching pier with special metal cramps, and the hollow cavities can then be filled with concrete for extra rigidity.

Another means of reinforcing an earth-retaining wall is to build it thicker at its base, stepping back the courses from the earth side to the final thickness at the top (see *Ready Reference*, page 80).

Long walls must also incorporate breaks in the bond called 'movement joints', which allow for seasonal expansion and contraction. Joints should run the full height of the wall and can be packed with a compressible material such as expanded polystyrene which can then be pointed with a weak mortar mix to conceal the gap. You should leave movement joints at every 3.6m (12ft) in brick walls and at every 1.8m (6ft) in block walls.

Drainage and damp-proofing

Because of their location, buried in the ground and holding back a large amount of earth, retaining walls are susceptible to dampness. It's vital, therefore, that you include adequate drainage in the structure so that the earth behind doesn't become waterlogged and heavy. In the long run this would weaken the wall and could even cause it to collapse. Freezing water trapped behind the wall could also cause the masonry to crack.

You can provide drainage in two areas: at the back of the wall and actually through its face. To drain the back you can set pipes of porous, unglazed terracotta or plastic – slightly sloping and surrounded by gravel to quicken the rate of drainage – behind the wall, just above the foundation. Take the pipe run along the wall to each end, where the rainwater can drain into a soakaway or other suitable drainage point.

To drain the retained bank through the wall you can simply leave 'weep holes' – open mortarless joints – every one metre (3ft) just above ground level, or set short lengths of 75mm (3in) diameter drainage pipe in the wall at these intervals, tilted slightly downwards to the outside.

In very wet areas, or where you're building a high wall, you'd be wise to dig out the earth behind the wall and make an infill trench of well-rammed broken bricks topped with gravel. This will help to relieve the pressure on the wall from expansion as the soil soaks up water in winter.

Where there's an excessive amount of water draining from your wall you should make a shallow gutter or gully at its base to carry the water to a suitable drainage point.

In addition to providing drainage for the bank you can further protect your wall from damp by applying two coats of bituminous paint to the back face, or by tacking on a sheet of thick 250 gauge polythene. Take care not to damage this membrane when you back-fill behind the wall.

It's not usual to incorporate a damp-proof course (dpc) in a garden wall, but for greater protection from rising damp you can bed a layer of slate between courses, or lay a course of water-resistant engineering bricks at this point instead. If the wall adjoins the house, and rises above the house damp course level for any reason, a vertical dpc must be included between house and wall to prevent any moisture from rising into the house structure.

Finally, you should bed sound copings of concrete or brick on top of your wall, to keep water out of the mortar joints. Precast concrete copings usually have a bevelled top for drainage, project beyond the face of the wall, and have channels called 'drip grooves' under the front edge to prevent water trickling back onto the surface of the wall. You can also buy a variety of special coping bricks with shaped edges for a softer, decorative effect.

Supporting the excavated earth

Where you're building your earth retaining wall in a steep bank, or in loosely-packed earth, you might have to construct a type of 'dam' from temporary timber struts and braces to shore up a series of vertical boards or planks called shuttering. This should be laid directly against the face of the soil to hold it in place while you can dig and lay your concrete strip foundations and build your wall.

On very high walls you might find it easiest to build the shuttering in stages as you excavate the site. You should leave about 300 to 600mm (1 to 2ft) between your proposed retaining wall and the face of the shuttering to allow you plenty of access when laying foundations and building the wall.

Once the foundations have been laid and you've completed the lower courses of bricks or blocks you should start to lay your drainage pipes. Set some actually in the wall, draining to the front, and lay others at the back of the wall, set in gravel or hardcore, to drain the sides.

Continue to build the wall in the normal way and, when it's completed, and you've pointed the joints, you should leave the structure for at least 24 hours to set before removing the shuttering.

Back-fill the wall with well-compacted soil or a porous filling (see *Ready Reference*) and top the wall with concrete or brick copings to complete the structure.

READY REFERENCE

BUILDING PIERS

Retaining walls over about 1.2m (4ft) high need supporting columns called 'piers' at each end, and at intermediate positions if they're very long. Piers should be:

- a brick or decorative block column bonded into a brickwork or blockwork wall (A)

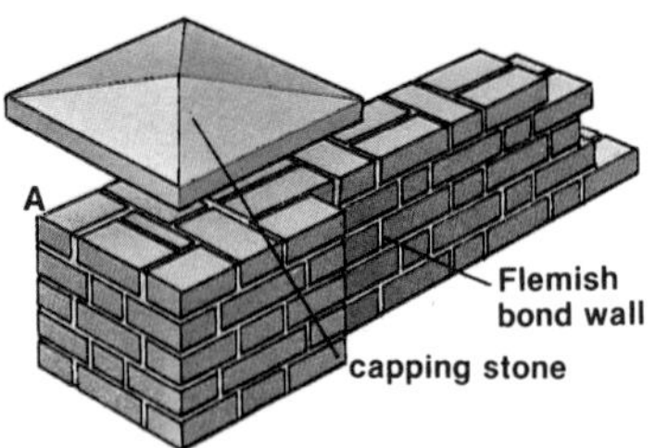

- a hollow concrete-filled blockwork column tied alongside a hollow block wall with metal cramps (B)
- a column of blocks tied to a solid block wall with galvanised expanded metal mesh (C).

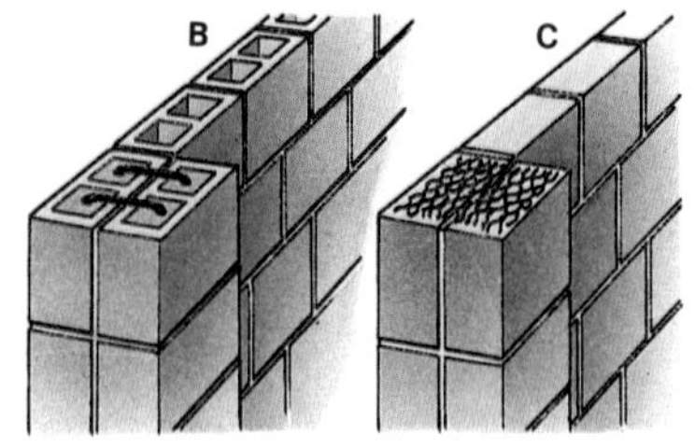

REINFORCING A LARGE WALL

High, heavy-duty walls will require reinforcement to hold back the weight of the earth. Hollow concrete blocks are the simplest to reinforce. To do this:

- set L-shaped steel rods (A) the height of the wall in a wide concrete strip foundation (B)
- slot hollow concrete blocks (C) onto the rods as you build the wall
- fill the cavities in the blocks with concrete (D).

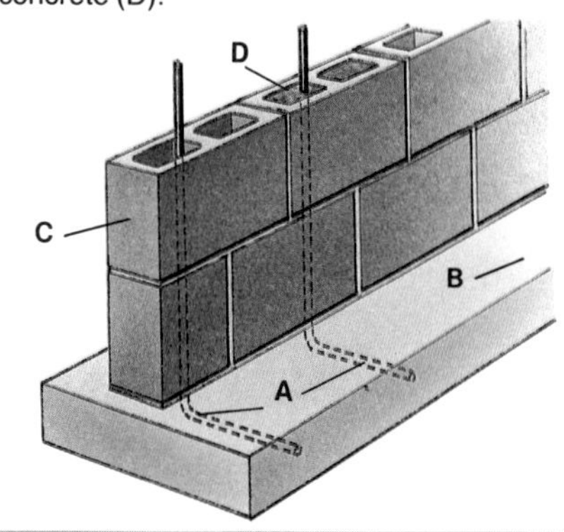

BUILDING ARCHES IN BRICKWORK

An arch can make a decorative feature of your door and window openings – or even give a grand treatment to your garden gate. Here's how to build a basic brick arch.

If you're making a large opening in a wall for a new door, window or serving hatch – or where you're knocking two rooms into one – you must include adequate support for the masonry above, and any load that bears on it. The usual way to span an opening such as this is to bridge it with a rigid horizontal beam called a lintel, or, for very large openings, a rolled steel joist, or RSJ. But this limits you to a square or rectangular opening, which you may not think is really suitable for a more decorative effect.

In the past, arches – although there were many complex, elegant variations on the basic shape – were used for more practical reasons: wider openings could be spanned than was possible with timber or stone lintels, and they were also used in conjunction with lintels or in long stretches of wall to relieve the pressures on the structures.

Nowadays, however, with the development of lightweight steel lintels, RSJs and reinforced concrete lintels – which can be used to span much wider openings – arches aren't a really practical or cost-effective proposition. Consequently they're used mainly for their decorative effect on smaller-scale structures.

How an arch works

An arch works in virtually the same way as a lintel, by transmitting the weight of the walling above, and its load, to solid masonry at each side of the opening.

The individual components of your arch – usually bricks, reconstituted stone or natural stone blocks – are laid on a curve, forming a compact, stable beam.

Types of arch

The type of arch you choose to build depends on whether you simply want a decorative effect or a load-bearing structure.

You can use arches on internal walls as conventional doorways, to create an open-plan scheme between two rooms, or as a serving hatch. On external walls you can form arches above doors and windows, and outdoors they can make an attractive feature of your garden gate, connected to your boundary walls or even to the house.

Decorative arches

If your arch is to be a purely aesthetic feature indoors you don't necessarily have to build a sturdy structure from masonry. Instead you can make a decorative arch to your own specification from plasterboard, hardboard or chipboard panels cut with a curved edge and fixed to the masonry at the sides of the opening underneath the lintel of a conventional doorway. You can use hardboard, which can be bent, for the underside of your arch. You don't need to alter the structure of the wall in any way.

To finish off your arch, if you're careful to conceal the join between it and the wall, you can simply decorate it with wallpaper or textured paint.

Prefabricated arch formers made of galvanised steel mesh offer a ready-made choice of arch profiles. They come in various widths of opening – and you can even buy simple corner pieces to turn an ordinary doorway into an arched opening. The preformed mesh frames are simply attached to the masonry, under the lintel, then they're plastered over to match the rest of the wall. You can even use them outdoors for a rendered arch finish.

Structural arches

However, where your arch is to form an integral and load-bearing part of a wall its construction is rather more complicated. Brick

READY REFERENCE

SIMPLE ARCH SHAPES

Medieval stonemasons arrived at the shape of the perfect arch by trial and error, and went on to develop many elegant and complicated variations on the basic shape. Three simple types are in common use today, and are easy to construct. They are:

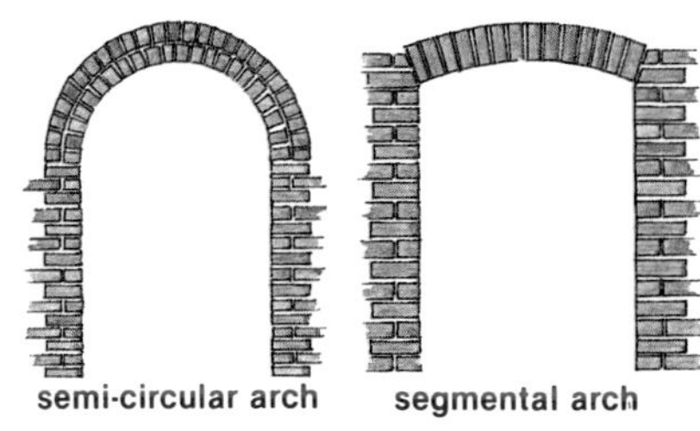

Flat arches are nowadays usually supported by a steel lintel and so are not truly load-bearing. On spans over about 1.8m (6ft) a flat arch should have a very slight upward curve to counter the optical illusion that it is sagging in the centre.

BUILDING INTEGRAL ARCHES

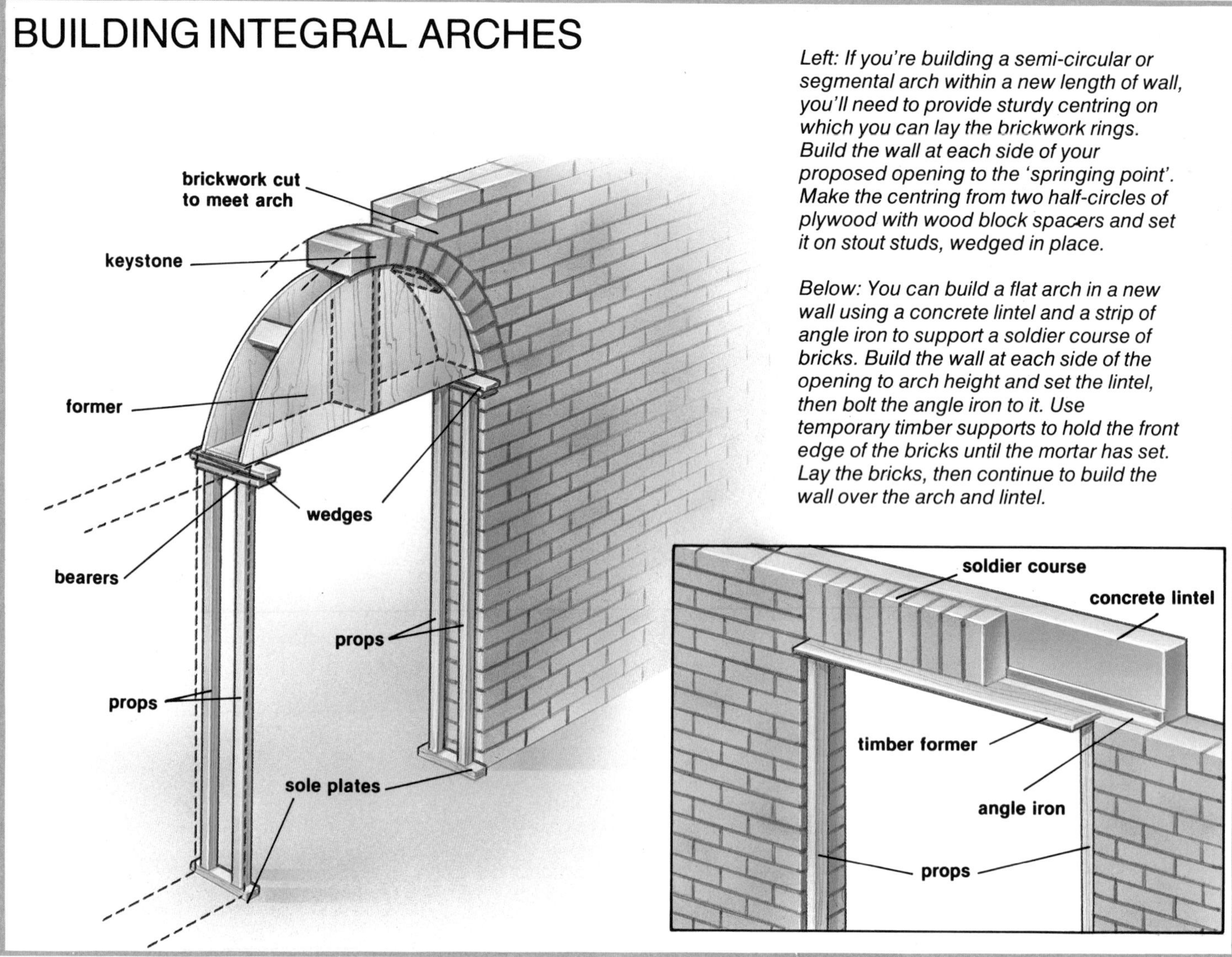

Left: If you're building a semi-circular or segmental arch within a new length of wall, you'll need to provide sturdy centring on which you can lay the brickwork rings. Build the wall at each side of your proposed opening to the 'springing point'. Make the centring from two half-circles of plywood with wood block spacers and set it on stout studs, wedged in place.

Below: You can build a flat arch in a new wall using a concrete lintel and a strip of angle iron to support a soldier course of bricks. Build the wall at each side of the opening to arch height and set the lintel, then bolt the angle iron to it. Use temporary timber supports to hold the front edge of the bricks until the mortar has set. Lay the bricks, then continue to build the wall over the arch and lintel.

is probably the best material to use for this type of arch; you can either leave it exposed as a feature or clad it with render outside or plaster inside for a smooth finish.

Arch profiles

There are various types of arch profiles you can make. One of the commonest is the 'flat' arch (see *Ready Reference*). It's suitable for narrow spans up to about 1.2m (4ft) and the bricks – usually special wedge-shaped types – are set to radiate from a central, vertical point, called the 'keystone'. You can make large flat arches by laying conventional-shaped bricks vertically in a 'soldier' course on a steel or concrete boot lintel (Chapter 5 pages 89 -91) or resting on a length of angle iron used in conjunction with a concrete lintel (see 'Building integral arches', above).

On openings wider than about 1.8m (6ft) the brickwork may look as if it's sagging fractionally and you can remedy this optical illusion by laying the bricks so that the centre of the arch is about 12mm (½in) higher than the ends. One way you can do this is to set the bricks on a curved strip of flat iron instead of the angle iron.

Simpler types of arch, which don't need the use of specially-shaped bricks, or the additional support of a lintel, are the 'semi-circular' (see *Ready Reference*) or 'segmental' types. Both of these arches form part of a circle. The centre of the circle in a semi-circular arch is on the imaginary line between the highest bricks on each side of the opening, from which the arch starts to curve inwards. Its diameter equals the distance between the sides of the opening. For a segmental arch the centre line of the much larger circle which the arch follows is some way below this point.

Because the bricks used for these arches are the conventional format the 'wedging' effect necessary to spread the load of the wall sideways is achieved by shaping the mortar joints between each brick. However, if you are contriving a solid wall above the arch, you'll have to cut the bricks at each side to fit the arch shape.

Building piers

The base of your arch, called the 'springing point', or the point at which it starts to curve inwards, must be supported on sound brickwork at each side, for it's here that the load of the structure is transferred.

If you're building an arch into an existing wall, or if you're building a new wall containing an arch, the load will be taken on solid bearings at each side. But if you're building a freestanding arch between two walls, such as a surround for a garden gate, you'll have to build separate supporting columns called 'piers' at each side. Build up your piers from 225mm (9in) thick brickwork on concrete foundations (see pages 30 -34) making sure that they're set perfectly vertical and that each course matches that of the opposite pier, (see step-by-step photographs on page 85).

BUILDING UP THE PIERS

1 *Mark guide lines on the foundations to indicate the line of the arch, and start to build up the first pier – in this case measuring 1½ x 1 bricks.*

2 *As the first pier rises, check at intervals with your spirit level that it is rising vertically. Tap any out-of-line bricks gently into place.*

3 *Measure out precisely the separation of the two piers, and start to build up the second pier. Check continuously that the two piers align accurately.*

4 *When you have laid six to eight courses in each pier, go back and point up the mortar joints while the mortar is still soft.*

5 *Continue to build up the piers one course at a time, checking at every stage that the courses are level and the pier separation is constant.*

6 *When the piers have reached the desired height – about 1.5m (5ft) for a garden arch – check the measurements accurately on each pier.*

MAKING THE FORMER

1 *Using the pier separation to give the diameter of the semi-circular former, draw the curve out on plywood and cut it out with a jig-saw.*

2 *Nail the first semi-circle to a stout piece of softwood just narrower than the wall thickness, and cut a number of spacers to the same length.*

3 *Nail on the second semi-circle, and then add the spacers at intervals round the edge of the former to hold it rigid when it's in place.*

BUILDING THE ARCH

1 *Set timber props at each side of the arch opening. If the brickwork will be continued above the arch, use two props and wedges at each side.*

2 *Position the former on top of the props, and use a spirit level to check that it is level and that both its faces are truly vertical.*

3 *Lay the first brick of the inner ring on top of one of the piers, and bed it down so its inner face fits tightly against the former.*

TIP

5 *Measure the curve length at each side and divide this by the brick width to indicate how many whole bricks will fill the ring. Mark their positions.*

6 *Add bricks one by one to each side of the ring, tapping them gently into place and adjusting the mortar thickness as necessary for even joints.*

7 *Continue adding bricks until you reach the top of the inner ring. Then butter mortar onto both faces of the keystone and tap it into place.*

9 *Point the mortar joins between the bricks in the inner ring, and then carefully spread a mortar bed 10mm (3/8in) thick on the top surface.*

10 *Build up the second ring in the same way as the first, trying to avoid aligning the mortar joins in the two rings. Add the second keystone.*

11 *Leave the former in place for at least 48 hours (and preferably longer) before carefully removing the props and allowing the former to drop out.*

4 *Use the spirit level to draw a true vertical line on the face of the former, passing through the centre point of the semi-circle.*

8 *With the first ring complete, use the spirit level to check that all the bricks are accurately aligned and that the face of the arch is vertical.*

12 *Trim away excess mortar from the underside of the arch, and point up the joints carefully to match those on the rest of the arch.*

When you reach the arch height you should leave the piers for about 24 hours so that the mortar sets before continuing.

Supporting the arch

Semi-circular or segmental arches are usually built on a timber former or support called 'centring'.

If you're building a simple freestanding arch between two piers you can make a fairly lightweight frame from two sheets of plywood cut to the profile you want for your arch (see *Ready Reference* and step-by-step photographs, page 85). Set the former perfectly level at the springing point on timber studs wedged against the piers at each side. You can then build your arch over the former.

If you're building an arch within an existing wall – or if you're building a new wall – you'll have to provide much sturdier centring (see page 84). You'll also need temporary support for the existing walling above the opening. You can do this by setting up adjustable metal props and timber needles which will support the masonry while you build the arch.

Building the arch

With your formwork in position you can start to lay the brickwork for your arch. The best way to do this accurately is to lay the bricks alternately from each side, finishing with the central, topmost, keystone brick.

So that you can keep the mortar joint thicknesses constant throughout the arch you'd be wise first to mark out the positions of each brick on the side of the plywood former as a guide to laying. When you're spacing out the bricks on your 'dry run', remember that the mortar joints will be thicker at the top than at the bottom and that the narrowest point shouldn't be less than 6mm (1/4in) thick.

Your arch can have one, two or three 'rings', or courses, of bricks. But you must ensure that as few vertical joints as possible coincide with those on adjoining courses, or this will weaken the arch. Point the joints as you go, while the mortar is still soft.

Finishing the arch

Once you've laid the rings of the arch you can fill in the wall surrounding it, unless your arch is freestanding, and has a curved top. Follow the bonding pattern used for the rest of the wall, cutting the bricks next to the arch to fit the curve.

When you've completed the arch leave the structure for about one week so the mortar hardens then remove the centring. You'll have to rake out the joints underneath the arch and repoint them to match the rest of the brickwork.

READY REFERENCE

TIP: MARKING CURVES

For a round arch, the diameter of the semi-circular former equals the pier separation. Use a pin, string and pencil to mark out the curve.

For a segmental arch, the curve centre is some distance from the edge of the former. Set the plywood from which the formwork will be cut on a flat floor, and mark out the curve as shown.

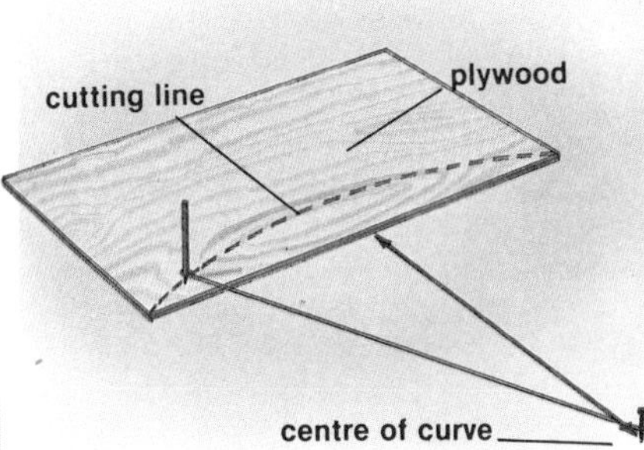

ARCHES IN CAVITY WALLS

If you are building an arch in a cavity wall, the bricks forming the arch must not bridge the cavity. For this reason arches in cavity walls are usually built with the bricks laid as stretchers rather than headers. Formwork is used as for arches in solid walls.

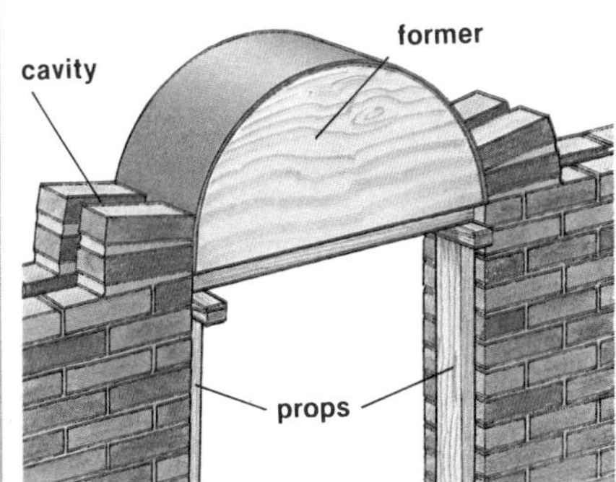

TIP: TILES FOR KEYSTONES

The dimensions of your arch may make it difficult to use a whole brick as the keystone without having very wide or very narrow mortar joints in the ring. In this case use pieces of flat roof tile – or even floor quarries – set in mortar instead.

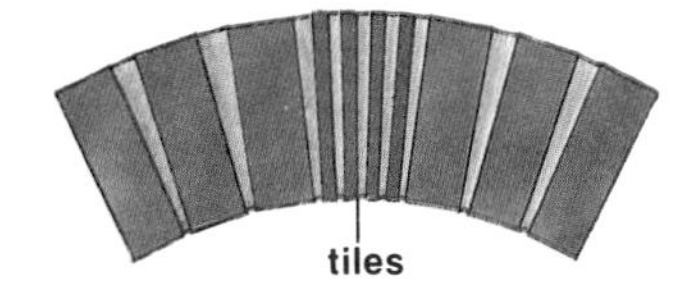

RENDERING EXTERIOR WALLS

Rendering gives exterior walls a good-looking and weather-resistant finish. Many different textures can be produced, depending on the way the render is applied.

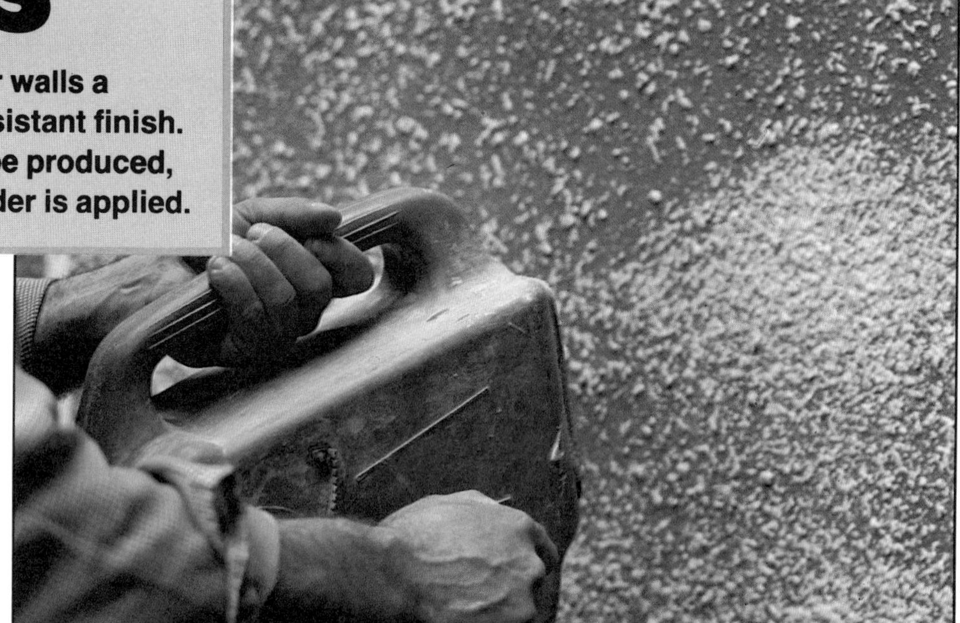

The range of colours and textures that can be produced simply by rendering a wall is quite extensive. For example, various colours and textures can be achieved by pressing small pebbles or crushed stone chippings into the surface of the rendering while it is still wet. Whites, greens, browns, greys and buffs are all possible. Many interesting textures are also produced by throwing the mortar on the wall – either by hand or machine – or by spreading it with a trowel and working the surface with hand tools.

In fact, a smooth surface is not really desirable at all, as it is prone to surface crazing and it encourages streaky stains. A more textured surface will mask minor cracks and any water running down the surface is divided and spread so streaks and patches are less likely – the rougher the texture the rendering has, the better.

The easiest way of getting different colours is to add pigments to the mix. However, for large areas it is difficult to keep a consistent colour for every mix so it may be more convenient to paint over a plain rendering with masonry paint. This reduces the risk of a patchy finish and the walls may be repainted at any time.

What surfaces can you render?

All walls can be rendered, but different surfaces need to be treated in different ways if the render is to adhere to the wall surface. If the wall has a rough texture it is said to have a good mechanical key, and the render will cling to it very easily. Also, if the wall absorbs water it will encourage a good bond with the render. Walls vary a lot and some may need special treatment to make sure the render will stick properly. These treatments may also limit the choice of finishes you can use.

Wall surfaces can be divided into five different types.

Dense, strong and smooth materials provide a poor key, and have little suction. These include dense, dry bricks, in situ concrete, dense concrete blocks and some close-textured lightweight concrete blocks.

A suitable key can be produced by chipping the surface to roughen it up, but this is very hard work over a large area. Alternatively, you could fit expanded metal lathing over the surface or apply a preparatory stipple or spatterdash coat of render (see below).

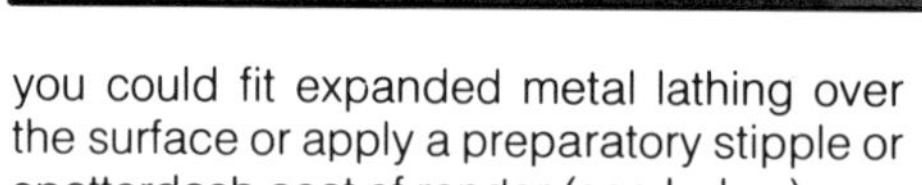

Moderately strong porous materials have a fairly good mechanical key and fairly high suction. These include most bricks in common use, and most lightweight concrete blocks. If the suction is too great or uneven then it may be necessary to apply a stipple coat first. Alternatively, all that may be needed is to rake out the mortar joints to provide a better key.

Moderately weak porous materials, such as some lightweight concrete blocks and some softer types of brick, usually have a very good key. The only problem may be excessive suction, but this can easily be overcome by damping the wall before you start rendering. However, this does not mean soaking or saturating the wall. Again, a spatterdash or stipple coat can be applied first to even out the suction.

Timber or steel surfaces do not provide a satisfactory bond, nor does crumbling brickwork or stonework, and you'll have to provide a separate support for the render. Here metal lathing is ideal. It must be made of galvanised or stainless steel or prevent corrosion, and the fixings should be of the same material. Portland cement actually gives protection against corrosion, so the first coat should be a cement-rich one, mix type I is suitable (see *Ready Reference*). Two top coats or render can then be applied to suit the final finish, giving at least 16mm (5/8in) of total cover.

Backgrounds containing soluble salts can sometimes damage the render. For example, some clay bricks like flettons and some stock bricks contain soluble sulphates which can be washed out into the render where they react with the cement, causing it to break up. This only occurs where excessive amounts of water get into the wall, but if there is a chance of this happening it is advisable to use a sulphate-resisting Portland cement for the rendering. In addition, ensure there is a protective coping to a chimney or rendered wall and that there is a waterproof membrane behind any retaining wall.

What is rendering made of?

Rendering is a mortar mix made of Portland cement, sand, some lime and possibly a few additives and pigments.

It is very important to use the correct type of sand, so make it clear to your supplier that you want a sand for rendering. It will be a type of sharp sand; soft sand and builder's sand for bricklaying mortar are not suitable.

The correct proportions for different mixes are shown in *Ready Reference*. But to save time and trouble in buying and mixing the materials, some part-mixed and completely pre-mixed products are available.

Ready-mixed dry materials for rendering are available in paper sacks, and all you need do is add water. The manufacturers have accurately batched the ingredients and they may have included additives to improve workability and setting. Don't make your own additions to the mix.

CHOOSING THE RIGHT MIX

The type of mortar mixes you use for the undercoats and top coat depends on several things such as the condition of the wall, the type of finish and the weather conditions the render is exposed to. See Ready Reference (right) for details of mixes I – IV.

Note that strong coats must not be laid over the weaker ones unless the strong coat is much thinner than the weaker one. Also, when render of the same strength is used for the top and undercoat the first coat should be thicker than the following one.

	SMOOTH, SCRAPED OR TEXTURED FINISH		ROUGH CAST OR PEBBLE DASH
Background to be rendered	**Sheltered conditions**	**Exposed conditions**	**For all conditions**
Strong and dense	II/III	II/III	II/II
Fairly strong, porous	III/IV	III/III	II/II
Fairly weak, porous	IV/IV	III/III	Not recommended
Metal lathing (requires two undercoats)	I/II/III	I/I/III	I/II/II

Masonry cement is a ready-mixed blend of ordinary Portland cement with additions to improve the mortar. It should be added to sand alone to produce a complete mortar. Don't add anything else.

Lime and sand mixtures can be supplied but make sure you order one specially for rendering. The supplier can also offer to include certain additives to improve the mix. These materials need to be measured out quite carefully so it is best left to the specialist supplier. The mixture is bought slightly damp and it should be kept in this state. You can store it for a week or more on a clean drained surface, but protect it with a tarpaulin or polythene sheet to prevent it getting wet or drying out too quickly.

When you come to use it it should be mixed with the correct amount of Portland cement – check with your supplier to find out what proportions you should use. Use either ordinary or sulphate-resisting Portland cement, but not masonry cement.

Fibre-reinforced rendering mortar is another ready-mixed product available in bags. Glass fibre strands are mixed in with the mortar and allow the mortar to be applied in a layer only 6mm (1/4in) thick. The fibres

APPLYING A STIPPLE COAT

1 *A stipple coat is used on hard, dense surfaces to provide a mechanical key for the render. Use a banister brush to scrub a thin layer of mortar into the wall first.*

2 *When you've covered a square metre (10sq ft) or so, load the brush and go over the same area with a stippling action. Leave for at least 7 days to harden.*

READY REFERENCE

THE TOOLS YOU'LL NEED

You will need some or all of the following tools:

- a hawk to hold the mortar up to the wall
- a plasterer's trowel – this is a rectangular steel trowel with a central handle
- a steel float for smooth textures
- a wood float for texturing the final coat or tamping pebbledash into place
- a scoop for appying spatterdash or pebbledash
- a comb – this consists of a wooden handle with protruding nails for scratching a key into the undercoat
- a screeding board for levelling the surface; this is usually 125mm (5in) wide, 15mm (1/2in) thick and about 1200 to 1500mm (4 to 5ft) long
- a spot board to hold the mixed mortar before use – approximately 700mm (30in) sq.
- a Tyrolean machine for applying a Tyrolean finish

A cement mixer is desirable for all but the smallest jobs; it can be hired by the day or week from plant hire shops.

You may also need a wheelbarrow, shovels, watering can, polythene sheeting, buckets, hammer and nails and possibly ladders or scaffolding or a platform tower.

MORTAR FOR RENDERING

Four different mortar mixes are suitable for rendering, ranging from Type I (which is the strongest) to Type IV (the weakest).

Type	cement:lime: sand	masonry cement: sand
I	1:0 to 1/4:3	–
II	1:1/2:4 1/2	1:3 1/2
III	1:1:6	1:5
IV	1:2:9	1:6 1/2

ESTIMATING QUANTITIES

In general you can assume that the total volume of the mix you'll need is the same as the volume of sand. The lime and cement fill the voids in the sand and only increase its volume by about 5 per cent.

- 1cu m (35cu ft) of sand will make enough mortar to give a 10mm (3/8in) layer over an area of 100sq m (1076sq ft).

For 1cu m (35cu ft) of sand you need to add the following quantities of cement and lime;

Type	cement (kg)	lime (kg)
I	500	50
II	350	70
III	250	100
IV	175	140

If you order ready-mixed lime and sand the supplier will advise you of the amount of cement you'll need.

APPLYING SMOOTH RENDERING

1 Prepare the wall by scraping off any old paint, hacking out the mortar joints and roughening up the bricks. Then nail up battens to divide the wall into bays.

2 Lay on a thin coat of render first and scratch it to provide a key for the second coat. This 'scratch' coat helps to even out suction in the background.

3 Leave the scratch coat to set for about a day; then lay on the second coat up to the level of the battens. Press it well onto the wall with the trowel.

4 Hold a long straightedge across the battens and move it upwards with sawing action to rule off the render. Fill in any hollows and rule off again.

5 Remove the battens, taking care not to disturb the render. Then fill the gap, pressing the mortar well in and smoothing over the join.

6 Leave to set for half an hour. Then pick up a small amount of mortar on a float and rub it onto the surface with a circular action to leave a smooth finish.

reinforce the rendering to prevent shrinkage and cracking and the finished surface has a good resistance to impact and wear. Again, follow the makers instructions and do not add your own additives.

Tyrolean ready-mixed finishing coating is available in a range of pale colours. It is made from white Portland cement, specially graded sands and colouring pigments, and just needs to be mixed with water before use. It is applied with a hand-operated machine to give a honeycombed texture over a backing coat of ordinary render. It can be applied direct to some surfaces, but joints in walling will tend to show through so a backing coat is usually preferable.

Resin-based high-build coatings are not really renderings but are more like thick paints. They can be used to finish a backing coat of render to give it colour, texture and weather-resistance. They are supplied in tubs ready for use and can be applied by brush, plain roller or textured roller. They can also be used indoors.

Choosing the right mix

Although you might choose a ready-mixed product for the top coat of rendering, there are many cases where you'll need a different mix for the undercoat, and you may need special mixes for the top coat too on some surfaces.

Mixes suitable for rendering can be made from cement, sand and lime or masonry cement and sand. The various mixes are grouped into four types (see *Ready Reference*). Type I is the strongest and Type IV the weakest. The stronger mixes are denser and more resistant to water absorption, but they are also more prone to shrinkage. As a render shrinks, stresses build up in the surface which may pull the render off the wall or cause it to break up a weak background. Shrinkage also causes cracks in the render. These cracks let rainwater into the wall and the dense render holds it there. More absorbent renders will soak up water over the whole surface, but readily allow it to dry out later. Cracks in absorbent renders are therefore not so serious.

The render does need a certain amount of strength to be resistant to frost damage, and the skill in rendering a wall lies partly in choosing the right strength mix for the job. The chart on page 157 gives a summary of the undercoat and top coats needed for different surfaces and weather conditions.

Preparing the surface

Before you start rendering you must prepare the wall properly. Any fixings such as rain-water pipe brackets or light fittings should be brought forward to the new surface level.

Short lengths of rigid plastic pipe are useful as spacers, with new, longer screws used to refix the items.

The top edges of the rendering must always be protected from the weather. So check that overhangs on window sills, copings and roofs are big enough to protect the edge completely. If the overhang is not enough you will have to extend it. This will most likely have to be done with the rougher finishes as they require a projection of 50 to 75mm (2 to 3in) beyond the finished surface.

You will have to fix beading at external corners and around window and door openings and at the bottom edge so you have a hard edge to work to (see *Ready Reference*).

Fix temporary battens to the wall to divide the area into manageable bays. Their thicknesses should be equal to the thickness of the render coat. There's no need to fix battens at the bottom edges since the beading will provide a firm edge. Finally, before the first coat is applied, dampen the area to even out the suction. Generally splashing with a brush is adequate but a garden spray might be useful. Keep your eye on areas that dry out and redampen them as necessary. But there should be no free water on the surface, and you should avoid overwetting it.

Applying stipple and spatterdash coats

Many surfaces benefit from a thin stipple or spatterdash coat before the main coats of render are applied. These coats will improve the key of a wall and they will even out the suction too. Before application, the wall must be brushed clean of any loose material and damped down if it is absorbent.

Spatterdash is a mix of 1 part Portland cement and 2 parts coarse sand, with just enough water to produce a thick slurry. It must be kept well stirred. All you have to do is pick up a small amount in a hand scoop and fling it on the surface, working from top to bottom. Build up a 3 to 5mm (1/8to 3/16in) layer and do not smooth or level it off in any way. Leave to dry for seven days.

A stipple coat is a mix of 1 part Portland cement and 1 1/2 parts sharp sand. Again it is mixed to a thick slurry but instead of using water, a mix of equal parts water and PVA building adhesive is used. This slurry is brushed vigorously onto the wall with a banister brush (dustpan brush) and then immediately stippled with it to provide the key.

Mixing the mortar

Unless you're rendering a very small area, you should try to hire a mixer. It will save you a lot of time and you will need all your energy for applying the render.

The materials for an undercoat or top coat should be mixed together dry until a uniform colour is obtained. Then add water a little at a time until just sufficient has been added to give a workable consistency. The mix must not be too wet or it will slump down the wall. Continue mixing for about two minutes after the water has been added, then turn it out onto a spot board and take to the work.

It is a good idea to cover the bulk of the mix with polythene to prevent it drying out. If the mix stiffens up, retemper it by turning with a shovel. Only add more water if absolutely necessary and then only with a watering can fitted with a rose. All mortar from a batch should be used within two hours of mixing.

Applying the undercoat

Select a mix suitable for your conditions and mix it up as described. Where local hollows require building up, it's best to fill them in first and allow the mortar to harden before carrying on.

The undercoat may be up to 15mm (5/8in) thick, but on dense backgrounds the *total* thickness of render should not exceed 15mm and in this case the undercoat will be much thinner.

Work with the hawk in one hand and the trowel in the other. Place about half a shovelful of mortar on the hawk and hold it to the wall. Then, starting at the top, press the mortar onto the surface a little at a time. When the hawk is empty smooth over the area before picking up some more mortar.

When one bay is covered, use a screeding board or long straightedge to rule off excess mortar. Then, when the coat has stiffened, comb the surface with scratches about 5mm (3/16in) deep. This forms a key for the following coats. Take care not to scratch right through the coat, and do not scratch at all if you're going to apply a Tyrolean finish.

Cover the surface with polythene to retain moisture while the mortar is setting – for about four days – then remove it to allow the mortar to dry out for another three days. You can then apply the next coat. Any fine cracks that may develop in this coat will not affect the final render.

Applying the top coat

How you apply the top coat depends on the type of finish you want. Smooth and scraped finishes are laid on with a trowel; with pebbledash the final coat is trowelled and the pebbles or stones are thrown onto it. Roughcast and Tyrolean are wet mixes thrown onto the wall either by hand or machine.

Whatever finish you choose it is essential to protect the surface while the mortar dries out. Any covering must be held clear of the surface, as uneven setting may cause a patchy appearance. Try not to let wind blow under the covering as that could increase evaporation and make the mortar dry out too quickly. If the surface does start to dry out early, careful spraying will help – mostly during the first three days.

READY REFERENCE

RENDERING EDGES AND CORNERS

Use expanded metal beading at edges and corners for a neat and strong finish:

- use stop bead around windows, doors and any other opening (A)
- use angle bead at internal corners (B).

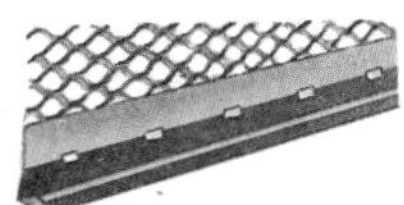

DEALING WITH THE DPC

Don't render over the dpc. It s best to stop the render above the dpc, in which case you can use a bell cast or flared render stop. The shape of the beading directs water away from the wall.

If the render is carried below the dpc, cut a slot in the render sloping down towards the ground to throw off the rainwater.

MAKE A NEAT JOINT

If you can t render the whole wall area in one go you must divide the wall into manageable sections. Plan the work so the joints follow a feature of the house; they will then be less noticeable.

A neat line for a joint can be made by cutting the mortar about half an hour after it has been laid. Use a straightedge and scrape off the mortar on the waste side.

TIP: DON'T USE TOO MUCH WATER

Most cracking in rendering is due to the use of too much water. Use only clean tap water and add the smallest amount that will make a workable mix.

FORMING ASHLAR JOINTING

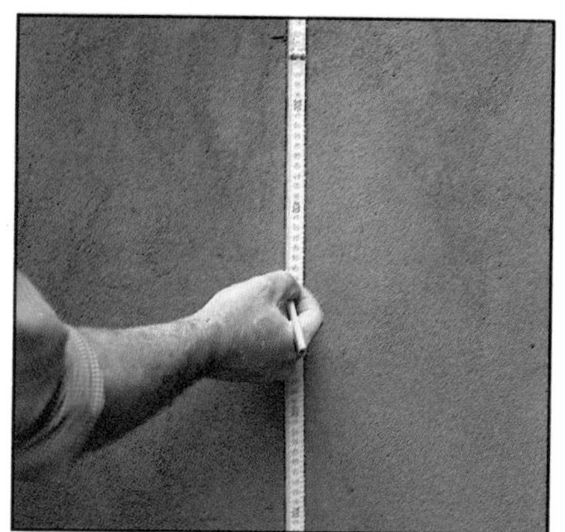

1 A very effective finish for smooth render is to score lines to imitate stone blocks. Mark out the position of the lines while the mortar is still wet.

2 Hold a straightedge on the marks, then draw in the lines. You can use a special tool like this, but a piece of bent tube or even an old nail will work just as well.

3 The vertical lines are marked in exactly the same way. The usual size for the 'blocks' is 230 to 400mm (9 to 16in) high and 460 to 800mm (18 to 32in) long.

APPLYING TYROLEAN FINISH

1 This finish uses a special mix which comes in several pastel shades. Mix it up to a fairly wet consistency and load the machine with 5 or 6 trowels full.

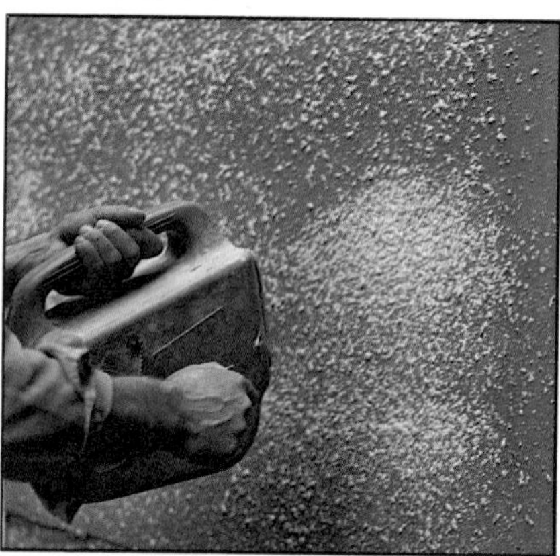

2 Turn the handle of the machine to flick the mixture at the wall (it might be best to practise first). Go over the whole wall first with just a light coverage.

3 Keep reloading the machine and go over the wall again until it is evenly covered and none of the background render shows through at any point.

Rough-cast

Rough-cast gives a very rugged-looking and rough-textured finish. Pebbles or crushed stones are mixed with the mortar which is then thrown at the wall as a wet dash. As the pebbles or stones are mixed in the mortar this is a very long-lasting and hard-wearing finish.

The mix consists of 1 part Portland cement, ½ part lime, 3 parts sharp sand and 1½ parts shingle or crushed stone. The maximum size of the coarse materials may vary from 5 to 15mm (3/16 to 5/8in). Add water until the mix has a fairly wet, plastic consistency.

The mix is taken to the work in a bucket and a hand scoop is used to fling the mix onto the wall. Great care must be taken to get an even texture and consistent thickness. If you make a mistake don't attempt to trowel the surface; instead, scrape off the mix while it is still wet and try again.

Pebbledashing

This is one of the most popular finishes and is produced by throwing small stones or pebbles onto a freshly-laid mortar coat. It is sometimes called dry dash. The undercoats for this finish should include a waterproofing additive (see pages 170-71).

The finishing mortar coat which receives the pebbles is called the butter coat. It is a mix of 1 part cement, 1 part lime and 5 parts sand, with enough water to make a slightly wetter mix than for smooth finished render. It should not dry out too quickly and to help this the background may be dampened to reduce its suction. The butter coat is laid 6 to 10mm (¼ to 3/8in) thick and a straightedge can be used to level off the surface.

The pebbles or stone chips must be washed and drained first. Place the stones in a bucket and fling them into the wet finish coat with a hand scoop. Try and obtain a close and even texture. When a large area has been covered, *lightly* tamp the surface with a clean dry wooden float to ensure a good bond.

A lot of the dry dash will fall off during the application, so spread a clean polythene sheet at the base of the wall to catch it. Have a sieve and watering can or hose ready to clean the stones so you can re-use them.

APPLYING ROUGHCAST RENDER

1 Roughcast consists of mortar mixed with crushed stones or pebbles. The wall should be smooth rendered, but left with a scratched finish.

2 Use a scoop or trowel to pick up a small amount of the mix. It's best to practise your technique on a piece of board before you start on the wall.

3 Fling the mix at the wall from a distance of about 500mm (20in). Start at the top of the wall and work down so any bits that fall won't spoil the work below.

PEBBLEDASHING

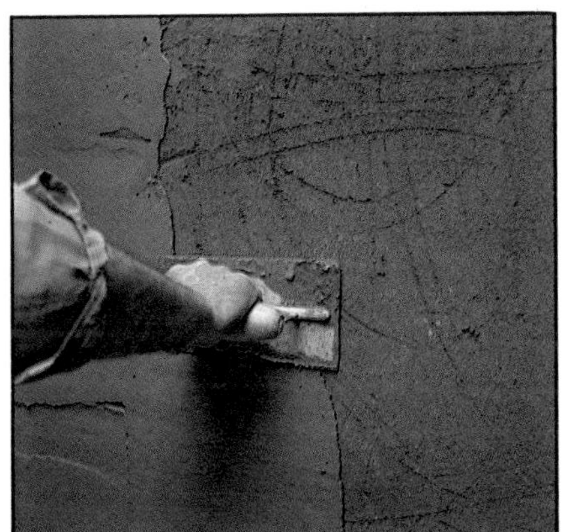

1 Make up a fairly wet mix of mortar and spread this onto the rendered wall. Don't cover too large an area at once or it will dry out before you can get to it.

2 Use only clean, washed stones or pebbles. Lay a sheet of polythene at the bottom to catch any pebbles that fall, as these can be washed for reuse.

3 Pick up some pebbles on the scoop and fling them at the wall, starting at the top and working down. Try to cover the wall with an even layer of stones.

Scraped finish

This is a fairly smooth-textured finish. The final coat of mortar is laid on with a trowel, levelled off with a straightedge, trowelled smooth with a steel float and allowed to harden for several hours. Then the surface is 'torn' or scraped with a suitable tool – an old tenon saw blade is ideal. This breaks up the cement-rich surface and drags out some of the coarser particles of sand. Take care to scrape the surface evenly, and when complete brush the surface with a soft brush to remove any dust and leave a crisp texture.

You can make a practically invisible joint with this finish. Work to an irregular but neatly cut edge and scrape to within 20mm (¾in) of it. The next day, lay the adjacent area tight to the cut edge and scrape the new area, working over the joint.

Tyrolean finish

This is a spattered finish applied with a hand-operated machine which flicks wet mortar onto a base coat. The result is an even, coarse, open honeycomb texture which is very attractive. The finish coat is made up from a prepacked mix which is available in a range of pale colours. These mixes vary so you should follow the makers' instruction regarding the amount of water you have to add. The hand spray machine can usually be hired from the suppliers of the Tyrolean mix or from plant hire firms, and they also supply instructions for its use.

Other textures

Smooth finishes can obviously be obtained by trowelling with a steel or wooden float. However, trowelling off like this will bring a lot of cement to the surface, and this is not at all desirable as it is likely to craze. A lambswool roller can be used to remove the cement and expose the sand.

A simple and very effective finish is to mark lines on the render to imitate blocks of masonry. Other textures are produced by working the surface with various hand tools. For example, you can use a banister brush, a stiff-bristled brush, a notched spatula or a wooden float. All sorts of swirls, whirls and random textures can be produced, but don't overdo it or the result may just look messy.

CHAPTER 5

ALTERATIONS

Once you've mastered the basic builder's skills, you can turn your attention to some more ambitious internal alterations. These may include replacing features such as doors and windows with new components, or even creating openings where none existed before. Fireplaces that have been gutted may need to be reinstated if you want to join the rush back to open fires; conversely, unwanted fire openings can be removed and blocked up to create much-needed living space. You can replace old, sagging ceilings with easy-to-handle plasterboard, and even sub-divide your living space exactly as you want it with partition walls.

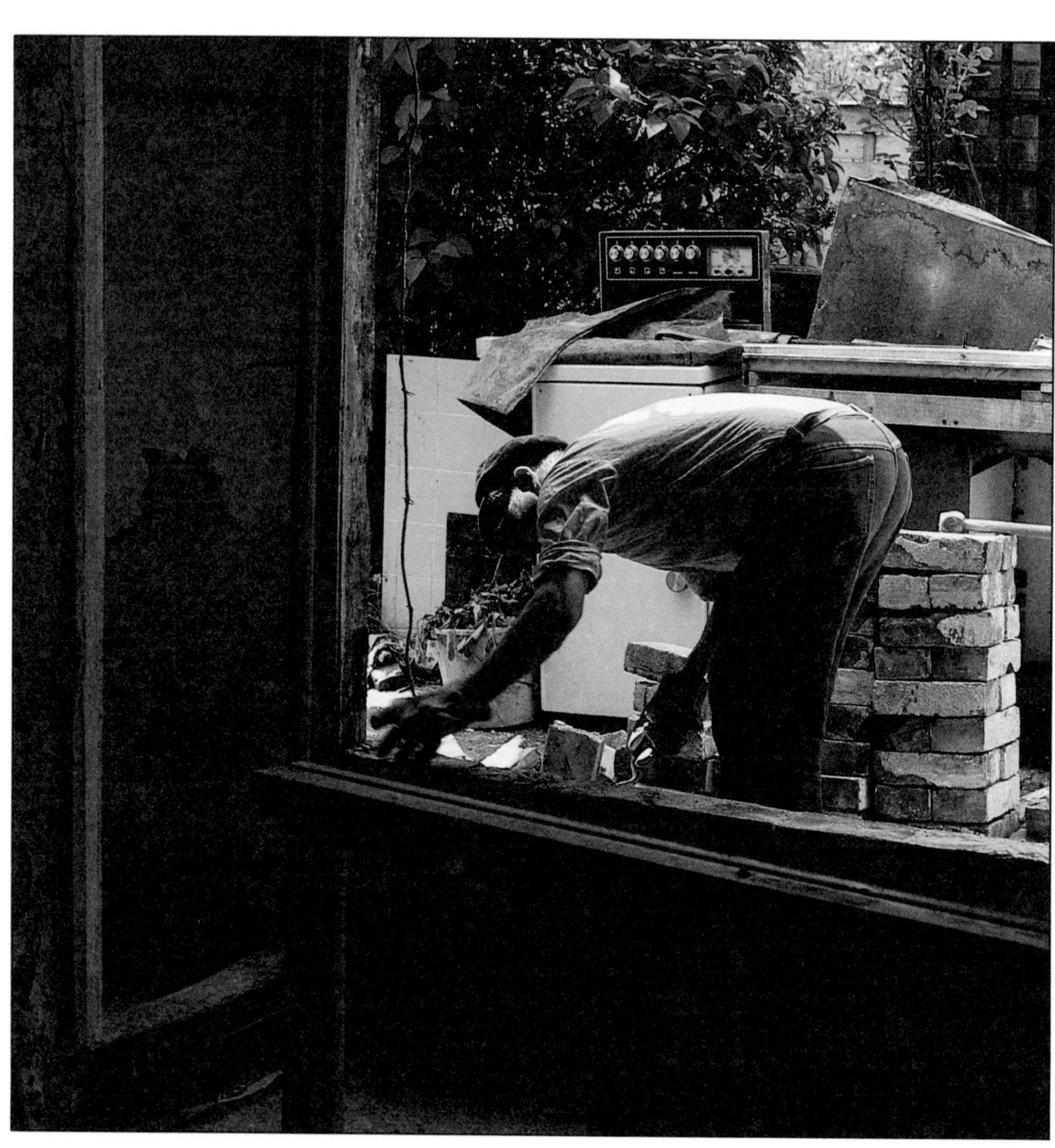

MAKING A NEW DOORWAY I – the lintel

Making a new doorway or window can dramatically change the way you plan your rooms. The first part of the job, described here, is to insert the lintel.

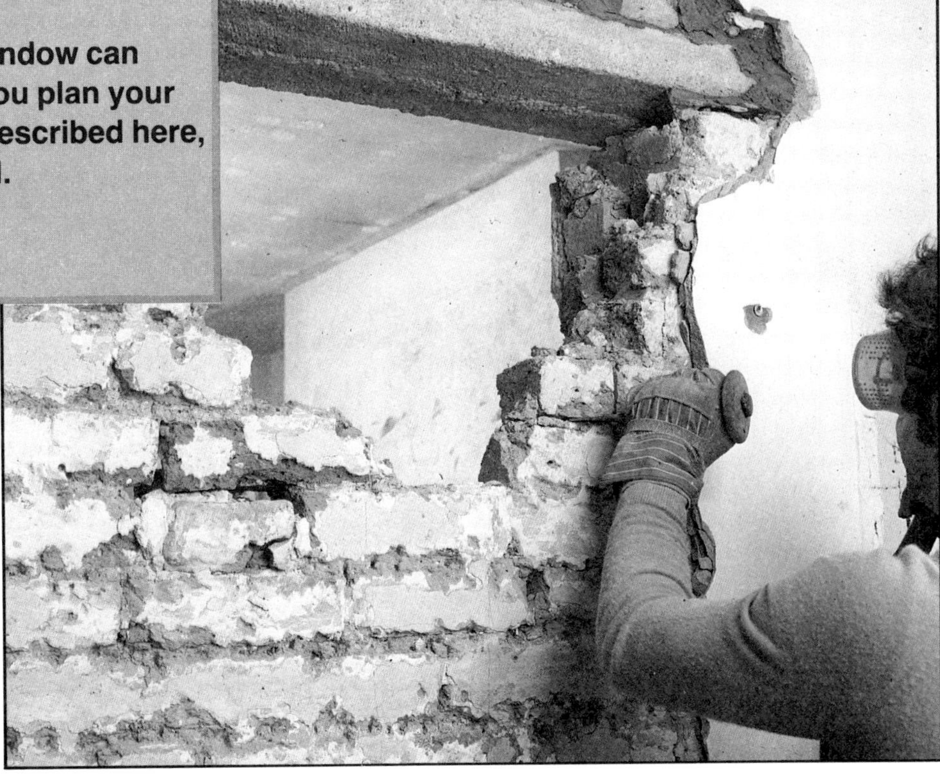

The way some of the rooms of your house are laid out may not be suitable for all the activities and storage needs of your household, particularly if you have growing children. You may also feel that the existing layout of your house severely cramps your style when it comes to designing decorative schemes for your home. The problem may be a purely practical one: the position of doors and windows may be an obstacle when it comes to placing seating, wall-mounted cupboards, bookcases and other furniture exactly where you would like them. Life can be made very much easier if there is an uninterrupted flow of traffic between two rooms, such as the kitchen and dining room, but this is often not the case in an existing plan.

You might be able to solve some of your problems by changing the function of your rooms; for example, converting a living room to a dining room. You could even alter the shape and size of the rooms – or create an entirely new room – by building a partition wall (see pages 137 -141). But, in addition, you may find it's necessary to reposition a door or window to improve an existing scheme or fit in with a new one.

It's a straightforward matter to block off a redundant window or doorway with bricks or plasterboard, but if you're making a new opening of this size in a wall you'll have to take into account some basic structural rules.

To cut out an area of wall you must support the masonry above, plus any additional load that bears upon it. The usual way to do this is to insert a rigid beam or 'lintel' made of wood, concrete or steel, directly above, which is recessed into the sides of the opening. Your choice of lintel depends on how your wall's been built and whether it's internal or external, load-bearing or non-load-bearing.

Planning the opening

Before you can make an opening in a wall – whether it's load-bearing or not – you must check with your local Building Department that what you plan is permissible under Building Regulations (see pages 185 -89).

It's not always feasible to change the position of a window or a door in an external wall: your new location might infringe local bye-laws or, especially in the case of a window, it might simply have a poor outlook. But you can usually make a new opening in an internal wall wherever you want.

The first job in planning your new opening is to decide exactly where you'd like it to be. Your decision will be influenced by where you'd like to put your larger pieces of furniture, and particularly those which will stand against the wall, such as a sideboard or bookshelves. If, for instance, you're going to make a serving hatch between kitchen and dining room you might want to position it directly above your sideboard to make serving easier. Work and traffic flows are also important aspects of room scheming, particularly when you're making a new doorway.

Although you have a lot of flexibility in positioning your opening, it's best to keep the top a reasonable distance below the ceiling so you'll have ample access when fitting the lintel and making good the wall around it. You should also avoid making your opening too close to the corner of the room because you'll weaken the structure of the wall; leave a margin of at least two bricks' length to be safe.

It's best to site your opening away from electrical socket outlets, wall switches or wall lights, but if that's not possible you'll have to reposition the wiring, accessories

READY REFERENCE

SUPPORTING THE WALLS

If you're making a door or window-sized opening in a wall you must:

- temporaily support the walling above the proposed opening, and any load bearing upon it, with adjustable metal props and timber 'needles' (A)
- install a lintel (B), resting on suitable bearings (C) above the opening to support the walling permanently.

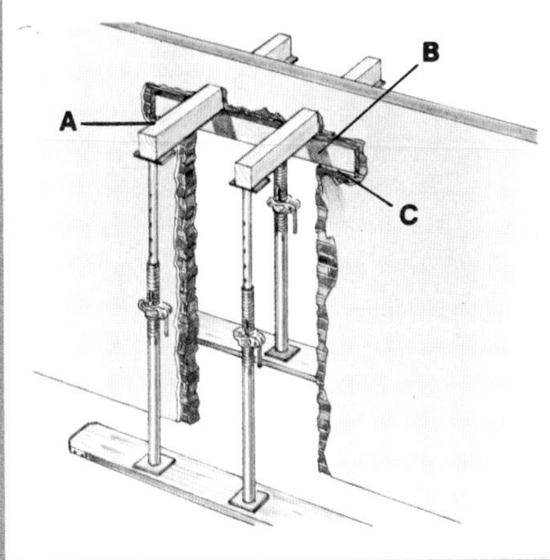

MARKING OUT A DOORWAY

1 *When you've decided where you want your doorway, mark one side on the wall then hold a spirit level against a timber straight edge at this point.*

2 *Adjust the straight edge so that it's vertical and draw along the edge using a pencil to mark one side of your proposed doorway.*

3 *Measure from your first guideline the width you want the doorway and mark the wall; extend the mark using the straight edge.*

6 *You'll also have to mark the lintel position on the wall: measure the length of the lintel and add on 25mm (1in) so it's easier to fit.*

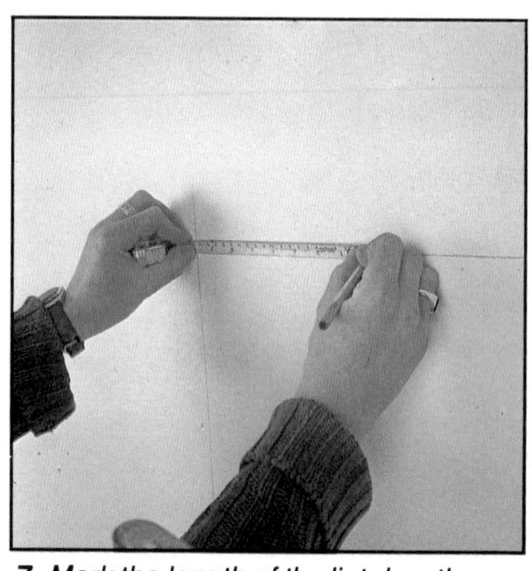

7 *Mark the length of the lintel on the wall; it should be central over the doorway and projecting beyond each side by about 150mm (6in).*

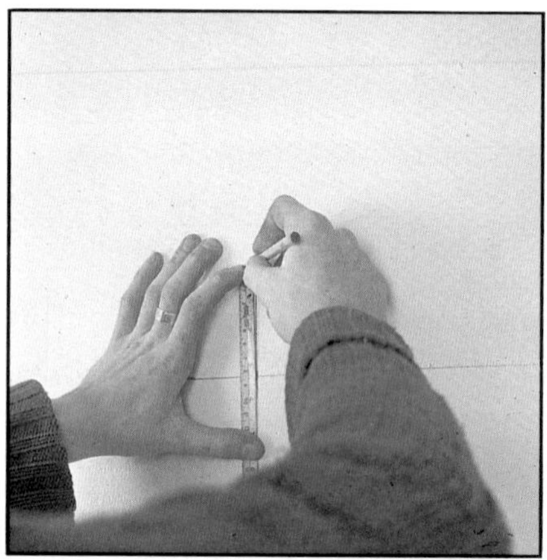

8 *Measure the depth of the lintel, add 25mm (1in) for fitting and transfer this dimension to the wall above the top of the door.*

and fittings. If there's enough slack in the cable you can simply move it further along the wall. Alternatively you'll have to extend or re-route the circuit to avoid the opening by joining in new lengths of cable.

You will also have to consider the dimensions of your opening and match them to the standard sizes of ready-made doors and window frames available from timber merchants unless you make the frames yourself to your own specifications.

Supporting the wall

It's possible to cut a door or window-sized hole in a non-load-bearing wall without supporting the wall above while you work. If it's properly bonded and the mortar joints are in good condition, the only area that's at risk from collapse is roughly in the shape of a 45° triangle directly above the opening. Without a lintel the bricks or blocks within this area would tend to fall out, forming a stepped 'arch'; the walling that's left would, however, be self-supporting – although weaker than it was – up to a span of about 1.2m (4ft).

So when you've marked out the position of your proposed opening it's quite safe to chop out the triangle of bricks or blocks above and insert your lintel. When that's in position, and you've filled in the triangle, you can remove the walling below to form your opening.

However, if you're going to cut an opening in a load-bearing wall you'll have to provide temporary supports not only for the walling above the opening but also for any load from floors or walls above, which bears upon the wall. The weight imposed upon a wall is quite considerable and if it's not adequately supported the brickwork or blockwork could 'drop'. This type of structural damage is impossible to repair and the only remedy is to demolish the wall and rebuild it. To support the walling you'll need four adjustable metal props and two stout timber battens called 'needles'. The props work rather like a car jack, and to fit them you'll have to cut two holes through the wall directly above the lintel's position at each end, and slot in the needles. You can then use the props to support each end of the needles.

Rigging up the supports is a two-man job: you'll need one person of each side of the wall to tighten up the props simultaneously so that the needles can be fixed level.

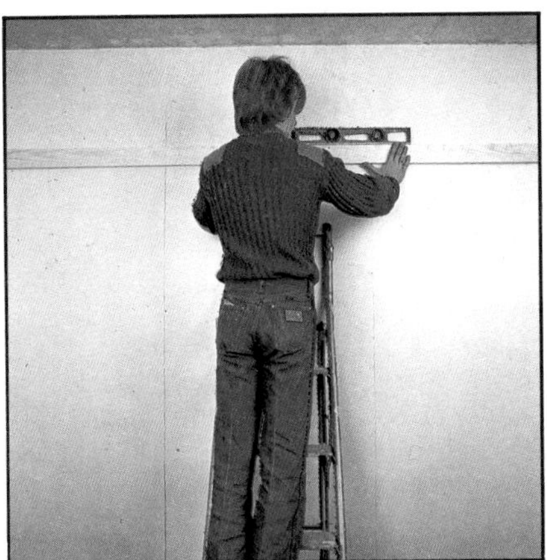

4 *You can mark the height of the doorway in the same way using your straight edge and spirit level to get it truly horizontal.*

5 *Once you're satisfied that the straight edge is level, remove the spirit level and scribe along the top to mark the top of the doorway.*

9 *Position your straight edge at this point using your spirit level then draw a line on the wall to mark the top of the lintel slot.*

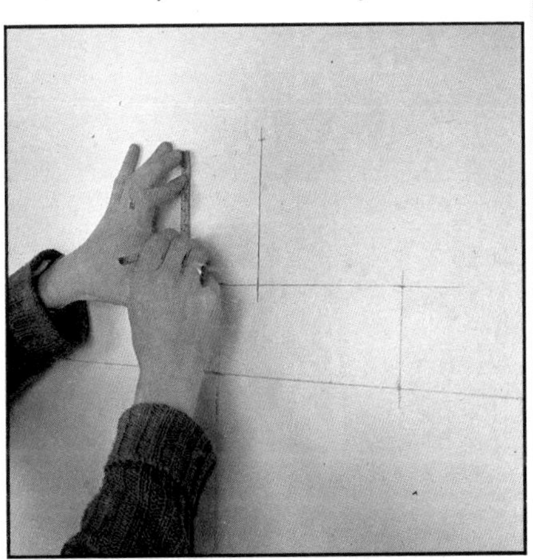

10 *Next you mark the slots for the needles directly above the lintel position, and centrally over the lines marking the sides of the door.*

If floor joists from the rooms above are resting on your wall you'd be wise to provide extra support for the ceilings at each side of the wall. Again you can use adjustable metal props with stout battens at top and bottom.

With the wall supported in this way you can safely chop out a slot in the brickwork or blockwork directly below the needles and insert your lintel.

It's possible to cut an opening in a cavity wall by more or less the same process but only the inner leaf – usually blockwork – is load-bearing. Cutting into a timber frame wall involves a slightly different process and will be dealt with in a later issue.

Marking out the opening

When you've selected the best location for your opening you'll have to mark out its shape on the wall. Take the overall measurements of your frame and add on about 25mm (1in) to give fitting tolerance; then transfer these dimensions to the wall. You might have to move your proposed opening along the wall fractionally so that both its edges line up with a vertical mortar joint; this keeps the number of bricks or blocks you have to cut to a minimum.

If the wall is plastered you could chip away a section about 500mm (1ft 8in) square from the centre of your proposed opening to reveal the bricks or blocks beneath. With this small area uncovered you'll be able to measure from a mortar joint to your line marking the perimeter of the opening. If the lines at each side don't correspond with a mortar joint you'll have to move them over until they do.

READY REFERENCE

MAKING BEARINGS

To enable a lintel to support a wall and distribute its loading both ends must rest on bearings. A bearing should be a slot 150 to 225mm (6 to 9in) wider than the opening plus 25mm (1in) at each end and on top to make fitting easier and to allow for slight settlement.
To bed the lintel evenly:

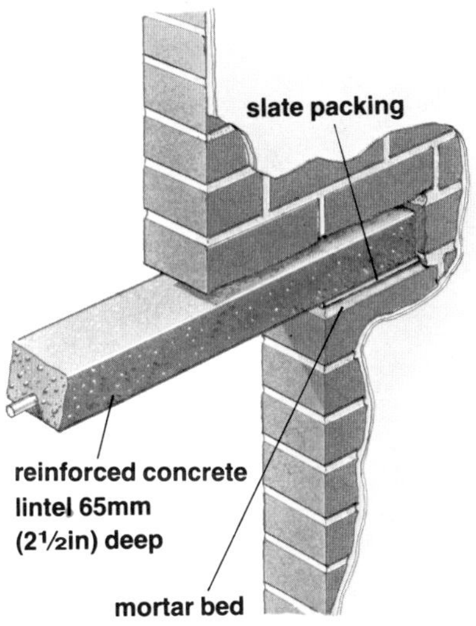

- trowel a mortar mix of three parts soft sand to one of Portland cement onto the bearing
- lift the lintel into place
- check the level of the lintel by holding a spirit level on its underside
- if it sits unevenly, pack out underneath with squares of slate
- when the lintel's level, fill the gap on top and at each end with mortar.

CHOOSING A LINTEL

If you're making an opening up to about 1.2m (4ft) wide in a properly bonded brick or block wall you'll need:
- a precast pre-stressed concrete lintel or
- a concrete lintel reinforced with metal rods or
- a lightweight galvanised steel lintel.

Special shaped lintels are made to cope with different types of walls and their loadings.

FULFILLING REQUIREMENTS

Any alteration or building work you carry out on the structure of your house must conform to a set of safety standards called Building Regulations. You should notify your local authority of your intentions, although approval might not be necessary for cutting a door- or window-sized hole.

CUTTING THE LINTEL SLOT

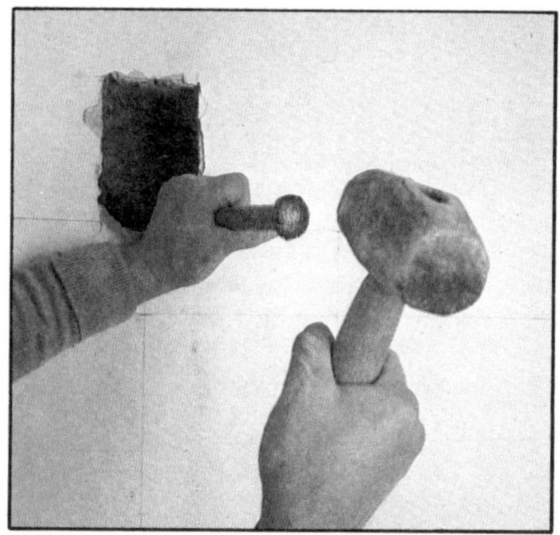

1 Start to chop the first needle hole, using a bolster chisel and club hammer. If you're cutting into a thick wall use a 100mm (4in) cold chisel.

2 When you've cut your first hole through the wall you can slot in your needle: It should protrude equally on both sides of the wall.

3 With one man at each side of the wall, start to tighten up the adjustable metal props under both ends of the needles.

4 Once you've rigged up the supports you can chop around the lintel guidelines then start to remove the plaster from within the rectangle.

5 Hack out the brickwork carefully; if the walling above drops, chop a margin of plaster around the lintel slot and remove the loose bricks.

6 Continue to chop out the bricks to form the slot for the lintel. Try to keep the top surface inside the slot, and the bearings, as level as possible.

When you're satisfied with the position of your opening you can draw its shape on both sides of the wall more accurately using a pencil held against a spirit level.

You'll also have to mark the position of the lintel on the wall directly above the line that marks the top of the opening. Make it wider than the opening by about 150 to 225mm (6 to 9in) to give a sufficiently sturdy bearing (see *Ready Reference*) plus an extra 25mm (1in) at each end and on top to make fitting the lintel easier.

Cutting the lintel slot

Before you can chop out any bricks or blocks you'll have to rig up your temporary supports for the walling.

If you're using a heavy concrete lintel you'd be wise to place it across two trestles in front of your proposed opening before you rig up the props and needles to save having to haul it in later.

Set up the adjustable props and needles first. If you're making a doorway or narrow window you'll need 100 x 75mm (4 x 3in) thick needles about 1800mm (6ft) long made of sawn timber. You'll have to chop out equivalent sized holes in the wall and feed the needles through. Support each end of the needles with an adjustable prop resting on a timber batten on the floor. The props mustn't be more than 600mm (2ft) from either side of the wall.

You may need to provide temporary supports for the ceiling, if it's necessary, by erecting three adjustable props between stout planks measuring about 100 x 50mm (4 x 2in) top and bottom and 1m (3ft) apart.

If you're working on a brick wall you can simply remove a single course of bricks in order to fit the lintel, but if the wall's made of blocks you'll have to remove a whole block and fill the gap above with a row of bricks. Start to cut out the bricks or blocks directly below the needles by chopping into the mortar joints with a bolster chisel and club hammer to loosen them.

When you've cut the slot for the lintel, trowel a bedding mortar mix of three parts soft sand to one of portland cement onto the bearings at each side and lift the lintel into place. If the walling and its load is to bear evenly on the lintel you'll have to make sure that it's bedded perfectly level. You can hold a spirit level below it and pack out where it rests on the bearings with squares of slate until it's level. Fill the 25mm (1in) fitting tolerance gap above and at each end of the lintel with mortar.

Wait for about 24 hours for the mortar to set before you move any of the bricks and blocks below the lintel.

See pages 100 -103 for how to cut out your opening and fit a lining frame for your decorative architrave and door.

FITTING THE LINTEL

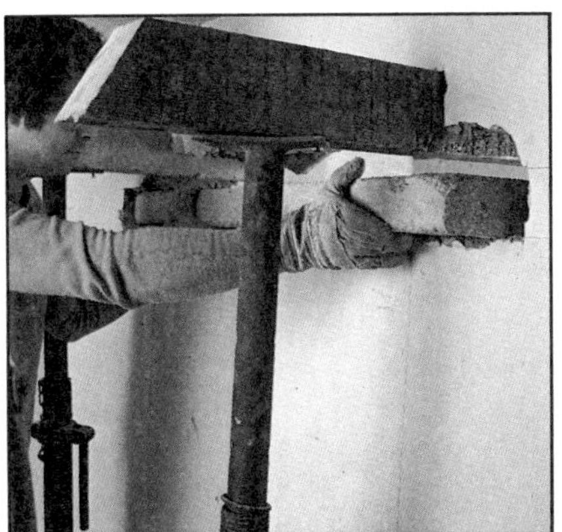

1 *Trowel mortar onto the bearing at each side of the doorway then lift the lintel into position and centre it carefully in its slot.*

2 *The lintel must sit firmly and perfectly level on its bearings; you can wedge small squares of slate underneath to pack it up.*

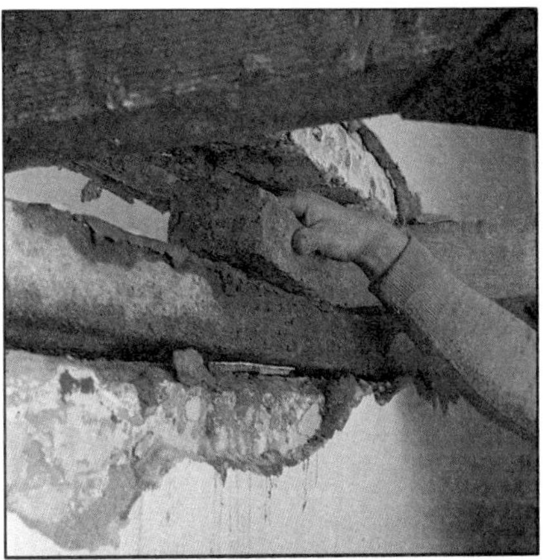

3 *When you're satisfied that the lintel is bedded evenly, start to replace the brickwork that's fallen out, bedding it in fresh mortar.*

4 *Continue to replace the bricks on top of the lintel – at both sides of the wall if it's 224mm (9in) thick – and copy the original bond.*

5 *You can remove the needles and props about 24 hours later, when the mortar has set, and then fill in the gaps left with more bricks.*

6 *Point the mortar joints flush with the face of the brickwork and leave them about 24 hours for the mortar to set before cutting the walling below.*

READY REFERENCE

TIMBER STUD PARTITIONS

If you're making an opening in a load-bearing timber stud partition:

- temporarily support the floor or ceiling above with adjustable props and planks
- remove the plasterboard or lath and plaster cladding from each side of the wall
- cut through one of the vertical studs at the height you want your opening
- notch a timber lintel into the studs at each side of the cut one
- notch a new timber stud into the sole plate and the new lintel to give the opening width you want
- skew-nail additional timber noggins between the new stud and one next to it
- if you've making a hatch, notch in a sill at the required height within the opening
- fix new cladding around the opening.

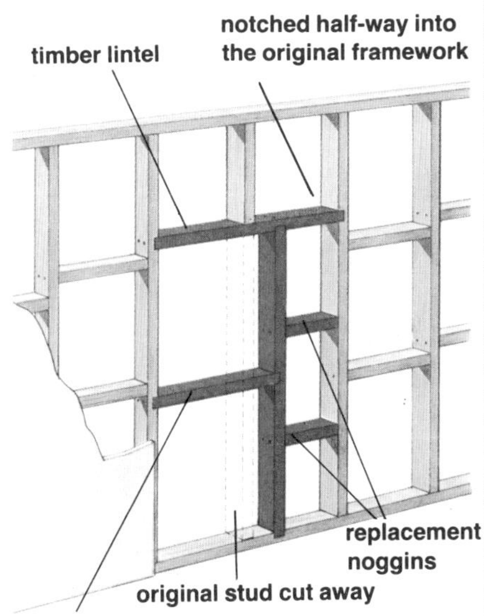

FIXING THE FRAME

To fix the frame in your opening:

- stand the frame temporarily in the opening
- cut recesses in the sides of the opening to take frame ties
- remove the frame and screw frame ties in position corresponding to the recesses
- return the frame to the opening and fill in the recesses over the ties with mortared in bricks
- fill the gap between the wall and frame with mortar and brick fillets before making good the wall.

MAKING A NEW DOORWAY II – the frame

You can change the layout of your rooms to suit your needs by making a new doorway. Part one of this job told you how to fit a lintel; the second part, described here, deals with cutting the opening and fitting a frame.

Making a new doorway or window in a masonry wall is a fairly straightforward job so long as you comply with some basic structural rules. These make sure that you carry out the work in safety and that your end result is sturdy, has no detrimental effect on the building, and is in keeping with the rest of your house.

The most important requirement is to provide some temporary means of support for the wall while you're cutting into it – usually adjustable metal props and stout beams called 'needles' – and the permanent support of a lintel above the opening.

If it's a non-load-bearing wall you'll only need to support the weight of masonry directly above the hole; if it's load-bearing however, you'll also have to support any load that bears on the wall higher up.

Pages 95 -99 cover in detail what's involved in the first part of cutting a hole in a wall: planning and siting your opening, marking out its shape on the wall and finally cutting a slot for, and installing, a pre-stressed concrete lintel.

Cutting the opening

Once your lintel is in place and the mortar on which it's bedded has set, the next stage of the job is to chop away the masonry below to form the opening, within the guidelines that mark its perimeter.

If you're making a doorway you'll have to prise away the skirting board first, using an old chisel and a mallet if necessary. Alternatively, you might be able to chisel away just the section of skirting within your proposed doorway, although you'll probably find it easier to cut it with a saw when it's been removed, and then to re-fix the cut sections to the wall.

Next you should chop along your guidelines using a bolster chisel and club hammer. This gives you a fairly clean, straight, cutting edge through the plaster surface and minimises the amount of making good that will be necessary later.

When you're outlined your proposed opening you can hack off the plaster within this area to reveal the brickwork or blockwork beneath. From this stage on the job's very messy, producing a large amount of rubble and clouds of choking dust. For this reason you'd be wise to remove all of the furniture from your room, or at least to cover it completely with dust sheets. You'll also have to remove or roll back your floorcovering and you might find it useful to lay down large sheets of heavy gauge polythene to collect the debris. Work with the window open and the doors into adjoining rooms closed to keep the dust to a minimum: you can also splash water onto the pile of rubble from time to time to help the dust to settle quickly.

Don't forget to wear old clothes, stout gloves and goggles to protect your eyes from flying fragments. You'll have to exert considerable pressure on your chisel when you're cutting masonry and it's easy for your hammer to slip off the top of the chisel onto your hand. It's a good idea to fit a special mushroom-shaped rubber sleeve, which has a guard at the top, onto the shank of your bolster chisel.

Start to remove the masonry just below the lintel. Cut into the mortar joints of the first brick or block, then lever it out with your chisel or wiggle it out by hand.

Once you're removed the first couple of bricks or blocks the rest shouldn't be too difficult to remove. The best way to work is to

CUTTING THE OPENING

1 *When you've fitted the lintel, chop down the pencil line marking the perimeter of the doorway, using a club hammer and cold chisel.*

2 *Remove the skirting then hack off the plaster within your guidelines. Keep your chisel almost parallel to the wall and fit a knuckle guard.*

3 *Start to chop out the brickwork directly below the lintel. Loosen the first brick by chopping around its mortar joints and wiggle it out.*

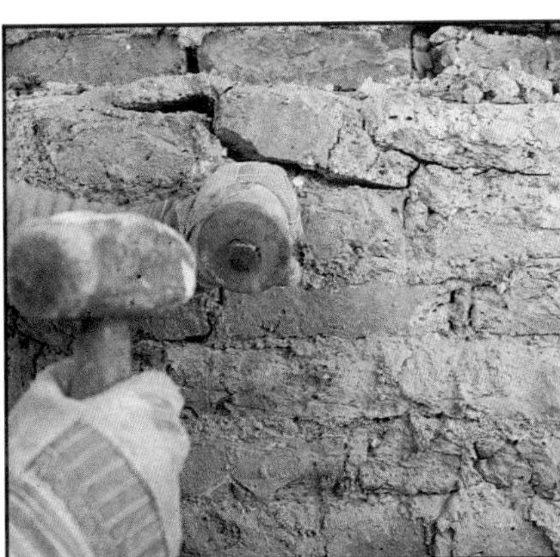

4 *If the wall is 225mm (9in) thick you'll have to remove one leaf at a time. Working down the wall, lever out individual bricks.*

5 *To make a clean cut at the edges of your doorway chop at right angles into the half- or three-quarter bricks protruding into the opening.*

6 *Lever up the brick you've just cut and remove it. It's a good idea to save as many of the whole bricks as you can for use in other jobs.*

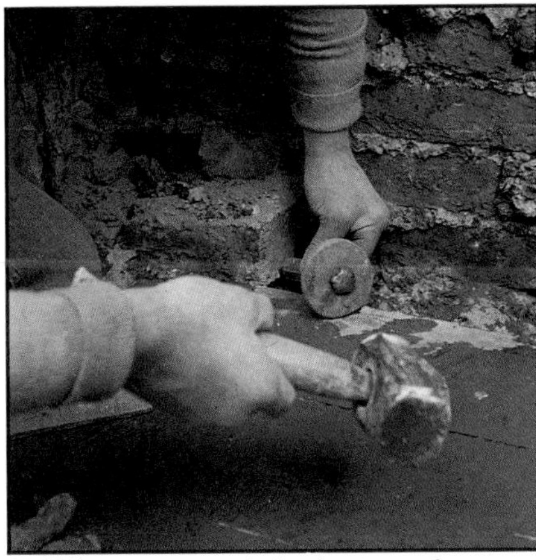

7 *When you reach the bottom of your doorway you'll have to remove the brickwork to just below floor level so you can continue the flooring.*

8 *Go around to the other side of the wall and start to chop away the second leaf of brickwork, working from the top down in the same way.*

9 *When you've cut out all of the walling within your guidelines, return to the edges and trim off any uneven areas of brickwork.*

FITTING THE FRAME

1 *Your lining frame is fixed inside the doorway with six metal frame ties located in 'pockets'. Decide on their positions and screw them in place.*

2 *Chop out pockets in the sides of the opening on the face of the wall where the frame is to be flush and lift the frame into place.*

3 *Hold your spirit level at the sides and underside of the frame to check that it's square and level: pack it out with timber wedges if it's not.*

4 *Trowel mortar onto the base of the pockets into which the frame ties are slotted and cut brick fillets to fit the holes. Butter the bricks with mortar and insert them in the holes.*

5 *Secure all of the six frame ties in their pockets with fillets of brick, pack out the gap between the frame and the wall, then trowel a fairly wet mortar mix over the surface.*

6 *Trowel out the mortar as smoothly as you can over all the areas of exposed brickwork. Don't make it flush with the wall surface; leave about 6mm (1/4in) for plastering.*

chop into the vertical joints first, followed by the horizontals you should then be able to lever the individual brick or block out. Some breakages are certain but you should try to keep whole bricks for use in other jobs.

Cut off the half or three-quarter size bricks that project into the opening on alternate courses as you come to them. Don't leave them until last to cut off or you could weaken their joints in the sides of the opening by hammering. It's always best to cut onto a solid surface. So chop into the wall opening at the side of the wall to ensure the brick splits vertically, then lever it out from below.

Your wall might be 225mm (9in) thick. Although this thickness of wall is usually external, you might come across one inside where an extension to the house has been added. One side – the original outer one – might be rendered and you'll find this difficult to remove without damaging the masonry below. You'll have to remove one half of the wall first, then go around to the other side and remove the remaining skin to form the opening.

On a solid floor you'll simply have to level off the surface but on a suspended timber floor you should remove the walling to just below the level of the floorboards. Don't remove too much walling or you risk destroying the damp-proof course. Fix timber battens to the sides of the joists and then screw a length of replacement floorboard or a panel or chipboard to them.

Where there's a difference in floor levels between the two rooms you've connected you'll have to make a step with floorboards or by laying concrete inside simple formwork (see *Ready Reference*).

When you've formed your opening you can tidy up the perimeter by filling any large cracks and voids at the edges with mortar. If you're making a doorway you'll have to fit a frame within the opening, to which you can attach a stop-bead, door and decorative architrave (see *Ready Reference*).

You can fix the lining frame to the sides of the opening with galvanised metal frame ties set into 'pockets' cut in the walling, or simply by screwing the frame to the masonry if it's fairly level.

You can fix a window frame with frame ties or by inserting timber wedges between the masonry and the frame and screwing through both.

With a 225mm (9in) thick wall, a door frame is set nearer the opening side of the doorway and the rest of the opening is plastered as a 'reveal': like the recess into which a window frame is fitted.

MAKING GOOD

1 To make good the area above your doorway spread a layer of Carlite Browning plaster onto the lintel and over the brickwork.

2 Make good the reveals flush with the frame: nail battens to the perimeter as thickness guides. Key the bricks with water for plastering.

3 Spread a layer of plaster onto the reveal and underside of the lintel using your steel trowel, filling any voids in the surface. Leave to stiffen.

4 When your first layer has stiffened you can spread on a second layer, bringing the surface of the reveal almost level with the thickness guide.

5 Fix a timber thickness guide to the face of the lintel and spread on a layer of plaster to the underside. Leave it to stiffen and apply another layer almost flush with the guide.

6 Use your trowel without any plaster to smooth the surface of the reveal. If you're going to fit a wide door stop to the reveal you needn't apply a thin coat of finishing plaster.

READY REFERENCE

MAKING A STEP

If there s a difference in floor levels between the two rooms you're connecting you'll have to make a step. You can:

- cast a simple concrete step or
- make a timber step

Ideally your door should open into the higher level, where you'd need only a small step within the opening (**A**)

If this isn't possible make a broader step projecting about 400mm (16in) into the lower level (**B**).

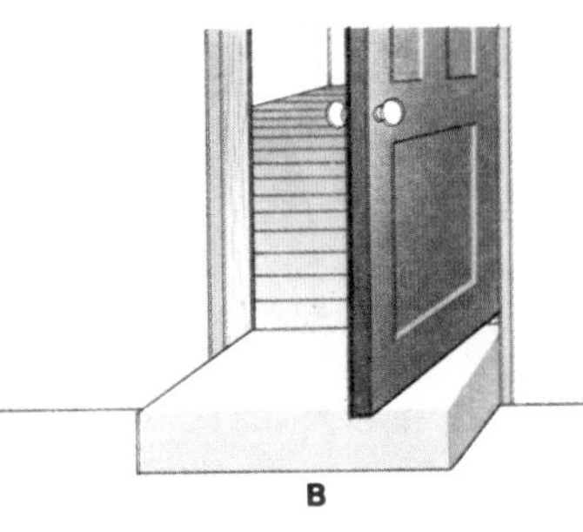

FINISHING THE DOORWAY

When you've cut your opening and fitted your lining frame you'll need to finish off the doorway. To do this:

- fix plain or moulded architrave (**A**) around the doorway over the join between frame (**B**) and wall with 38mm (1½in) oval nails
- fix a stop bead (**C**) to the inside of the frame with 25mm (1in) oval nails
- hang your door (**D**).

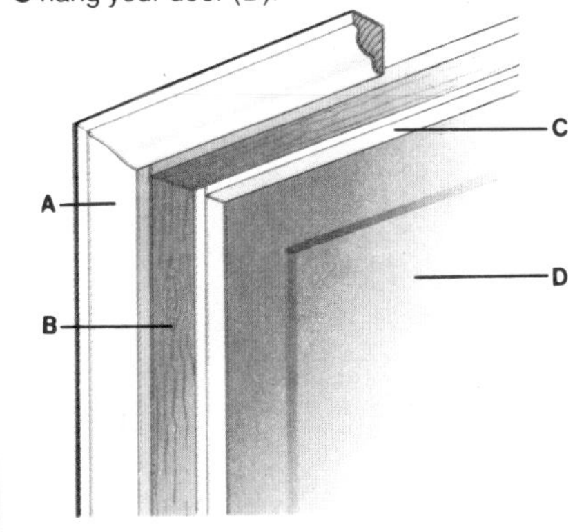

INSTALLING PATIO DOORS I – preparation

Aluminium-framed patio doors are the ideal replacement for old French windows. They offer greater insulation and are quick and easy to install. Part one of this article describes how to prepare the opening.

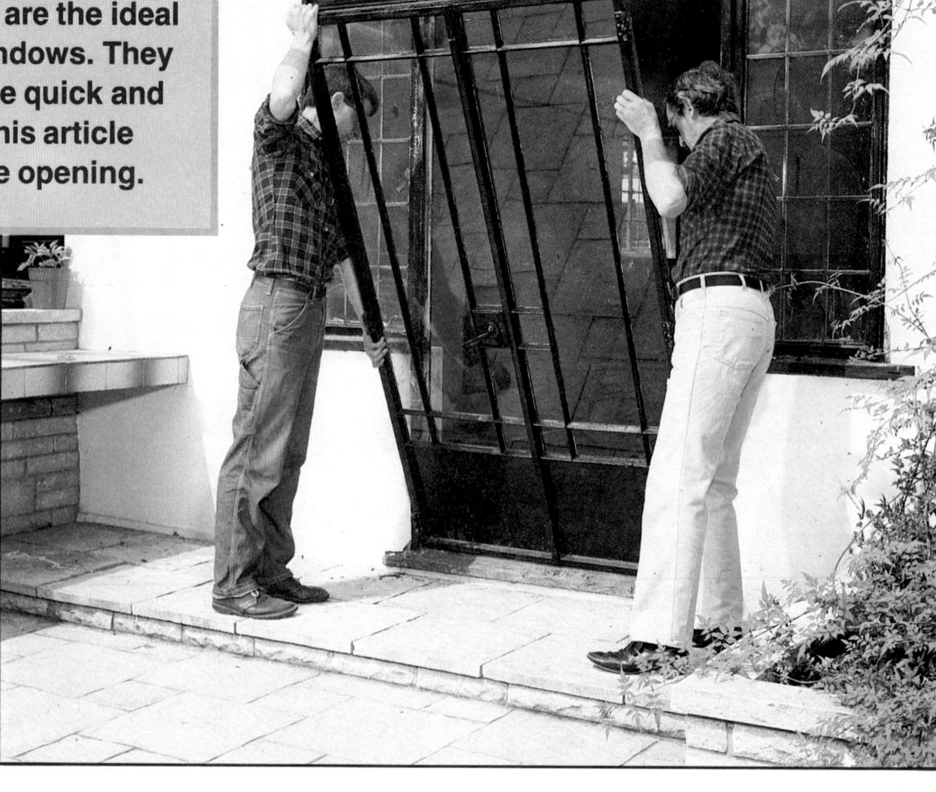

Traditional French windows or exterior glazed doors may provide convenient access to your garden – and they certainly admit more light to your room than a conventional window – but they're a notoriously easy way in for burglars. This is partly because they open outwards and so are difficult to look successfully, and partly because they usually consist of many small panes of glass, which don't offer much security. They tend to admit draughts, too, and if they're old and in poor condition they may well leak and rattle in bad weather.

Commonly, the frames used to be made of wood, which rots, or steel, which is prone to rust. Nowadays, however, patio doors with aluminium – or even plastic – frames, double glazing, and space-saving sliding action are made as replacements. The most readily available type will have a protective decorative grey, satin or 'anodised' finish, or a factory-applied coloured surface, normally just black or white.

You needn't, of course, stop at substituting aluminium doors for existing French windows or glazed doors: you can also make an entirely new opening in your wall to take your new frame, so long as you provide the necessary structural support for the walling above and any load that bears upon it by installing a new lintel to bridge the opening.

Although standard-sized doors are available from all the major door manufacturers, it's also possible for you to have them made to your own specification: this is especially useful if you're unable to – or you simply don't want to – modify the size of your existing opening to take an 'off-the-shelf' frame.

How the doors are made

Although there are small differences between models in the way modern patio doors are put together, they all consist of the same basic components (see *Ready Reference*). These are usually a main aluminium or UPVC frame screwed to a hardwood subframe or surround with an integral sill which is, in turn, screwed to the masonry at the perimeter of your opening.

The individual panels are supplied ready-made for you to insert in the frame. They have nylon or stainless steel rollers at the bottom, which you simply clip onto metal runners at the base of the outer frame. The top of the doors locates in a channel. All meeting faces – between adjoining doors and the frame – are sealed with weatherstrips to keep out draughts and to prevent moisture from penetrating.

The double-glazed panes usually comprise two factory-sealed sheets of 6mm (¼in) thick laminated or toughened glass – sometimes tinted – with a 12mm (½in) gap between giving an overall thickness of 24mm (1in). Some doors may even have a gap between panes of 20mm (¾in) for even greater heat and sound insulation.

Although aluminium is durable and never needs decorating, it's not a good insulator, so the hollow frames usually incorporate a thermal barrier, which stops condensation forming on the inside of the frame in cold weather.

If there aren't any other openable windows in your room, you'd be wise to install a ventilator into your new door frame; this is usually an unobtrusive panel with sliding vents, and a fly-screen which you fix at the top of the doors, just under the main frame. They allow fresh air to circulate in the room without you having to leave the door open.

Aluminium patio doors are also much more secure than conventional glazed doors; they incorporate tough locks and must have an 'anti-lift' device, without which any determined burglar would lift out a sliding panel easily.

Choosing patio doors

Visit the showrooms of local suppliers to see what styles are available, and to obtain price quotations. Your final choice should largely depend on the finish of other door or window frames in your house, especially those on the same elevation as your new doors. You should take care to maintain the overall appearance of your property; patio doors with, for example, a silvery-grey finish may look totally out of place in a house with other frames made of timber and painted white.

Aluminium patio doors are commonly made of two, three or four separate panels, and there are various combinations of positions you can have for the fixed and opening parts (see the *Ready Reference* column on page 110). Your choice depends on the position of furniture inside the house and access to your garden.

Ordering the doors

To calculate the size of the doors and frame you'll need, first measure accurately the height and width of your existing opening. But if you're altering the opening size, or your

REMOVING THE OLD FRAME

1 *To fit a new frame the same size as the old one, chop away some render at each side to expose the bricks, then measure for the new frame from here.*

2 *A metal unit has doors hinged to a surround. Unscrew from the timber frame and use a wood block and hammer to break the bond of the putty.*

3 *Get someone to help you to lift the doors away; then remove the smaller side windows and fanlights – if any – in the same way.*

4 *To remove the timber subframe, cut the verticals at an angle, about 150mm (6in) above the sills; then lever them away with a bolster.*

5 *You should then be able to pull away the main vertical frame jambs. They're held by cut nails; you'll need to exert some pressure to remove them.*

6 *The head piece of the timber frame is held by cut nails also, and you should remove this in two parts, by cutting it in half at the centre.*

READY REFERENCE

PATIO DOOR FORMAT

An aluminium-framed patio door usually consists of:

- a timber subframe with integral sill (A)
- an aluminium outer frame, incorporating the sliding track (B)
- one or more fixed double-glazed panels (C)
- one or more sliding double-glazed panels (D)
- a threshold strip (E)
- a head infill (F)
- a security lock (G)
- door stops (H)
- an anti-lift device (I)
- a ventilator hook (J)

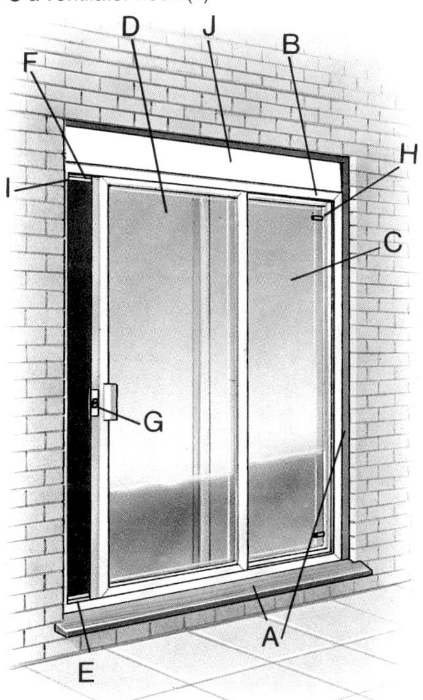

TIP: CUT BACK THE RENDER

You may find that the external render has been carried over the timber frame of your original patio doors at the sides. Where you're fitting new doors of the same size, cut back a little render at top and bottom to reveal the brick edge so you can measure the opening.

CHANGING THE OPENING SIZE

If you're enlarging the width or height of your opening, or you're fitting a new patio door where none existed previously, you'll have to fit a new lintel above to bear the weight of the walling and any load it supports.

original doors are the 'keyhole' style with windows at each side incorporated into the main frame, you'll simply need the overall dimensions of your proposed doors, including the hardwood subframe.

Measure at two points in each direction, and if there's a variation use the smaller one when ordering. You can always fill in any slight gap between the frame and the masonry – indeed, some 'fitting tolerance' is preferable – but if the frame is even fractionally too big you'd have to enlarge the opening.

On rendered walls you'll have to chop out a small area of the cladding at the top and bottom on each side of the existing frames to expose the brickwork so you can measure accurately. When measuring the height, take your reading from the top of the existing frame to the underside of the sill.

The two dimensions you'll end up with – which, in effect, give you the overall size of the opening – are all your supplier will need.

Removing the old frame

If you're simply replacing existing patio doors with new ones of the same size you'll probably be able to carry out the work in one day. But if you have to enlarge – or even decrease – the size of the opening you should tackle the preparation on one day and fit the new doors on the following day.

Before you start work, clear the room of all furniture and carpets, or stack it out of the way at the opposite end to the doors and cover it with dust sheets.

The first job is to dismantle the old doors and windows. If your original doors are wooden, you simple unscrew them from their main frame at the hinges to remove them. Repeat this procedure for any fanlights and side windows.

Often you'll find an old metal door and window unit will be set into a rebated timber frame, and the doors will be hinged to their own metal surround, which is screwed to the woodwork. To remove these you'll have to locate the screws, which will probably be covered by layers of paint.

With all of the screws removed you can take off the doors. They may still be held by bedding putty; if so, use a small wooden block and a hammer from the inside to break the bond so you can lift the doors free of the metal frame.

If there's a fixed light above either the door or opening windows it will probably be connected to the metal frame below by bolts and to the surrounding timber by screws. Use an old screwdriver or a glazier's hacking knife to chop out the putty covering the bolt and screw heads in the fixed light and unscrew them.

With all of the glazed sections removed, you'll be left with just the wooden framework to tackle. Cut through each of the vertical rebated sections that may separate the doors from the window about 150mm (6in) above the window sills, and then lever the lower portions away from the walls and from the old sills. Then pull away the top sections from the headpiece of the frame above.

With your club hammer and cold chisel, chop back any rendering that may cover the outer frame so that it's flush with the brickwork (see *Ready Reference*). Next you can cut through the frame jambs about 150mm (6in) above the sills, making your saw cuts at an angle so you can remove the pieces easily (see *Ready Reference*). Use a crowbar to lever out the lower pieces of frame and the sills. Remove the upper sections of the vertical jambs similarly; they'll be held with cut nails driven into the brickwork and you'll need to exert quite a lot of pressure to lever them free.

Remove any nail heads that are left projecting from the wall, then cut the frame head in two and lever it away.

PREPARING THE OPENING

If you're fitting new aluminium patio doors that are the same overall size as your original French windows all you have to do, having removed the old frame, is to tidy up the opening to receive your new frame. But if you're going to extend the size of your opening to take higher, broader doors, you'll have to fit a new lintel above to take the weight of the walling and any load bearing upon it. On solid 225mm (9in) walls you could use a prestressed concrete or hollow steel lintel; on cavity walls you could use a concrete lintel for the inner leaf and a steel support lintel for the outer leaf.

1 *Use a timber straight edge against a spirit level to mark the sides of the opening to be cut on the projecting part of a keyhole-style window and door unit.*

4 *Start to hack off the render from the masonry; remove individual bricks from the top. Force the chisel under a brick to break the mortar bond.*

5 *Cut down onto any bricks that protrude into the opening to give as clean a break as possible, to avoid substantial making-good later.*

Preparing the opening

If you're going to install a new frame the same size as the original you can simply tidy up the masonry at the perimeter of the opening, ready to accept your new timber subframe (see part 2 on pages 108-112). But if you're enlarging the opening you'll need to fit a new, wider lintel above the first (see *Ready Reference*).

Where you just have to cut back the masonry of a keyhole-style opening (see step-by-step photographs on this page), or when you've fitted your new lintel, you should mark a pencil line down the outside wall using a spirit level and a long timber straight edge to indicate the sides of the new opening. Repeat this procedure on the inside wall, then chop down the lines with your bolster chisel to give a clean cutting edge.

Before you start to chop out the masonry, you'd be wise to fix up a dust sheet or a

2 *Do the same at the other side of the opening, so that you're outlining a square or rectangular opening; mark the inside face of the wall, too.*

TIP

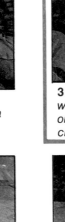

3 *Chop down your pencil guidelines with a club hammer and bolster chisel, on both faces of the wall, to give a clean cutting edge at the perimeter.*

6 *When you reach the bottom, cut out the bricks to just below floor level so that you can set your new subframe flush with the floorboards.*

7 *You can make good any damage to the corners of the reveals by fixing metal angle bead onto the masonry, and then spreading on a layer of render.*

polythene sheet across the opening, but tacking it to the inside, to contain the considerable amount of dust thrown up during hammering. You should wear goggles as protection against flying fragments.

Using a club hammer and bolster chisel – and starting at the top of the wall – hack off the render on the outside, followed by the plaster on the inside, at each side of the opening to reveal the brickwork.

When all of the render and plaster has been removed, start to chop out the bricks individually. Cut through the half bricks that protrude into the opening at the perimeter line from above.

When you've removed all the masonry to the perimeter of your opening take out the old door sill and give the mortar beneath and to the sides of it a thorough brushing with a stiff-bristled brush, ready to take the wider hardwood sill of your new door. The floor of your room may be higher than the ground level just outside the opening, in which case you'll have to lay a low plinth of bricks so that the new sill is at floor level.

Cutting out a new opening is relatively straightforward in a solid 225mm (9in) wall: you can simply cut the sides as straight and flat as possible or, for neatness, 'tooth-out' the bricks, removing the cut bricks, and fill in with half bricks placed flat side out. But if you've a cavity wall you'll have to close off the gap (see *Ready Reference*). It's also vital that you incorporate a damp-proof course around the opening to stop any moisture bridging the cavity and causing damp on the inside face of the wall.

This completes the preparation of your opening. See part 2 of this job on pages 108-112 for details of how to fit a timber subframe, followed by your new aluminium patio doors.

READY REFERENCE

CLOSING A CAVITY WALL

If you have a cavity wall with an inner section of bricks or blocks and an outer one of bricks, you'll have to close off the gap between the two leaves and install a vertical dpc before you can fit your new frame. To do this:

- tooth out half- or cut bricks at the perimeter of your opening and replace them with whole bricks laid header-on to form the reveals
- add cut bricks to close the gap
- insert a strip of bituminous felt between the two leaves as a dpc.

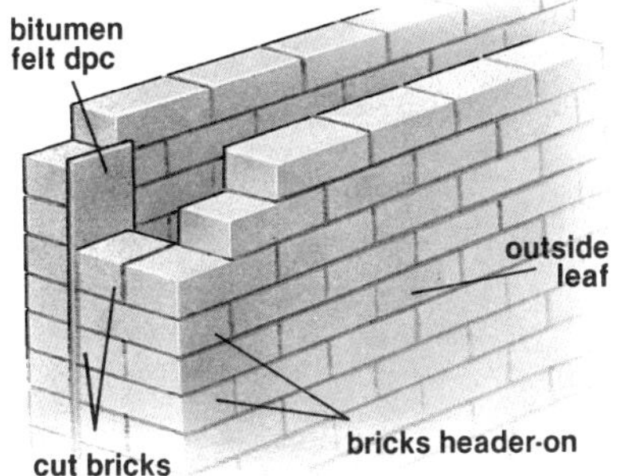

BUILDING A PLINTH

If the floor of your room is more than about one brick's depth above the ground outside you should lay a low plinth of bricks across the base of the opening so that you can set your new hardwood timber sill at floor level.

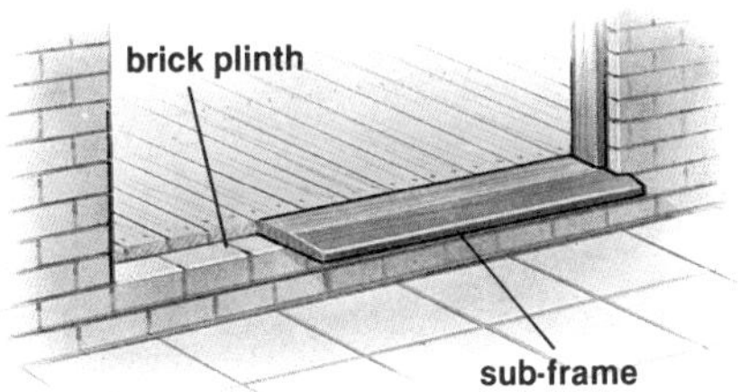

TIP: REMOVING THE OLD FRAME

To remove your old door frame, cut through the jambs about 150mm (6in) above and below the sills and lever out the sections.

INSTALLING PATIO DOORS II – installation

New patio doors with durable aluminium or plastic frames usually come in kit form for easy assembly. Part one of this article described how to prepare the opening; here's how to fit the doors.

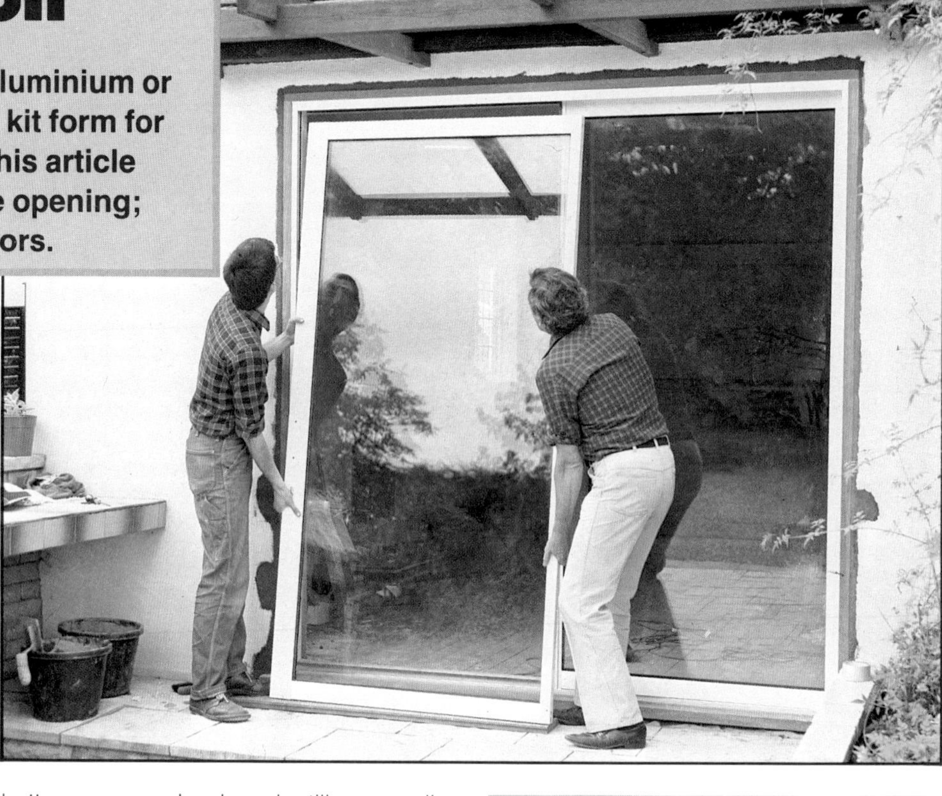

Aluminium or plastic-framed patio doors have many benefits over traditional French windows and exterior glazed doors. They're double-glazed, with efficient weatherproofing, operate with a smooth, space-saving sliding action, and admit more light to your room. Also, they're quick and easy to install, and they have an attractive finish which means you never need to decorate them.

Although they're popular as replacements, you can, of course, fit them where there was no opening previously, as long as you install the necessary support for the wall by installing a new lintel.

When you've removed your old frame (see part one of this job on pages 104 - 107) or you've increased the size of the opening, you can fit your new frame and doors.

Parts of the frame

Your new doors will probably have two, three or four separate panels, depending on the width of your opening, but it's not usual to have all of them sliding – which you have fixed and which opening mainly depends on access to and from your garden and where your furniture is placed indoors.

Although it's possible to have doors made to your own size specifications – and, indeed, containing as many panels as you want – there's a wide range of standard sizes available, which should suit most locations.

When your new patio doors have been delivered you should make sure that all the components have been supplied, by checking them off against the manufacturer's instruction booklet, which should also be supplied with the kit.

Although the individual parts may vary from one maker to another, all of the doors consist of the same basic components (see the *Ready Reference* column on page 105. Basically the installation consists of a hardwood subframe and a metal outer frame into which the doors slot, although plastic-framed doors may not have a subframe.

Your kit should contain the double-glazed door panels themselves, already assembled and with nylon or stainless steel rollers at the bottom; a new hardwood still, generally made from prepared 150x50mm (6x2in) Brazilian mahogany, with two integral jambs that form the sides of the subframe, and a head piece of 100x50mm (4x2in) hardwood, which forms the top of the subframe. These four pieces of timber should be in good, clean condition and they're usually assembled with mortise and tenon joints. The frame may have projecting 'horns' at the sill and head, which protect the corners during storage and transit. You can saw off the head horns when you're ready to fit the frame but you should leave the sill horns in place to make the frame more rigid when fitted. The frame may also have a prepared rebate into which the aluminium frame will fit.

This aluminium frame will probably come in four separate pieces, comprising two jambs, one sill and one frame head. With these there'll also be two short lengths of aluminium, which you use at the door threshold and between the fixed door and locking jamb of the frame head.

Your kit may also include a ventilator unit, which is fixed to the outer metal frame at the top, and ensures a constant flow of air into the room. The security lock should already be fitted to the opening door, and lengths of weatherstrip should be fitted in channels to the frame sections at all meeting faces. All of the screws, brackets and various other miscellaneous items used to assemble the

READY REFERENCE

PATIO DOOR COMPONENTS

An aluminium-framed patio door unit usually consists of: a timber subframe with integral sill; an aluminium outer frame, including the sliding track; one or more sliding double-glazed panels; a threshold strip; a head infill; a security lock; door stops; an anti-lift device, and a ventilator hood.

FIXING THE SUBFRAME

To fix the hardwood subframe inside the opening:

- drill holes for the fixing screws through the frame jambs and into solid masonry
- push special long plastic wall plugs, which come complete with their own screw, through the clearance holes and into the masonry, and drive them fully home to secure the frame.

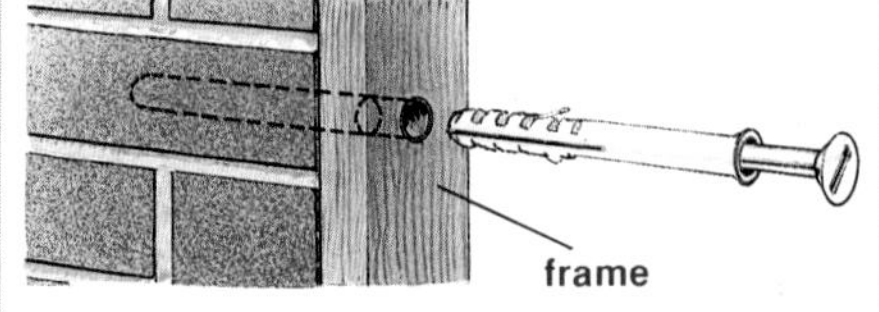

frame will also be provided with the kit.

Other materials you'll need – but which may not be supplied with the doors – are sealants and mastics for weatherproofing. They usually come in tubes for use with special applicator guns. You'll also need some help in fitting the doors – they're very heavy to lift and you'll also need some assistance when setting the doors on their tracks.

Assembling the timber subframe

The first step you'll probably have to make is to assemble the timber subframe, using a little wood glue on each joint. Make sure the tenons are fully seated in their mortises and check that the corners are absolutely at 90° with a try square.

Saw off the head horns, then offer up the assembled frame to your prepared opening to check that it will fit; you'll have to cut out some of the masonry at each side to accommodate the projecting sill horns and remove any brickwork that prevents easy insertion of the frame. Once you're satisfied that the frame will fit snugly, remove it and lay it flat on the ground.

Assembling the aluminium frame

Next, identify each component of the aluminium outer frame to make sure you know how it fits together, then assemble it. It's best if you do this on soft ground – your lawn, for example – or on an old sheet or blanket laid on the ground, so that you don't scratch the surface finish of the aluminium. Take special care to ensure that the side jambs are located properly, and that the locking jamb is on the correct side.

Apply a bead of weatherproofing mastic to the sill and head joins, assemble them and secure with self-tapping screws, through the pre-drilled holes in the frame. Check that the joints are tight, then wipe off any excess mastic. Lift the assembled frame and slot it into the wooden subframe, which is still lying flat, to check it fits. Once you're satisfied that the fit is perfect, remove the aluminium frame and leave it lying flat on soft ground.

Fitting the frames

You can now fit the hardwood subframe into the opening from the outside. If the subsequent components are to fit properly, and the doors to slide without binding, the subframe must be set perfectly square and true – both vertically and horizontally – within the opening. You'll probably have to wedge the frame to get a perfect fit.

Pack out under the sill first with small wooden offcuts to jack up the frame until it touches the top of the opening. Use your spirit level continually on the sill and adjust the wedges until the sill is perfectly horizontal.

Move on to the jambs next, using wedges to adjust and hold the frame in position. Use your spirit level on both the sides and faces of the jambs to check the plumb, then recheck the sill, to make sure that your adjustments to the jambs haven't altered the horizontal level. A long metal straight edge held against the face of each jamb will indicate whether or not the wood is flat and straight: if it's at all bowed you can correct this when you fix the frame finally.

To check that the frame is seated truly square you must measure the diagonals accurately. You'll need someone's help to hold one end of your tape measure at a top corner while you stretch the other end to the bottom diagonal corner. Repeat this procedure for the other diagonal. If there's any variation at all between the two measurements you'll have to re-adjust the sill or the head pieces and possibly re-align the jambs. Only when the diagonals measure precisely the same and the jambs are exactly vertical should you fix the frame permanently.

To do this, drill clearance holes to take the fixing screws through the jambs and into the solid masonry. Follow a line that will eventually be covered by the aluminium outer frame to conceal the screws. Make a minimum of four fixings on the jamb adjacent to the fixed door, and six on the locking jamb side for additional security in the area of the lock.

Push wallplugs, which have their own built-in screws, through the holes you've drilled in the jambs and into the masonry (see *Ready Reference*); then drive them fully home.

It's possible that by screwing the frame in place you may have secured it in a warped position, so you should check again with a straight edge that the jambs are flat; loosening or tightening the screws or

FITTING THE TIMBER SUBFRAME

1 *Once you've removed your old frame, assemble the jambs, head and sill of the timber subframe and position it in the prepared opening.*

2 *Place your spirit level on the sill to check it's level. Leave the 'horns' protruding at the sides of the sill and build them into the masonry.*

3 *Use timber offcuts to wedge up the frame at the sill. Check again with your spirit level until you're sure that it's perfectly horizontal.*

4 *Measure both diagonals of the frame by stretching a tape measure from one corner to another. If they're the same, the frame's square; adjust if necessary.*

FIXING THE ALUMINIUM FRAME

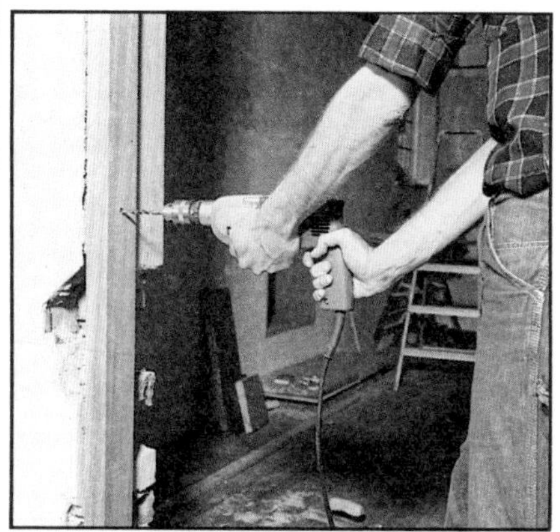

5 *When you've adjusted the frame so it's square, drill clearance holes through the jambs into the masonry to take the fixing screws and plugs.*

6 *You can then assemble the outer aluminium frame, on soft ground so you don't scratch the finish. Apply non-setting mastic to the meeting faces.*

7 *Fix the sill and head pieces to the jambs and secure them at the corners with self-tapping screws, inserted through their pre-drilled holes.*

8 *You'll need help to position the assembled aluminium frame within the subframe. Place the sill first then swing it up into the opening.*

adjusting the wedges should remove any bowing in the frame. Finally, fit more wedges under the sill to keep it perfectly flat.

The aluminium door frame can now be fixed in place. If the wooden subframe has a rebate you should apply a bead of non-setting mastic to the rebate before you push the frame into place firmly. Offer up the base of the metal frame to the subframe sill and press it firmly against the back of the rebate.

If there's no rebate on the subframe, you should fit the aluminium frame flush to the inside edge of the timber one.

Screw the frame – which will probably be pre-drilled – to the subframe using the screws provided; these will probably be countersunk for the sill and round or pan-head types for the head and jambs.

Check again with your straight edge and spirit level to make sure the entire assembly is square and level.

Installing a fixed panel

The next stage is to insert the fixed panel into the frame from the outside. First of all you'll probably have to attach a fixing bracket at the top, just behind the interlocking edge. Rest the panel the correct way round on the block of wood alongside the sill to protect its bottom edge, while you lift it into place.

With your helper's assistance, lift up the panel and slot its edge into the head frame, then swing it down and push it tightly against the sill upstand. Slide the panel along until it's firmly against the jamb, then secure it at the top and bottom using the screws and brackets provided. You'll have to drill holes in the head and sill to take the self-tapping screws, so make sure you've positioned the panel accurately in the frame.

The threshold strip, which stretches from the fixed door across to the locking jamb, must be cut to the precise length required; it

READY REFERENCE

COMMON DOOR ARRANGEMENTS

The position of doors and fixed panels depends mainly on the siting of furniture in your room and on access to your garden. Common arrangements of panels are:

for a two-panel door:

- the opening panel on the right-hand or left-hand side

for a three-panel door:

- the opening panel in the centre, sliding to the left or right

for a four-panel door:

- the opening panels in the centre, sliding outwards.

MAKING FRAMES WATERPROOF

If your timber subframe has a rebate:

- apply non-setting mastic to the rebate
- offer the base of the aluminium outer frame up to the timber sill
- swing the frame upright and press it firmly against the back of the rebate.

If your subframe has no rebate:

- position the metal frame flush with the subframe's inside edge
- seal the junction between metal frame and subframe with mastic.

TIP:FIT A VENTILATOR

If there aren't any other opening windows in your room, you should fit a special ventilator unit on top of the metal outer frame. This will introduce a gentle supply of air to the room; most types include a fly-screen.

FITTING THE DOORS

1 *Check that the metal frame is positioned perfectly squarely in the subframe, then screw it to the timber at the sill, head and jambs.*

2 *Lift the fixed panel into the metal frame, swing it onto the sill and secure with its brackets; fix the threshold and head infill.*

3 *Reset the sliding door on a timber batten to protect the rollers, lift it into the head channel and swing it down to locate on its track.*

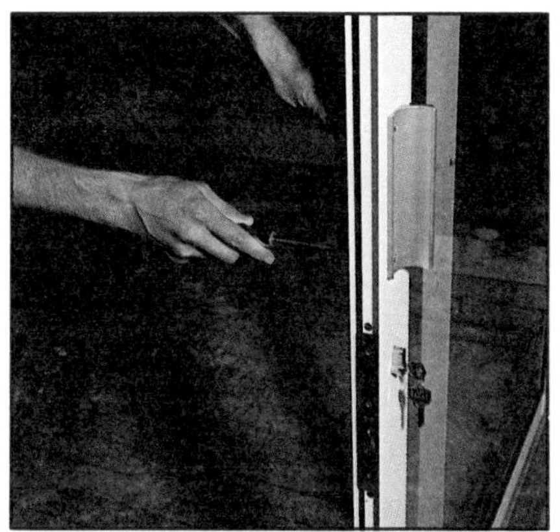

4 *When you've adjusted the roller height so the door fits tightly and squarely in the frame, attach the handles and the locking mechanism.*

5 *Screw the rubber-tipped door stops to the outer frame jamb, positioning them about 150mm (6in) from the top and the bottom.*

6 *To make good any gaps between the subframe and the masonry, first cover the outer face of the frame with masking tape; leave a 25mm (1in) gap.*

is then slotted and screwed into position to cover the bottom fixing bracket. The head infill strip, which fills the gap at the top of the doorway, is similarly cut and fitted; it conceals the top fixing bracket. Before you insert either of these infill pieces you should check that both have weatherstrips firmly fixed in their channels.

Fitting a sliding panel

You can fix the sliding door next, by lifting it up into the frame head and swinging the bottom inwards until its rollers rest on the track, which is an integral part of the metal sill. Don't move the door until you're sure that it's sitting correctly on the track and that there's no contact between the door frame and the sill, since this could scratch the sill surface.

You'll need to adjust the height of the rollers to enable the door to close tightly and evenly against the frame. The usual way to do this is to insert a screwdriver – usually a Pozidriv type – into a hole at the base of the door to reach the mechanism. Your helper must lift the door slightly with an improvised lever such as a flat length of metal or a small crowbar. Insert it beneath its bottom edge as you make the adjustment or you could damage the screw threads.

First of all lower the rollers, which raises the entire door higher into the frame. This makes it difficult for a burglar to force out the door; it must be lifted free of its rollers before it can be removed.

Once you've aligned the sliding door to your satisfaction, you can screw on the handles and then adjust the lock keep to suit the height of the security latch; there'll probably be some plastic infill strips, which you can cut to size and fit at each side of the keep. This is because the final position of the lock and keep depends on the extent of the roller adjustments you've had to make to locate the door accurately.

Rubber buffers will probably be supplied with your kit of components, too, and these are commonly screwed to the jamb of the aluminium outer frame, against which the sliding door opens.

The 'anti-lift' device – normally nothing more than a simple plastic block – is fitted to the frame head adjacent to the locking jamb, so that when the door is closed its locking stile only just clears the block; it's invisible from outside and prevents the door from being lifted out of the frame.

Weatherproofing the door

Although at this stage your new door is fully operational, you'll still need to make it completely weathertight. Apply a bead of silicone rubber sealant to the small gap

MAKING GOOD

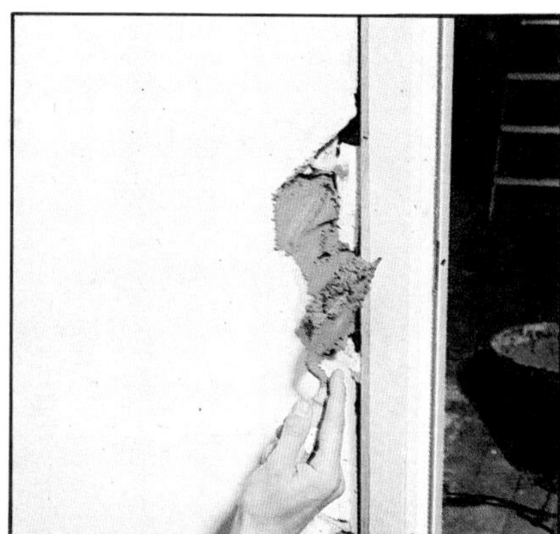

7 Trowel mortar into the gaps at the perimeter of the opening, taking it onto the frame by about 25mm (1in), then remove the masking tape.

8 Weatherproof the small gap between the metal frame and the timber subframe on the inside face with a silicone rubber sealant.

9 Make a 'pugging tool' from two offcuts of timber nailed together in a T-shape so that you can more easily push mortar under the sill.

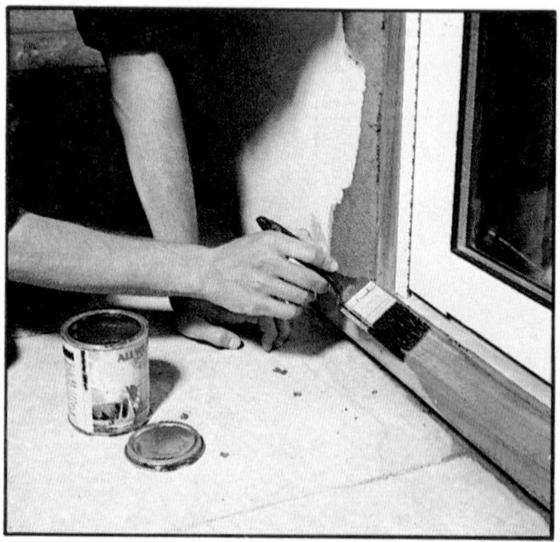

10 When you've made good the sides of the opening inside and out, apply three coats of exterior varnish to the exposed timber subframe.

between the wood subframe and the aluminium outer frame on the inside of the installation, then immediately clean off any excess which may have spread onto either frame, with a clean, damp cloth, to prevent smears from spoiling the woodwork.

Do the same on the outside of the installation, but use an exterior frame sealant in a colour that matches either the wood or the aluminium.

Any gap between the external brickwork and the wooden subframe must also be made good. You can fill large gaps with fillets of brick or stone embedded in, and covered by, mortar. Press the mortar well into gaps all round the frame, overlapping the subframe by about 25mm (1in). You can protect the rest of the timber face from smears with masking tape, as mortar will stain the wood quite easily.

In areas where access isn't easy, such as under the sill, make a small 'pugging stick' – simply two pieces of timber nailed together in a T-shape – to help you push the mortar into place.

Making good the frame

The interior plasterwork around your new door must also be made good. Where you've had to cut away sections of plaster at the reveals in order to fit the frame you'll have to replaster complete corners using expanded metal angle bead which you attach to the exposed masonry and plaster over (see Chapter 6 pages 154 -157).

Clean the frame thoroughly, then varnish the exposed timber frame, using at least three coats of top-quality exterior varnish. The timber can, of course, be painted, if this is more in keeping with the rest of your house. You can likewise paint the interior woodwork to match the decor of your room.

READY REFERENCE

ADJUSTING THE ROLLERS

You'll have to adjust the rollers on the bottom of the sliding panel so that the door will close tightly and squarely against the frame. To do this:

- get a helper to lift the door slightly using a piece of flat metal or a slim crowbar beneath its bottom edge
- insert your screwdriver – usually a Pozidriv type – into the hole at the base of the door to reach the mechanism

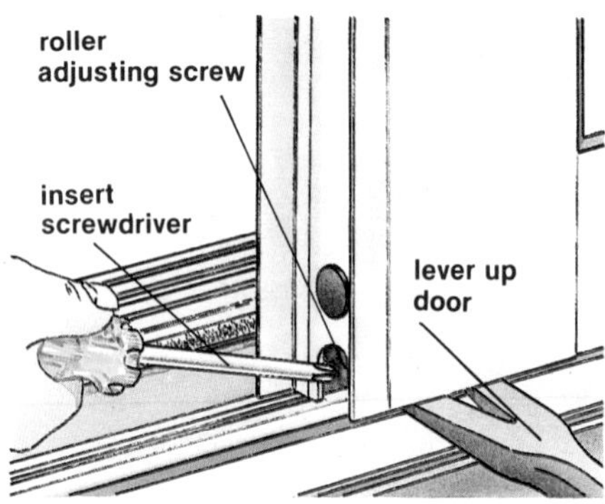

- lower the rollers to raise the door; raise them to lower the door.

TIP:FIT GOOD LOCKS

Glazed doors are particularly vulnerable to forced entry by burglars, so it's vital that you make them secure. To do this:

- fit the stile lock provided so that it's fully engaged in the locking jamb
- fit an 'anti-lift' device to prevent the door being lifted out of the frame even if locked; it's usually a small block screwed to the head piece, snug against the locking jamb on a two-panel door, and allows the locking stile to just clear the block

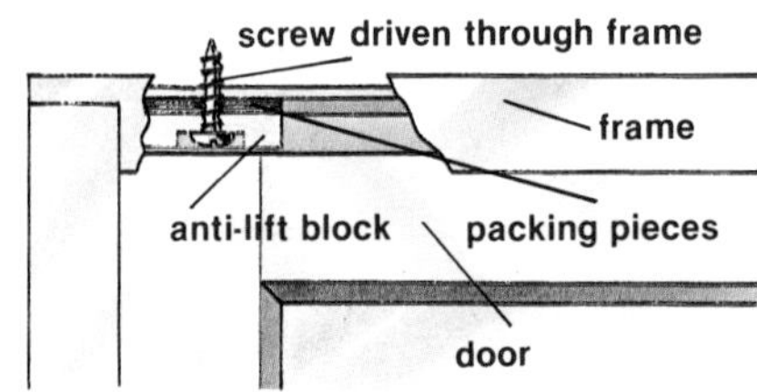

- fit an optional key-operated security bolt at the top of the locking jamb.

PREVENTING CONDENSATION

Aluminium isn't a good insulator and, although actual heat loss through the frame would be minimal, condensation would form and stream down the inside of the frames during cold weather. To avoid this, choose doors incorporating 'thermal breaks' – non-metallic sections that stop the transfer of heat and so allow the inside frames to stay warm.

FITTING A NEW WINDOW I – the opening

As well as admitting light and keeping the elements at bay, windows also perform an important decorative role. If they're in poor condition – or if you don't like their style – why not replace them?

There are numerous reasons why you may decide to remove or replace a window or, indeed, to install a new one. The most common one, however, is likely to be age. Whether your windows are made of wood or metal, they're likely to deteriorate unless you maintain them regularly. It's possible to make minor repairs to the frames by tightening up loose joints, replacing sash cords or even renewing damaged sections of timber with new wood. But if the damage is widespread, the only solution to the problem is to fit a new window.

Equally, your reason for replacing a window may be aesthetic rather than practical. Previous renovations might have included the fitting of new windows that you think are incompatible with the original style and age of the property.

Another possibility is that you feel there aren't sufficient openings to give adequate ventilation, or that the existing ones are in the wrong places. You may also want to make adjustments to the position of your windows within an area of wall to fit in with a new decorative scheme.

If you're replacing your existing window with one of the same size you won't have to worry about providing any support for the opening while you remove the old frame and fit your new one. If you're increasing the width of the frame or making an entirely new window opening, you'll have to install a rigid beam or lintel to support the walling you've removed. Look back at Making a doorway (pages 95 -103) for how to cut a hole in a solid brick wall ready to take your frame.

Planning permission

Before you go ahead with your plans to replace your windows you must contact your local Building Inspector with details of what you intend to do. Local planning bye-laws may dictate the type of window you can fit. At all events, you'll have to adhere to the Building Regulations, which ensure that you carry out the work safely and that the structure, when complete, is sound.

Choosing a replacement window

It's important to choose a window that's compatible with the style and age of your house; there are many styles of traditional and modern windows, made in a variety of materials, commonly wood or steel, and nowadays aluminium and sometimes even plastic. Some of these materials, especially if they are badly maintained, will only have a limited life. Anodized aluminium, for example, although it requires little decoration, will oxidise in seaside areas, where there's a salty atmosphere.

Steel windows, although tough and long-lasting, are prone to rust and, like plastic and aluminium, don't always lend themselves to older-style homes, although you can conceal their starkness by painting them. Wood, although it can rot, comes in the widest range of styles and can be painted to match other woodwork in your choice of colour.

Plastic window frames are much more durable than the steel types. They require little maintenance, but have the disadvantage that they don't take paint well, so you're limited to the factory-made colour – usually white. Unless you're making new timber frames to your own specification, you'll be limited in your choice of window size and style by the standard dimensions available. These range from 600 to 2400mm (2 to 8ft) wide and from 450 to 2100mm (18in to 7ft) in height. If you can't find a standard frame size or style that fits your opening – and it's not possible for

READY REFERENCE

WINDOW FORMAT

A basic window frame consists of:

- a horizontal sill at the bottom (A)
- vertical jambs at each side (B)
- a horizontal head at the top (C)

In addition they may include:

- a cross piece called a transom (D), forming fixed or opening lights (E)
- vertical struts called mullions (F), forming fixed panes or opening sashes (G)
- glazing bars (H) dividing large panes into smaller units.

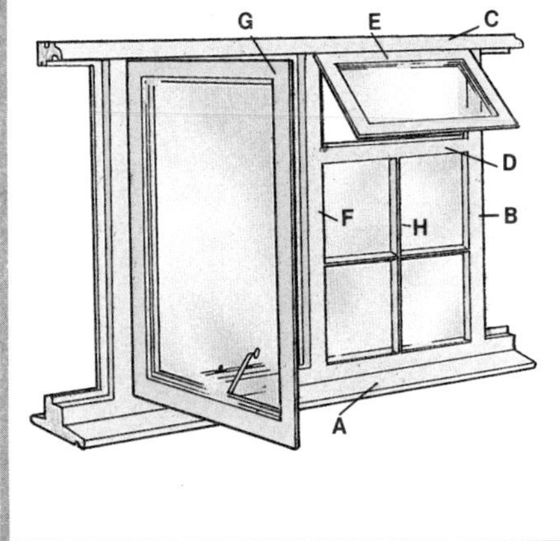

REMOVING THE OLD FRAME

1 *Before you can remove your old window you'll have to support the wall above and its loading, with metal props and a timber needle.*

2 *With the masonry suitably propped, you can safely chop out the lintel or brick soldier arch from above the opening, saving any whole bricks.*

3 *Release the fixing nails, screws or bolts holding the frame and use two bolster chisels or a crowbar to lever it out and lift it clear.*

4 *Once you've removed the window frame you can mark out the shape and position of the new lintel on the inside wall then chop out the bricks.*

you to enlarge or decrease the size of the opening – you can either adapt a standard frame of similar size by dismantling it and cutting it down to the required size, or make a frame specially designed to fit the opening.

How a window is made

A basic, single-pane window frame usually consists of a horizontal 'sill' at the bottom, vertical 'jambs' forming the sides and a 'head' at the top. The jambs usually have tenons at each end, which fit into mortises in the head and sill. In addition, separate smaller panes can be made at the top by adding a cross-piece called a 'transom'. The small panes, called 'lights', can be either fixed or hinged in their own frames. In the same way the large, main frame can be divided into smaller fixed panes, or opening 'casements' or 'sashes', by adding intermediate vertical struts called 'mullions'. The fixed lights and sashes can be further divided into smaller panes with 'glazing bars'. The separate frames can be hinged at the side, hinged at the top, pivoted horizontally or vertically, or can slide vertically within the main frame.

The most common type of window frame is called the 'standard casement' (see *Ready Reference*). It usually consists of a vertically-hinged casement, a fixed pane next to it, and a top-hung light, although there are many variations on this theme.

How frames are fixed

Window frames are fixed to the masonry within the opening in a variety of ways. Wooden frames can be secured with galvanised metal 'frame ties' (see *Ready Reference*) screwed to the frame and embedded in the mortar between the bricks or blocks of the wall. Alternatively screws or nails are driven through the frame and into wooden plugs set in the mortar joints. Sometimes screws are simply driven into plugged holes in the masonry; and you may even find frames that are simply wedged into the opening with timber wedges. New wooden frames sometimes have extra long sills and heads: the projecting 'horns' are intended to protect the corners of the frame from damage in transit but you can build them into the masonry for a really tough fixing. Commonly, the head horns are cut off, and the sill horns set into the masonry.

Metal frames are either fitted direct to the masonry by metal 'lugs' inserted in the mortar joints, or to a timber subframe, which in turn is screwed or nailed to the masonry (see *Ready Reference*). To remove the former type you'll have to locate and reveal the bolts that hold the frame in place; they'll either be buried in the putty retaining a fixed pane or inside the jamb of a hinged one. If they're rusted and you can't unscrew them, you can chip away some of the masonry at the side of the frame, insert a hacksaw blade and saw them in half to release the frame. When the metal frame is fixed to a timber subframe, screws are used, without lugs.

Cutting wooden frames to size

Most timber merchants stock ready-moulded timber sections for the construction of window frames but this is a far more costly and time-consuming method of making a window than buying a ready-made unit. Although it's possible that your choice of frame won't fit inside your opening, you can adapt the size of the frame or the size of the opening.

The best way to alter the size of a wooden window frame is to dismantle it first. If your frame is too wide you can trim a small amount of timber from the head and sill with a tenon saw and re-cut the mortises in each ready for reassembly.

If the frame is too deep for the opening you can cut a little bit off the jambs but you'll have to re-cut the tenons.

Such alterations can only be contemplated on simple frames with large glass panes; they are far too complicated to try on Georgian-style windows.

Removing the frame

Before you remove your old window frame you'd be wise to unscrew any hinged sashes within the frame, to make the unit lighter to carry. You could also consider removing any fixed panes of glass in case you accidentally drop the frame on removal. If you're careful you may even be able to save the glass for use elsewhere. Chop around the putty with a hacking knife, remove the glazing sprigs or nails holding in the glass, then lift out the pane and set it aside.

The screws holding a timber frame in place will probably be covered by layers of

INSERTING THE LINTEL

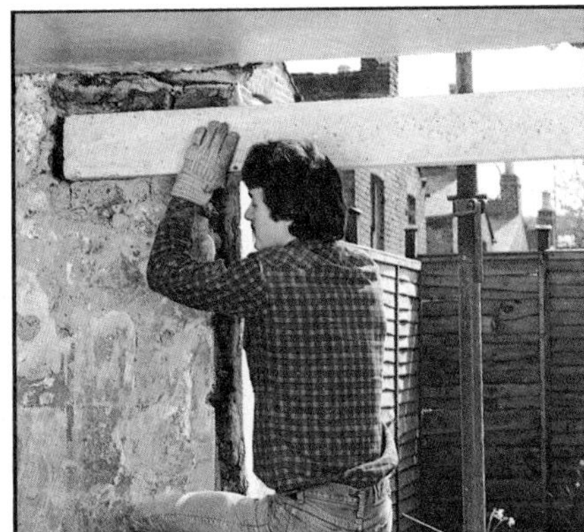

1 *Cut the slot for the new lintel at each side of the opening on the inner leaf and, with some help, lift the lintel onto the bearings.*

2 *While your helper steadies the lintel, lever up each end with a chisel and pack in tile or slate bearings together with mortar to set it level.*

3 *Check under the lintel with your spirit level to make sure it's bedded evenly on its bearings, and pack out with slate if it needs adjustment.*

4 *Leave the lintel for 24 hours for the mortar to set. Then draw a line down the wall against your spirit level to mark the span of the opening.*

5 *Chop down your guidelines with a club hammer and bolster chisel to give a clean edge, then hack off the plastic up to the line to reveal the brickwork beneath.*

6 *Starting directly under the lintel, chop out the inner leaf of bricks up to your guideline, keeping the edge as clean and smooth as possible, so that the frame will fit snugly.*

READY REFERENCE

FRAME FIXING

Wooden window frames are usually fixed within the opening with:

- galvanised metal frame ties screwed to the frame and embedded in the mortar joints of the wall (A)

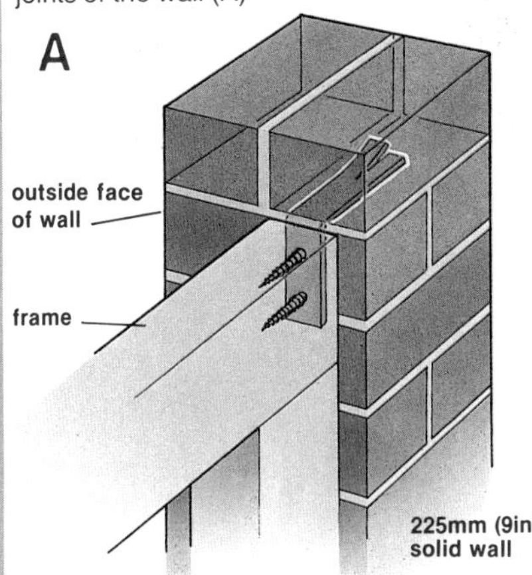

- screws or nails driven through the frame into wooden plugs in the mortar joints
- the horns built into the brickwork (B)
- screws driven into plugged holes in the masonry

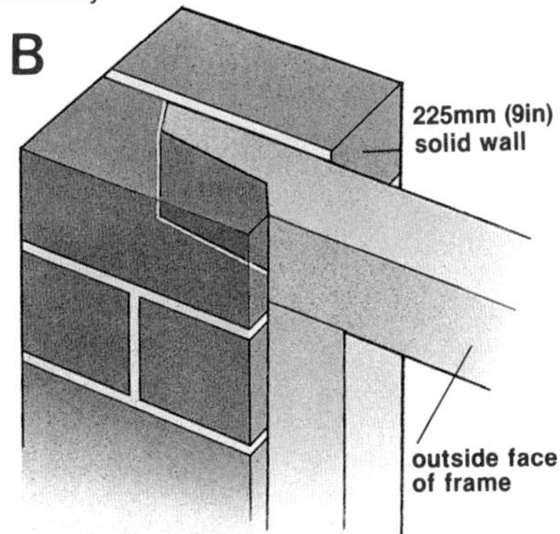

Metal frames are usually fixed with:

- screws to a timber frame, which is nailed or screwed to the masonry
- metal lugs inserted into the mortar joints and bolted to the frame (C).

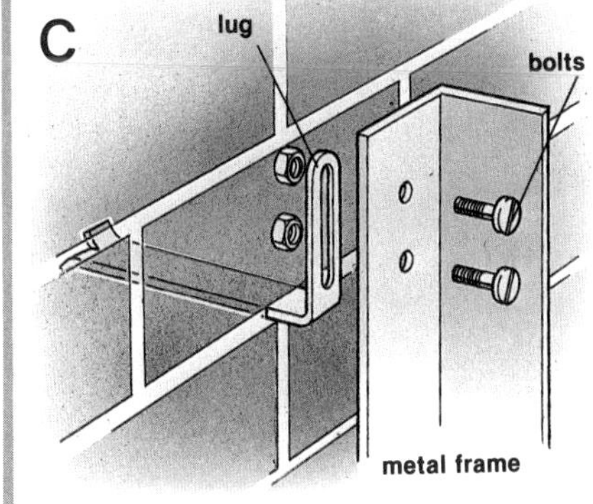

MAKING THE OPENING WIDER

1 *Brick up on top of the lintel and around the needle, following the original brickwork bond both for a blended look and for strength.*

2 *Measure the width of your new opening, transfer this dimension to the outer face of the wall and then scribe a line down the wall.*

3 *Starting at the top, just under the lintel, chop out the bricks individually up to your scribed guideline, saving the whole bricks.*

4 *Work your way down the window sides, toothing out the brickwork as you go by alternately removing the half or header bricks.*

5 *If there's a concrete sub-sill in the opening, you can remove it after you've cut back the sides by breaking its bond with the bricks.*

6 *Clear the opening of debris then begin bricking up the sides with cut bricks to give a square, clean surface for the frame.*

7 *Build up from the base of the opening, then mortar cut bricks into the toothed recesses so that they are flush with your scribed guidelines. Fit the cut ends first.*

8 *Check with your spirit level as you work that the sides of your new window opening are truly vertical, or you'll find that your new frame probably won't fit perfectly squarely.*

9 *Lay bricks to bring the base of the opening to within a brick's depth of the new frame position, copying the brick bond originally used both for strength and a matching look.*

paint and their heads may even be countersunk and concealed with filler or dowels: scrape away the paint to locate them. You may be able to chisel out the dowels or filler to reach the screw heads but if they're awkward to withdraw – or if nails have been used – simply insert a hacksaw blade between the frame and wall and saw through them.

Use a couple of bolster chisels – or even a crowbar – to lever the old frame from the opening – you can work from the outside or the inside, but you'd be wise to enlist someone to support the frame while you lever it out and, if it's large, to lift it away.

If you're not intending to save the frame for use elsewhere you can saw through the sill, transom and head near each corner and remove it in pieces.

Clear away any loose mortar, projecting nails and plugs from the opening. If there's a separate concrete or tiled sub-sill you can remove this by breaking the mortar bond with the masonry and levering it free.

Where possible you should try to have the new frame ready to fit as soon as you've removed the old one, but if you're going to alter the size of the opening you won't be able to complete the job in one day. Cover the opening with a large sheet of heavy gauge polythene held in place with timber battens lightly pinned to the wall with masonry nails to keep out draughts and rain.

FINISHING THE OPENING

1 *If you want to copy a soldier arch used on the original outer wall you'll have to make up a timber former to support the bricks.*

2 *Once the arch is completed you can brick up over it and around the needle, using the bricks you've saved to match the rest of the wall.*

3 *Wait for about a day for the mortar to set before removing the props, but leave the arch former until you're ready to fit the frame.*

4 *Fill in the hole where the needle was inserted and repoint the mortar joints throughout your new brickwork to match the rest of the wall.*

Altering the opening size

You can fit a deeper window frame in your opening without rigging up any temporary structural support for the walling, but if you're intending to increase its width or height you'll have to prop up the masonry with adjustable metal props and stout timber 'needles' (see pages 98 -99) while you reposition the lintel or fit a new, wider one to support the wall and its loading.

When you've positioned the props you can mark out where the lintel is to be fitted, cut a slot for it and then fit the lintel in place. Once it's installed, you can safely chop back the reveals to the new window width.

It's much easier to increase the depth of the opening (see *Ready Reference*): you can simply cut out the brickwork at the bottom to the new depth.

Your new depth of opening may not conveniently fall at an exact number of bricks, so you'll have to lay a course of split bricks to bring the opening to the correct level.

Making the opening narrower (see *Ready Reference*) is also possible, but you'll need a supply of bricks that match the existing masonry for filling in, or the finished job will look patchy; unless, of course, the exterior walls are going to be rendered. Continue the brick bond in your narrower opening by carefully knocking out all the half bricks down the sides and saving them. Mortar into both sides of your 'toothed' wall the required number of whole or cut bricks you'll need to decrease the width of the opening then use the half bricks to fill in between. Any cut bricks at the reveals should be mortared in, cut end first, so you have a neat, square edge within the opening.

When you're increasing the width of your opening you should also 'tooth out' the reveals to fill in with cut bricks.

Closing a cavity wall

A solid 225mm (9in) wall doesn't present many problems when you're replacing a window: all you have to do is increase or decrease the size of the opening, continuing the masonry bond originally used, and then fit your new frame from the inside. But if you have cavity walls you'll have to close off the gap between the two leaves and insert a damp-proof course (dpc) to prevent the cavity being breached by moisture.

You must also fit a vertical dpc at the sides of the opening. Chop out the half bricks on alternate courses of the inner leaf and lightly tack a strip of bituminous felt to the outer leaf. Mortar nearly whole bricks into the recesses left by the removal of the half bricks; position them header-on, at right angles to the inner leaf so they butt up to the felt strip and close the cavities. Fill in the gaps on alternate courses at the reveals with fillers of brick. Allow the felt strip to project into the reveal so it can butt up to the window frame or locate in a narrow groove in the jamb. You'll also have to fit a strip of dpc along the bottom of the opening separating the two leaves. Your lintel at the top will effectively seal the cavity.

If you are replacing an existing sash window with a purpose-built modern one, don't bother to close the cavity; the space will house the sliding sash weights.

FITTING A NEW WINDOW II – the frame

An old, rotten window frame not only looks unattractive but also admits rain and draughts to your house. Part one of this job dealt with removing the old frame and preparing the opening; part two describes how to fit the new frame.

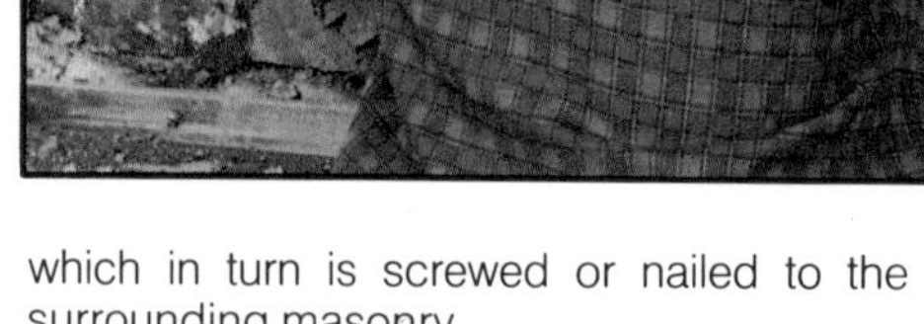

Windows take quite a battering from the elements, so it's hardly surprising that they need regular maintenance to keep them in working order and looking good. But if they've been neglected and deterioration has become too widespread for repairs to be made successfully, you will have to consider replacing them.

Choosing a replacement window

It's not necessary for you to fit exactly the same type of frame as the one removed. There's a vast selection of different ready-made styles to choose from, featuring numerous combinations of fixed and opening panes. If you can't find a suitable off-the-shelf frame, you could even make one to your own specification.

There's no reason why you shouldn't make your new window larger or smaller than the original, but if you're making structural changes such as this you must inform your local Building Inspector both to prevent infringement of local planning bye-laws and to ensure that you carry out the work safely.

To increase the width of an opening you'll have to install a new, longer lintel to support the masonry above, but if you simply want to make a narrower window you can 'tooth-in' matching brickwork at the reveals without disturbing the original lintel. Making the height of the window larger or smaller is best done at sill level to avoid the aggravating job of having to move the lintel.

Any work that involves disturbing the structure of the walls will need the temporary support of adjustable props.

How frames are fixed

Once you've adapted the size – or even the shape – of your opening to accept the frame you should leave the structure for about 24 hours so that any new mortar sets. You can then remove any props or arch formers (see Installing a window 1) and fit the frame into place.

There are a variety of ways you can secure your frame within the opening. A steel or aluminium frame, for instance, can be either bolted to metal lugs inserted in the mortar joints, or first screwed to a timber subframe which in turn is screwed or nailed to the surrounding masonry.

A timber frame can be secured with galvanised metal 'frame ties', which are screwed to the jambs (sides) of the frame and embedded in the mortar joints. Another method of fixing a timber frame is to drive screws or nails through the jambs into wood wedges set in the mortar courses of the brickwork. But the simplest method of all is to screw the frame into plugged holes drilled in the masonry, although this type of fixing isn't as tough and long-lasting as the other methods.

Some new frames have projecting 'horns' at each side of the sill and head piece. Although they're intended to protect the corners while in storage or transit, you can build them into the masonry at top and bottom for a really tough fixing. It's quite common for the head horns to be cut off flush with the jambs and for the sill horns alone to be built into the masonry.

Positioning the frame

Whichever method of fixing you choose, it's imperative that the frame sits perfectly squarely in the opening. If it's crooked or at all warped, for instance, you could find that the opening sashes tend to bind in their frames and a fixed pane of glass, if it's not bedded flat in its rebate, could shatter at even the slightest vibration.

Warping can occur if the lintel and its load bears directly on the frame (this is sometimes caused by settlement of the masonry), so it's usual to leave a slight gap of about 3mm (1/8in) between the lintel and the top of the frame. You can seal this gap with a flexible, non-setting mastic applied by a special 'gun' (see the *Ready Reference* on page 121). It's intended to compensate for any slight movement in the structure; if you were to fill the gap with mortar, the filling could crack and eventually allow moisture to seep inside.

The position of your frame in the opening depends on whether you want to fit a sub-sill or whether the sill built into the frame is going to rest on the masonry at the base of the opening. If you want to make a concrete sub-sill you can cast one in formwork when you've fitted the frame, which you'll have to wedge up to the required height. You can, on the other hand, fit a timber sub-sill. Thirdly, you can simply bed the frame's integral timber sill on a mortar bed.

If you're installing the frame in a cavity, wall, you'll have to fit a vertical damp-proof course (dpc) between the two leaves of the closed-off cavity and set them in a groove in the jambs to prevent moisture seeping through; you'll also need to fit a dpc beneath the sill.

You may also want to locate the frame within the opening so that it sits flush with either the inside or outside face of the wall, or recessed from each face to leave 'reveals' on each side (see *Ready Reference*).

When you've decided where and how you're

WINDOW AND SILL CONSTRUCTION

*A standard casement window consists of: a head **(1)**; a sill **(2)**; jambs **(3)**; a mullion **(4)** to divide the frame into two panes; a transom **(5)** forming a small top-hinged light **(6)** at the top. There may also be a side-hinged casement **(7)**. On the inside there's a window board **(8)** which may be timber or tiles set in mortar.*

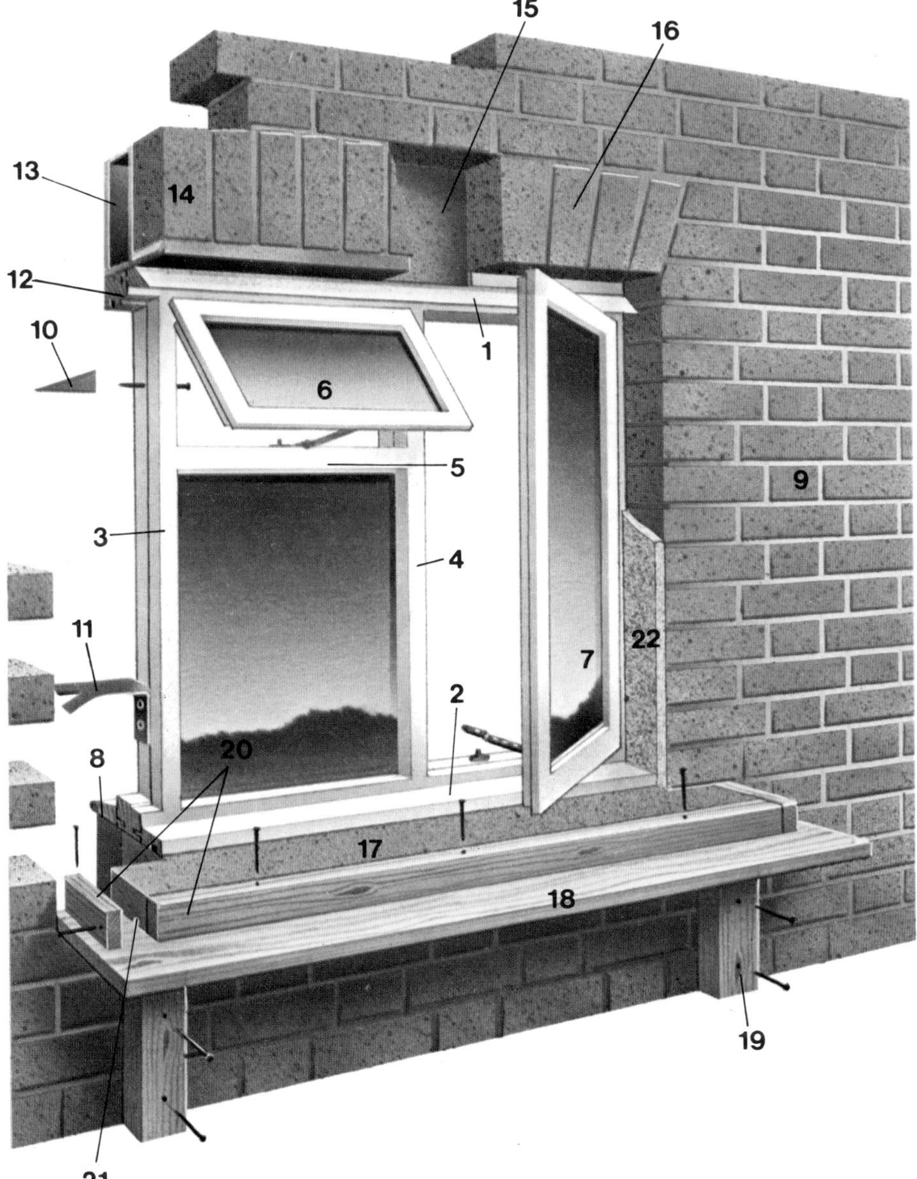

*The frame is fixed into a cavity or solid brick wall **(9)** with nails driven into wood wedges **(10)** hammered into the mortar joints, or with metal frame ties **(11)** screwed to the frame and set in the joints. Or, horns **(12)** can be built into the masonry. The masonry above the opening must be supported on a beam: either a hollow steel lintel **(13)** with a ledge to take a course of bricks standing upright **(14)**, or a concrete lintel **(15)** with a curving soldier arch **(16)**.*

*There may be a concrete sub-sill under the frame **(17)**, cast in formwork. A plank **(18)** supported on studs **(19)** makes the base, three battens nailed on top, the front edge and sides **(20)**. A length of sash cord set in the concrete **(21)** makes a drip groove. The reveals **(22)** can be rendered.*

READY REFERENCE

POSITIONING THE FRAME

The position of your window frame within the opening depends on whether there's a sub-sill.

If there isn't a sub-sill:

- place the frame near the outside face of the wall with its integral sill overhanging the brickwork by about 25mm (1in), leaving reveals for plastering on the inside of the opening.

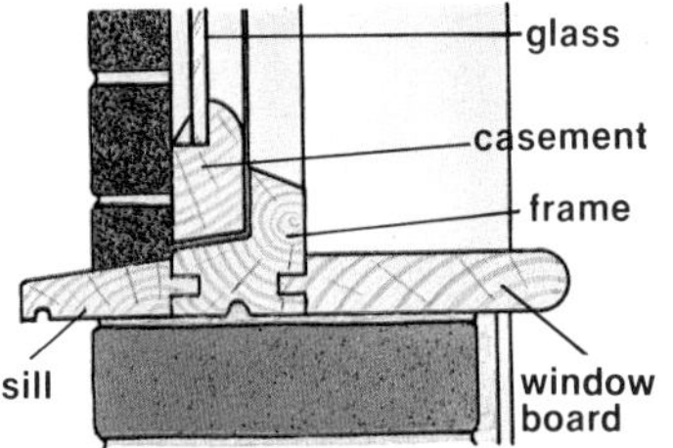

If there is a sub-sill:

- place the frame flush with the inside face of the wall, reducing the amount of plastering necessary or

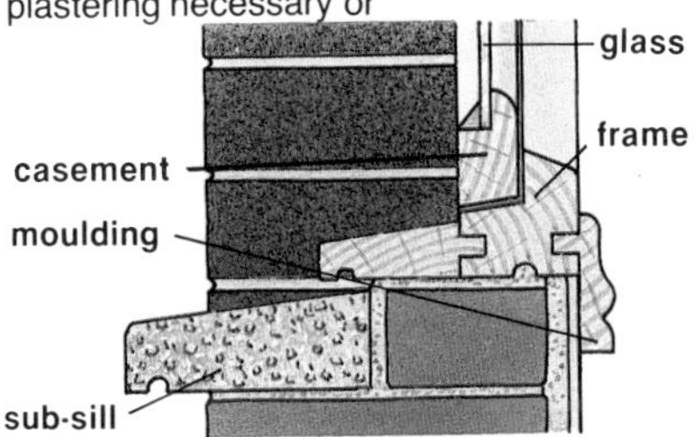

- place it centrally within the opening, leaving reveals on both sides of the window.

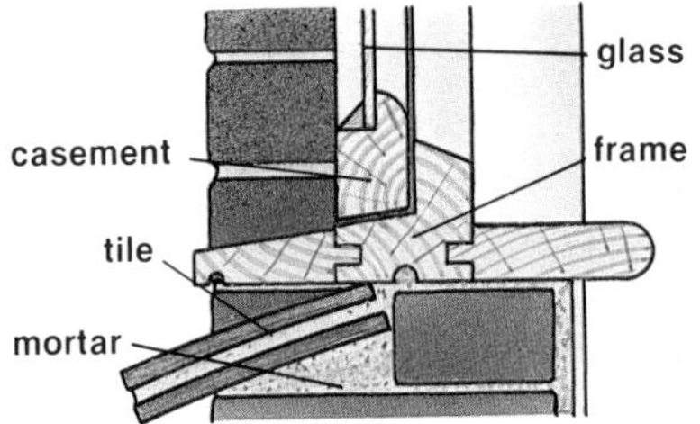

GLASS FOR GLAZING

Glass is sold by thickness and it's important that you buy the correct type for a particular window size. If you choose a thin pane for a large window there's a danger of it flexing and breaking too easily. You should use:

- 3mm glass for panes up to 1sq m (11sq ft) in area
- 4mm glass for larger panes up to 2.1m (7ft) long
- 6mm glass for large panes longer than 2.1m (7ft)

FITTING A TIMBER FRAME

1 *Where a timber arch former has been used you will have to remove it before you can fit your new frame within the window opening.*

2 *Decide into which masonry joints your timber fixing wedges are to be fitted, and hack out a slot using a club hammer and bolster chisel.*

3 *Cut your wedges to shape using a wood chisel, then hammer them firmly into their slots so that they just protrude into the opening.*

4 *Lift the frame into the opening and find its approximate position. In this case it should be sitting almost flush with the inside face of the wall.*

5 *Use timber blocks under the sill to wedge the frame in place beneath the lintel. Leave a 3mm (1/8in) gap for sealing with flexible mastic.*

6 *Place a spirit level on the sill to check that the frame is sitting horizontally in the opening; adjust the timber packing if necessary.*

7 *Check with the spirit level held against the face of the jamb that the frame is square within the opening. Adjust by tapping gently with the handle of a club hammer.*

8 *Hammer 100mm (4in) cut nails through the jambs into each wedge, punching in their heads. Recheck the level of the frame in case your hammering has dislodged it.*

9 *Once you're satisfied that the frame is fixed accurately and securely, brick up underneath. Use bricks split along their length if the depth is less than a whole brick.*

going to fix your new frame, offer it up to the opening to make sure it will fit.

Fitting a timber frame

Before you finally fit a timber frame you should treat all its surfaces – especially those that will be inaccessible when the frame's fitted – with a coat of wood primer. A new, off-the-shelf frame will probably come ready-primed – it's usually pink in colour – but you'd be wise to give it another treatment in case the factory-applied coat has been damaged in transit.

If you're going to secure the frame by its horns you should chop out slots for them in the masonry at the sides of the opening using a club hammer and bolster chisel. Be very careful, when cutting the holes for the horns, that you don't weaken the bearings, which support the lintel and its load. If you did and the masonry 'dropped', you wouldn't be able to return it to normal without substantial rebuilding.

To screw the frame into wallplugs you must drill the holes in the sides of the opening – three at each side – and insert the plugs. As your frame must be perfectly level and the fixing holes accurately positioned you'd be wise to drill the jambs of the frame first, wedge it temporarily in the opening level and square and mark the wall through the holes with your drill bit.

Remove the frame, drill the holes in the masonry and plug them, then return the frame to the opening, set it level and square again and drive in the fixing screws. Check that the frame is level and square and adjust it if necessary by re-drilling the fixing holes.

To fix the frame to timber wedges you can chop out the mortar joints at two points on each side of the opening and hammer in triangular-shaped timber wedges, cut with a chisel from a length of 50x25mm (2x1in) softwood. Don't cut them too thin or they'll simply split and the fixing nails won't hold securely.

Lift the frame – unglazed – into the opening and temporarily wedge it in place with blocks of wood while you ensure that it is level and square. The jambs should be tight against the triangular wedges; if they're not you'll have to adjust them.

Hold your spirit level against the jambs to make sure that the frame is sitting truly vertical and adjust it if necessary, by tapping gently with the handle of your club hammer. When you're satisfied that it's straight, transfer the spirit level to the top of the sill to check that the frame is truly horizontal. Insert more packing underneath, or remove the existing packing, to set the level. Don't forget to leave a small gap beneath the lintel at this stage. Re-check the vertical once more in case your horizontal adjustments have knocked the frame out of square again, then nail or screw it to the wedges. Check again with your spirit level that the frame hasn't moved; if it has you may have to release some or all of the fixings and adjust the position of the frame. Once the frame is securely and accurately fixed in the opening you can remove the temporary timber packing.

If you've set the frame above the base of the opening, intending to install a concrete or timber sub-sill in the gap, you'll have to brick up under the frame on the inside face of the wall. You may have to insert bricks split along their length if the gap is less than a course deep.

Mortar in bricks to cover the horns of the frame at the top and bottom, and point in the gap between the jambs and the masonry on the inside of the window frame with mortar, covering the triangular timber wedges. Leave a gap of about 3mm (1/8in) so that you can spread on a layer of finishing plaster flush with the rest of the wall.

Making good the walls

Leave the window for about 24 hours for the mortar to set before making good the wall round the perimeter of the opening with plaster. Where you've left reveals inside the room you can apply a layer of Carlite Browning plaster to the brickwork, followed by a thin layer of Carlite Finish plaster (see Chapter 6 pages 148 -153), which you can polish to match the rest of the wall.

You can form the corners of the reveals by attaching expanded metal angle beads to the brickwork; they have a 'nosing' at the angle which serves as a guide to spreading on the plaster to the correct thickness. Alternatively you can tack thin timber battens temporarily to the wall at each side of the corner so they project slightly, to serve as thickness guides. Plastering angles and reveals will be dealt with in more detail in a later issue.

The brickwork reveals outside the window can be either left exposed as a feature or rendered with a mortar mix. You can apply the render to the correct thickness by using timber battens as thickness guides in the same way as you plaster internal reveals. Dampen the brickwork with clean water splashed on from a brush to provide a key for the mortar, then trowel it on and smooth it flush with the thickness guide. Leave it to set for about 24 hours before removing the battens and making good the edges of the screed with more mortar.

Fitting a metal frame

Hold the frame against the opening and mark the wall at the sides where the metal fixing lugs are to be recessed. Remove the frame and chisel holes in the masonry with a club hammer and cold chisel to correspond with the lugs. Return the frame to the opening, with the lugs bolted on loosely, and wedge it in

READY REFERENCE

WEATHERPROOFING WINDOWS

You must seal the joint between the window frame and the masonry to prevent moisture penetrating to the inside. To do this:

- fill the gap between the frame and the masonry with mortar to within about 10mm (3/8in) of the outside edge of the frame
- when the mortar has set fill the remaining gap with non-setting mastic applied by a special gun (A). The flexible mastic will accommodate any settlement or other movement in the frame or wall
- fill the 3mm (1/8in) gap between the frame and the lintel with mastic also.

FITTING THE WINDOW BOARD

To make the shelf or 'window board' on the inner face of the sill:

- screw a length of 18mm (3/4in) thick timber to plugged holes in the masonry at the base of the opening, overlapping the face of the wall by about 25mm (1in)
- fit a ready-made window board to an off-the-shelf frame notched to take it (A)

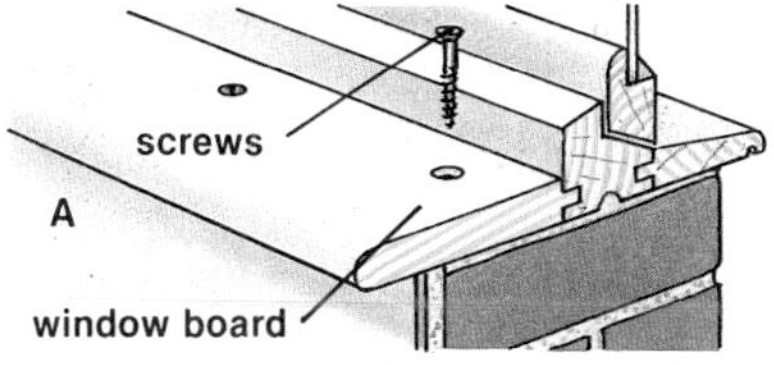

- set ceramic or quarry tiles on a mortar bed, sloping the shelf downwards for a decorative effect (B).

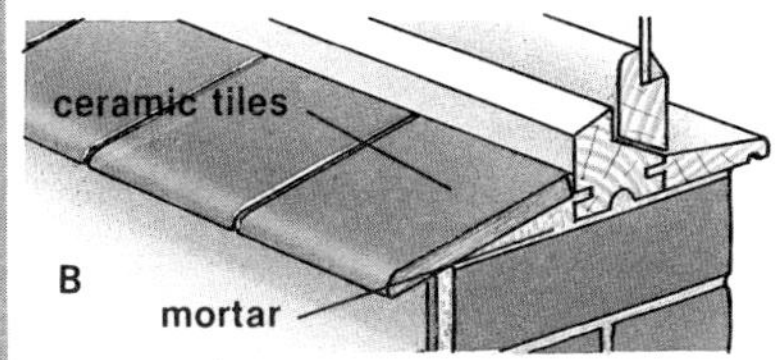

place, square and upright. Push the lugs in their slots and mortar around them using a mortar mix of one part cement to three of sand. Leave the mortar for about 48 hours to set before tightening up the bolts. You can then remove the packing and make good the perimeter of the opening wedges.

An easier, and a much more efficient method, is to screw the frame to a timber subframe, which you fix to the masonry. When you've fitted the frame, seal the joint between it and the subframe with a non-setting mastic.

Making a concrete sub-sill

To set a concrete sub-sill on site, you'll have to erect timber formwork against the wall to make a trough. You'll need a length of board to form the base of the sill, which you can wedge in place on stout timber studs, resting on the ground and attached to the wall with masonry nails. Nail a long batten to the base board to form the front edge of the sill and two shorter ones at right angles to form the ends. You could also use slightly tapered size battens and a thinner nosing button to give the sill a slight shape to the front.

An easy way to make a drip groove under the front edge of the sill, to prevent rainwater from trickling between the bottom of the sill and the wall, is to place a length of sash cord in the base of the trough in the required position, and trowel in the concrete on top. When the concrete has set and the formwork has been removed you'll be able pull the cord free, leaving a half-round groove the length of the sill.

Fill the trough with a concrete mix of four parts sand to one part cement, making sure you don't leave any air pockets, and then smooth the top with a steel float flush with the thickness battens. Leave the formwork in place for about 24 hours for the concrete to harden.

Sealing and glazing the window

Fill the gap between the frame and the brickwork with mortar to within about 10mm (3/8in) of the outside edge of the frame then, when the mortar has set, apply a flexible non-setting mastic over it and between the lintel and the frame (see *Ready Reference*).

You can glaze your new window as soon as possible after it's fitted. But if you have to leave it overnight you can tack a sheet of heavy gauge polythene over the frame on the outside to prevent rain or wind from getting in. What type of glass you use really depends on the type of frame and what the room it's in is used for. But remember to measure accurately the sizes which you require. Leave the newly glazed window for about a fortnight for the putty to dry before applying primer, undercoat and top coats of paint.

If you are installing sealed-unit double glazing with proprietary gaskets, make sure these are correctly positioned.

MAKING GOOD THE WINDOW

1 *Trowel mortar into the gap between jambs and masonry, covering the wedges. If it shrinks on setting you may have to add more mortar.*

2 *If you've made a brick soldier arch over the outside of the opening there may be a curved area of lintel visible, which you can render.*

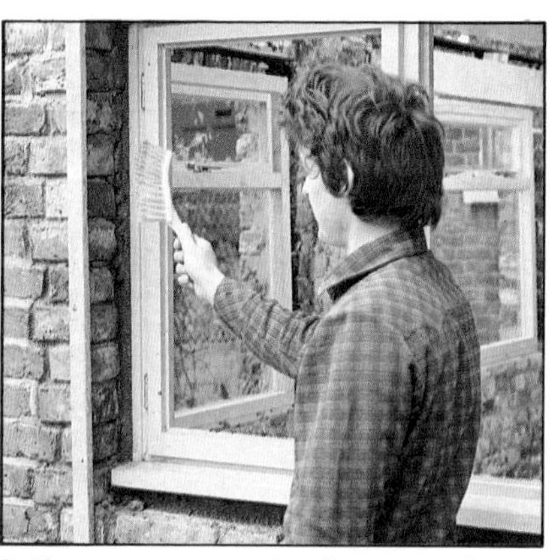

3 *If you want to render the external reveals, tack battens to them; allow them to protrude into the opening by about 6mm (1/4in). Wet the brick surface.*

4 *Trowel mortar onto the reveals, smoothing it flush with the edge of the timber batten. Remove the batten after leaving for 24 hours.*

5 *Set up timber formwork against the wall to cast a sub-sill. Nail a shelf on studs to form the base with battens nailed on top for the sides. Trowel in mortar and smooth out.*

6 *While the sub-sill is setting, glaze the window. Bed the panes in linseed oil putty, then retain them with glazing sprigs or panel pins and seal with triangular fillets of putty.*

FITTING A NEW FIREBACK

If you're renovating an open fireplace one of the jobs you're likely to come across is replacing an old, cracked fireback. This is essential if you want your fire to be safe, and although it is rather a messy job it is not at all difficult to do.

Until quite recently, all houses were built with an open fireplace in at least one room, and houses built in the 19th century are likely to have one in every room. Few people nowadays actually rely on open fires to heat their homes or their water and many fireplaces were removed or covered up, especially in the '60s and '70s. Gas fires and central heating were obviously cleaner and more convenient. However, it is becoming more and more popular to open up a fireplace again, especially in the living room.

When you uncover the old fireplace, you may be lucky to find both the surround and fire opening in good condition. But often the surround was removed before the opening was concealed, and the fireback may be damaged or partly missing. You'll probably have to have the chimney swept and clean up all the accumulated soot and dust before you can check what condition it's in.

Of course, you may have been using your fireplace for many years. But after so much intense heat from the fire as well as occasional knocks from the poker, the fireback will start to deteriorate and may need replacing.

The fireback is the part most likely to need attention in any case. Most firebacks are made of refractory concrete or fireclay, and their function is to provide a safe and convenient enclosure for the fire, helping to direct the heat into the room and the smoke up the chimney. But with age, cracks will start to form and will eventually open up so the fireplace becomes unsafe to use.

Repairing small cracks

If there are just a few cracks in the fireback it is relatively easy to repair them and there'll be no need to replace it entirely. You'll need some fire cement and a small trowel.

If the fire has been used recently you should leave it for two days to cool down completely. Then brush off all the soot and dust and rake out the cracks with the point of the trowel, undercutting them slightly. Brush out all the debris from the cracks and soak the area with plenty of clean water. The water helps the fire cement to adhere to the cracks by stopping it drying out too quickly. Fill in the cracks, using the trowel to work the fire cement well in. Then smooth off the surface by drawing a wet brush over the crack.

If the cracks are large, or part of the fireback has come loose, then a repair like this will not be safe and you'll have to replace the whole fireback.

Removing the old fireback

Although this is a heavy and dirty job, it doesn't require any great skills and you shouldn't have any problems with it. Before you start work, lay down dust sheets around the fire and, if you've a tiled hearth, lay some sheets of cardboard over the tiles first to protect them from accidental damage.

Now you can remove the fire grate. If this is an old-fashioned one it will be free-standing and can merely be lifted clear. A modern appliance will be fixed by screws driven through its base into plugs inserted in the back hearth; the gap at its sides and base will be pointed with fire cement. Withdraw the screws in the normal way (leaving the plugs intact) and break the concrete pointing by tapping it with a hammer and cold chisel.

Under the fire cement at the sides you will come across short lengths of rope. These are made of asbestos and you should keep them as they will be needed when the fire is re-installed. Remember always to treat asbestos with respect, for if you create and inhale any asbestos dust there is a possible health hazard.

You can now take out the old fireback. As it fits behind the fire surround you must break it up to remove it – use a club hammer and cold chisel, or even a crowbar. If the fire surround is missing then you can probably wrench the fireback forwards without needing to break it up. Behind the fireback you will find a mass of solid rubble – packed in to help dissipate the heat from the fireback – and most of this should be cleared out. It is not very hard and will break up easily with a cold chisel and hammer. When this is all removed you'll be left with a brick-lined recess known as the builders' opening.

Ordering a new fireback

New firebacks are available from builders' merchants and fireplace specialists. Measure the size of the old one and order a new one of the same size. The part you need to measure is shown in *Ready Reference*.

A fireback for a normal-sized fire opening has to be installed in two pieces so that it can expand under heat without cracking. However, it will normally be delivered to you in one piece. About halfway up you will see a break line. Tap gently along this line with a

REMOVING THE OLD FIREBACK

1 *Start to knock out the old fireback at one of the top rear corners using a club hammer and cold chisel. It should break up quite easily.*

2 *If a large crack forms, follow along it with the hammer and chisel – you'll often be able to remove quite large pieces of fireback in one go.*

3 *There may be a line of fire cement along the front edges of the fireback. Break this off carefully without damaging the fire surround.*

4 *Often the rubble behind the fireback will be loose and can easily be removed. If it's solid you'll have to break it up first with a club hammer and bolster.*

hammer and bolster until it splits in two. Some suppliers call this a one-piece fireback while others call it a two-piece fireback, so be sure to describe exactly what you want when you're ordering. Large fire openings may need a four-piece or six-piece fireback.

Installing the fireback

The heat thrown out by an ordinary domestic fire is intense and you must take steps to counteract this. Obviously, you must use non-combustible materials throughout, but that's not all. The heat will also cause the materials to expand. As different materials expand at different rates they could cause a lot of strain if they were allowed to press against each other. So various 'buffers' are installed in a fireplace to prevent the expansion causing any damage.

The first step is to fit lengths of asbestos rope behind the top and sides of the fire opening. If the original rope is in a good enough condition you can use that. The lower half of the fireback is then manoeuvred into position and pulled forward so it lightly compresses the rope at the front of the opening. The area behind the fireback will be filled in, but first you need to fit another sort of buffer. This one is made from two sheets of corrugated cardboard which should be cut to the shape of the fireback and then tucked in place behind it. The idea is that once the fire is lit this cardboard will char away under the intense heat, leaving a gap for expansion of the infill and fireback. Often the paper stays intact – you might have come across some when you took out the old fireback – but that doesn't matter.

The whole area around the rear of the fireback – both the back and sides – must be filled in with a weak mortar mix. A suitable mix can be made from four parts vermiculite

READY REFERENCE

PARTS OF A FIREPLACE

Here are some of the specialist terms used to describe a fireplace:

- the decorative part fixed to the wall, made of wood, marble, metal or tiled concrete, is called the **fire surround**
- the opening at the centre where the fire burns is the **fire opening**
- the floor in front of the fire – often tiled – is the **superimposed hearth**
- below the superimposed hearth is the **constructional hearth**
- the concrete under the fire is the **back hearth**
- behind the opening is the **fireback.**

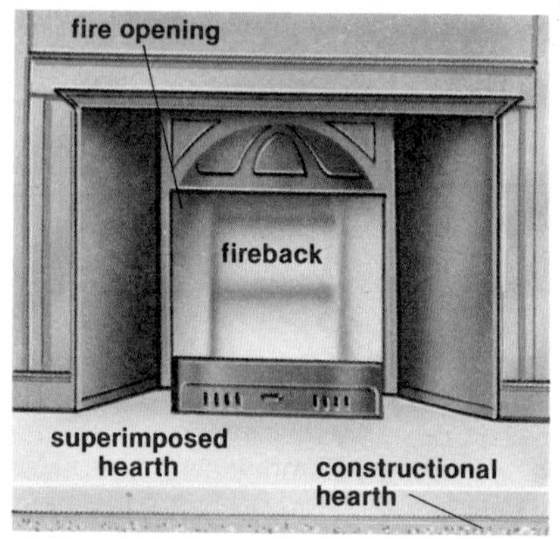

MEASURING UP FOR A NEW FIREBACK

You need to measure the old fireback to find out what size to order. Standard sizes are 16 and 18in (406 and 457mm), but larger sizes are available. The part you need to measure is the inside front opening (A).

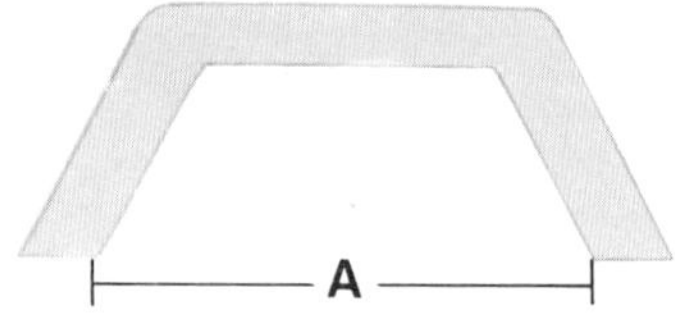

TIP: FIT A THROAT RESTRICTOR

A throat restrictor will help cut down on the amount of fuel used by reducing the air flow up the chimney. It comes in the same sizes as the fireback and is quite easily fitted at the same time.

CONSTRUCTION OF A FIREBACK

This illustration shows what goes on behind the fireback. The corrugated cardboard and asbestos rope act as buffers to protect the fireback against expansion. Then the whole area behind the fireback is filled with vermiculite mortar and rubble. Extra mortar is used to form the flaunching at the back and sides.

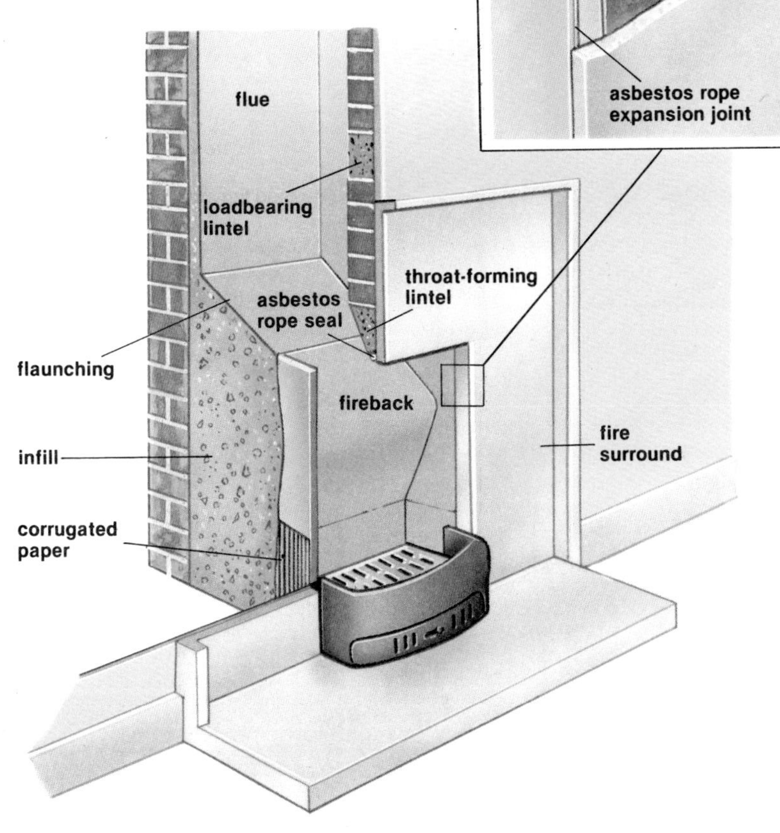

(an inert filler available from builders' merchants) to one part lime or cement; or nine parts sand to one part lime and one part cement. Do not make the mix too wet; it should be damp rather than soggy. The infill should be brought to the top of the fireback and tamped down so it is truly solid. You can, if you wish, put old bricks or bits of the old fireback in the middle of the infill to save you mixing too much material. This type of mortar mix is ideal, as it has good heat insulation qualities and will not expand too much.

You can now fit the top half of the fireback. Sit it squarely on top of the lower half, bedding it onto a layer of fine cement and levering it into place if necessary with a bolster. Using the same mix, fill in behind this section – there's no need for any corrugated cardboard – and tamp it down until the filling is level with the top of the fireback.

Just behind the surround (and over the front of the fire opening) is a lintel, called the throat-forming lintel. If you peer inside you should be able to see that the centre section of the lintel slopes upwards. You must now add extra infill mortar to the top of the fireback so that its slope is parallel to the slope of the lintel. Together, these two slopes form an opening known as the throat, which conducts the combustion gases and smoke up to the flue. It will normally be about 110mm (4½in) wide. The flaunching, as the sloping section is called, should be trowelled smooth as a rough surface can cause friction that would impede the flow of gases. Fill in the sides too to avoid any ledges where soot could accumulate. Finally fill in the join between the sides of the fireback and the fire surround with some fire cement and smooth off with a damp brush.

Replacing the grate

With the fireback in place you can now put back the grate. If it's a free-standing one you simply stand it in place. If it's an old-fashioned one, this is the time to think of replacing it with a modern fitted one which will perform much more economically and be easier to control.

Whether it's a new or existing modern fire the installation procedure is much the same – except that a new fire will need new plugs in the back hearth to take the fixing screws. Note that the plugs should be made of some non-combustible material – metal for instance. To fix the grate in place, drive the fixing screws into the plugs. Then push asbestos rope into the gap between the sides of the appliance and the fire opening. You can hold the rope in place with a smear of fire cement; then make the fixing permanent by pointing lightly with more fire cement.

Leave the installation for a few days to allow the various mortars to dry out and set properly before lighting a fire, and make only small fires for the first few days.

Fitting a throat restrictor

While you are fitting a new fireback it's worth considering adding a throat restrictor to the fireplace. This is a device for reducing the size of the throat, and it has several advantages. As smoke and gases rise up the chimney, they draw with them air from the room, and this air has to be replaced with air from outside the room. This is a cause of the notorious draughts that used to be associated with open fires. However, modern appliances with their restricted air intake have largely overcome this problem. In fact, the air changes they do cause are often beneficial in that they give good ventilation to the room and help to overcome condensation. Even so, the new air that is drawn into the room has to be warmed up, and that means extra work for the fire – and extra fuel burned. A throat restrictor, by reducing the amount of air flowing up the chimney, cuts down on the number of air changes, and this means the fire doesn't have to work so hard.

The restrictor is a metal device that is placed on top of the fireback and held in place by the flaunching. Once the flaunching has set and the fire has been fitted you can adjust the restrictor. Light the fire and get it going properly, then slowly close the restrictor until you reach the point when puffs of smoke start to billow into the room. You have then closed it just too far, so open it up slightly until the smoke flows properly upwards; that is the correct setting. You will probably find that you have reduced the throat aperture from something like 110mm (4½in) to about 20mm (¾in). From time to time the gap will need to be adjusted to suit varying weather conditions and your heating needs.

FITTING A NEW FIREBACK

1 *In most cases a one-piece fireback will be supplied, but if this is difficult to fit in the opening it can be split along the break line.*

2 *You need to tap along the line very gently for quite some time before the two halves will separate. If you try to hurry it will crack in the wrong place.*

3 *Make up sufficient mortar for the infill and use some of this along the bottom of the opening to bed in the bottom section of the fireback.*

4 *Position the fireback on the mortar, tapping it into place until it is central in the opening. The sides must be vertical and the top horizontal.*

5 *Push one or two layers of corrugated cardboard behind the fireback. This chars away when the fire is lit, leaving a gap for expansion.*

6 *Using a weak mortar mix fill in behind the fireback, tamping it down until it is level with the top. You can fill up part of the space with rubble.*

7 *Trowel a layer of fire cement along the break line and bed the top half of the fireback in place. Make sure the two halves are lined up.*

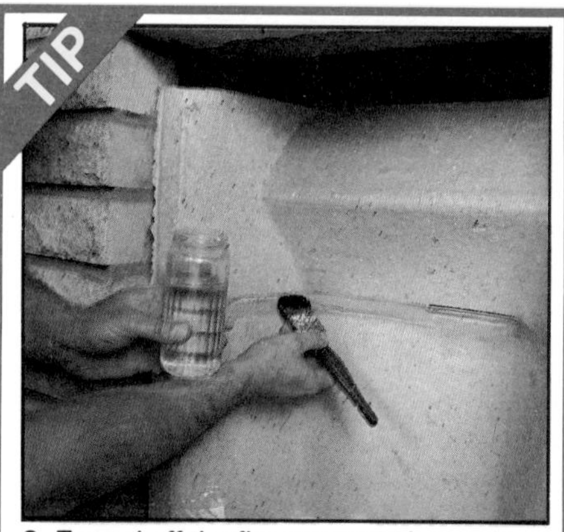

8 *Trowel off the fire cement, then go over the surface again with a wet brush for a really smooth finish. Point the sides at the same time.*

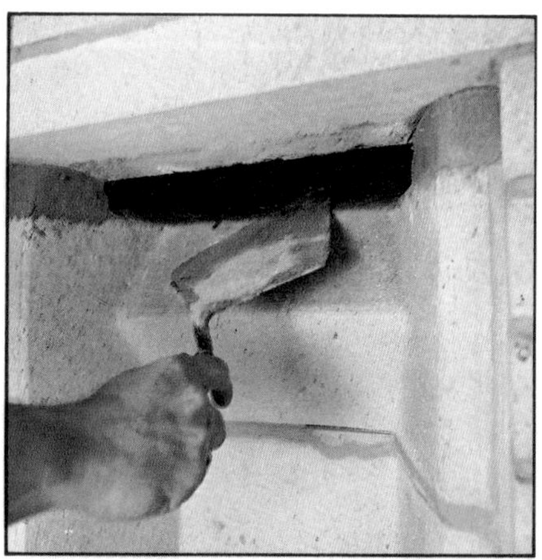

9 *Finally, fill in behind the top section and, using the same mortar, build up the flaunching at the back so its slope matches that of the throat-forming lintel.*

REMOVING A FIREPLACE

If you've a drab, old fire surround that detracts from your decor, or a fireplace made redundant by central heating, you can remove it and block off the opening, as long as you ventilate the flue.

Although natural fire with a decorative surround can be an attractive focal point in a room, if you live in a centrally heated house you may feel that obsolete fireplace takes up valuable wall space and detracts from the rest of your decor; it can also create difficulties in placing furniture exactly where you'd like it.

Many modern homes are designed and built without chimney breasts or fires, so the problem doesn't arise, but if you live in an older house with fireplaces in every room you may want to remove one (or more) of them. It's possible to completely remove the chimney breast. But, although this can give you a considerable amount of extra space, it'll tend to make your room look 'boxy'; and you'll also lose the useful alcove space at each side of the chimney breast, which is ideal for fixing shelves.

The best solution is to remove the fireplace and its surround and block up the hole. It isn't a difficult job but it can be rather messy, and you'll probably need some help to lift the heavy hearth and surround.

Surrounds and hearths

There are many types of modern and older-style fireplaces, although they're usually fixed to the wall in a similar way. The opening, known as the 'builder's recess', might have been converted to take a gas or an electric fire but once this and its wiring or pipework is removed, blocking it off is just the same as for a 'real' fire.

Your fire surround may be one of the decorative cast-iron types and will have four of six integral 'lugs' at the sides through which are inserted the screws that fix it to the wall. You may also find a separate cast-iron section framing the opening and secured to the main surround by nuts and bolts.

This type of surround is very heavy, so you'd be wise to dismantle the separate section before you try to remove it from the room.

Surrounds are also made of brickwork or stonework; they're built like a wall flat against the chimney breast, and sometimes they're bonded to the wall. You can remove either type by dismantling them piece by piece.

Timber surrounds, whether elaborately moulded, carved, or simple hardwood frames, are usually screwed to timber battens fixed to the wall, sometimes as a separate mantelshelf and side pieces. They often have a central tiled area, which may be fixed directly to the wall or stuck on a slab and secured to the wall with metal lugs and screws. Cast-iron and wooden surrounds can be valuable and some types are much sought-after, so you may think yours is worth selling. You'll have to take great care when removing it so you don't damage it.

Concrete slab surrounds are usually clad with ceramic tiles and have wire mesh or steel rod reinforcement, and fixing lugs set in the concrete at the sides.

Most decorative or 'superimposed' hearths are made of reinforced concrete, often clad with tiles, or stonework. The decorative hearth is set in mortar on a concrete slab called the 'constructional' hearth, which is set flush with the floor surface. It forms the base of the fireplace opening and also extends into the room – often with an asbestos tape or rope expansion joint where it protrudes from the cavity – to provide a barrier between the heat of the fire and the combustible material of the floor.

Inset tiled hearths are mostly found in upstairs rooms. You probably won't need to remove the tiles if your floorcovering conceals them, but you may have to lay a concrete screed over them or glue hardboard on top to bring the surface level with the floor.

Firebacks and lintels

There will be a fireclay fireback cemented inside the cavity, which protects the brickwork of the chimney breast from the heat of the fire. You can remove this by chopping out the mortar which secures it, but if on the other hand you think you may want to restore the fire to working order later, leave it in place.

The builder's recess itself is formed in the chimney breast and has a concrete lintel or iron bar and brick arch above to support the walling above the opening.

Removing your fireplace can be a messy job, so before you start, it's wise to have the chimney swept: any vibrations could bring

HOW FIREPLACES ARE BUILT

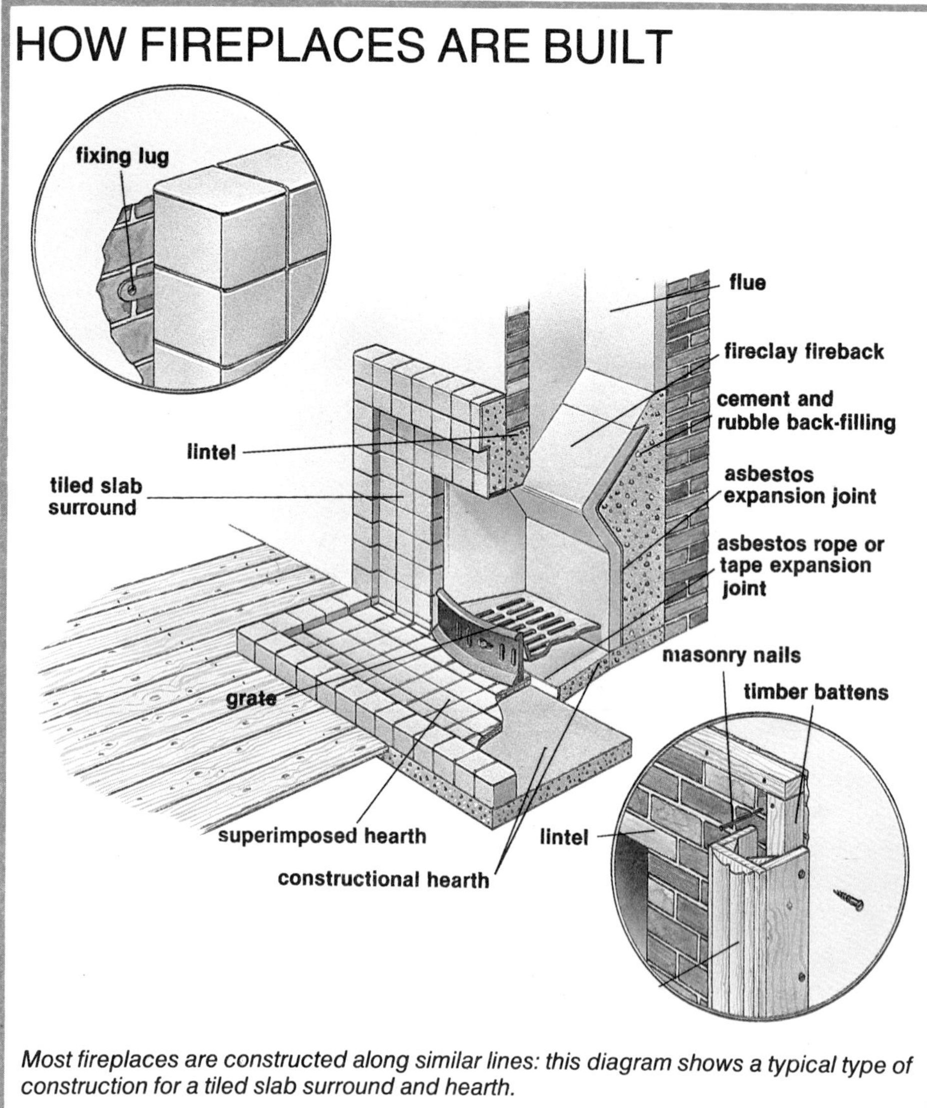

Most fireplaces are constructed along similar lines: this diagram shows a typical type of construction for a tiled slab surround and hearth.

soot down. Clear the room of as much furniture as you can, or group it in the centre of the room and cover it with dust sheets. Roll back the floorcovering from around the hearth and lay a large sheet of heavy gauge polythene on the floor as protection against dust and debris: don't use newspaper as it tears too easily. You'll also need something in which to collect the debris, such as a large metal bucket or even a wheelbarrow.

Removing the hearth

The decorative hearth is usually laid after the surround has been fixed to the wall, so this is the first thing you should remove. If, on the other hand, the surround has been laid on top of the hearth you'll have to remove this first. A little investigation will show you what the first move should be.

Use a club hammer and bolster chisel to loosen the mortar bond between the two hearths and ram a crowbar or garden spade underneath to prise them apart. It's a good idea to wedge offcuts of wood under the hearth to give you more leverage. If you find the hearth too heavy to lift – or if it's firmly bedded in mortar – you may be able to smash it into smaller, more manageable, pieces with a sledge hammer, although you'll have to cut the metal reinforcement with a hacksaw. Don't forget to wear goggles as protection against flying fragments. A stonework hearth can be removed in the same way.

When you've lifted the hearth clear you can make good the surface of the constructional hearth by filling any cracks or voids with ready-mix mortar. If your floor's solid you can simply level off the hearth area to floor level by laying a screed of concrete, but if the floor's timber, and you want to continue the floorboards up to the wall, you'll have to chisel away the constructional hearth to just below the floor level with your club hammer and bolster chisel, then fit a

READY REFERENCE

CAPPING THE CHIMNEY

To prevent moisture forming in the disused flue, where it could cause dampness on the face of the chimney breast:

- fit a metal cowl (**A**) to the chimney pot (**B**) to ventilate the flue and prevent rain getting in or
- remove the chimney pot and bed a half-round tile (**C**) over the opening to the flue.

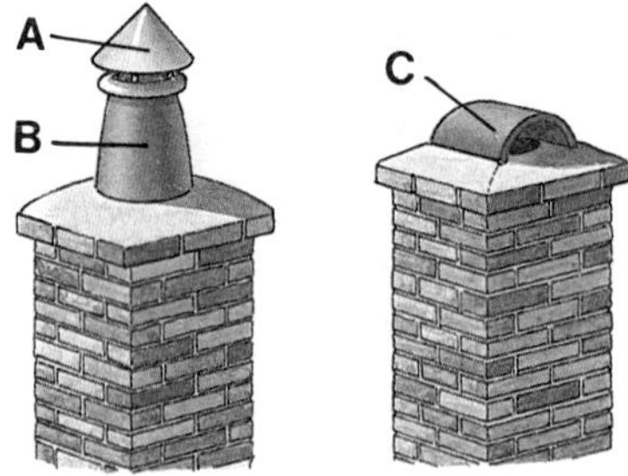

CONTINUING THE FLOOR

If you've got a timber floor you want to feature, you'll have to continue the floorboards over the hearth area. To do this:

- chisel the constructional hearth (**A**) to just below floor level
- nail a new joist (**B**) across the hearth
- lay short lengths of floorboard
- fit new skirting across the chimney breast.

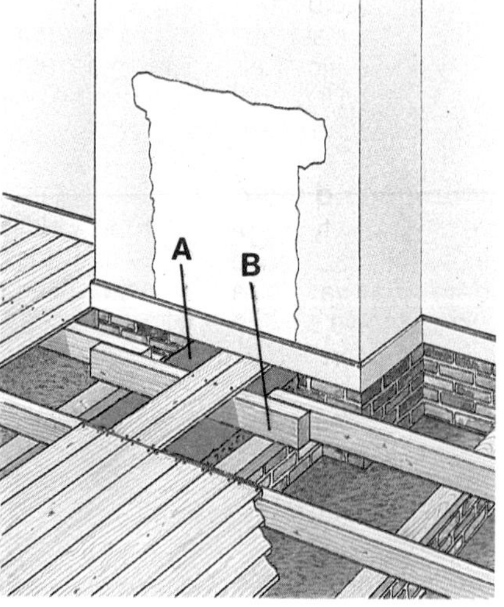

MAKING A FEATURE

If you want to make a feature of your fireplace opening:

- remove the fireback and rubble behind it by breaking it with a hammer and chisel
- fit a ventilated chipboard roof just above the opening within the chimney
- mount shelves on battens inside the opening for books, ornaments or the TV.

REMOVING THE HEARTH

1 *Not all fireplaces have slab hearths and tiled surrounds. Some of the older types have a recessed hearth, and a decorative inner surround.*

2 *If your tiled hearth is set flush with the floorboards you needn't completely remove it; lift the tiles using a club hammer and bolster chisel.*

3 *Your surround might have a separate mantelshelf simply bedded in mortar; break the bond by lifting the front edge.*

4 *If the mantelshelf is marble or metal it'll be very heavy. Get someone to help you lift it off the columns and carry it from the room.*

5 *If the side columns are fixed to the wall by lugs, unscrew these first and pull the columns free; they may simply rest against the wall.*

6 *When you've removed both columns you should be able to pull away the heavy cast iron centrepiece, which may have tiled panels.*

new joist on which you can lay new floorboards.

Removing the surround

Your cast-iron or tiled slab surround will probably be fixed to the brickwork with screws inserted through its side lugs into wall plugs. Small surrounds usually have only two lugs, one at each side, about 75mm (3in) down from the top; large ones may have two at each side and maybe two at the back of the mantelshelf.

The lugs are usually screwed directly to the brickwork and buried in the plaster surface, unless your surround is a later addition. To avoid having to remove too much plaster, strip off a margin of your wallcovering from the perimeter of the surround; if it's not apparent where the lugs are located, tap the surface with your knuckles until you hear a change of sound. When you've located all the lugs, chip away about 50mm (2in) of plaster at each point to reveal them, using your club hammer and cold chisel.

Undo the fixing screws using a screwdriver: it's a good idea to squirt a little penetrating oil on them first, and leave it to soak in. If they're still difficult to remove you can drill out their heads using a drill bit the size of the screw shank, or saw them off with a hacksaw.

Wedge a crowbar behind the edge of the surround near the top lugs and insert an offcut of timber behind it to protect the wall and give better leverage. Lever the surround away from the wall slowly and lower it gently to the floor: you may need someone to help you at this stage.

If the surround is too heavy to remove in one piece – and you don't intend to sell it – you can break it up with a sledge hammer, even if it's made of cast iron. Cover it with sacking first to contain any flying fragments, and wear goggles as a precaution.

You can dismantle a stone or brick surround piece by piece, and if you're careful to keep the bricks intact you may be able to use them elsewhere. In which case it's useful to number each piece in relation to its position on the wall.

As you remove the bricks or stone blocks you may find metal 'ties' enbedded in the mortar joints, which hold the surround against the wall, and a steel plate supporting the walling above the opening.

Timber surrounds are screwed at the sides to a timber frame fixed to the chimney breast. The screws are usually countersunk about 6mm (¼in) and concealed with filler or dowels, glued in and smoothed flush with the surround. If you can't locate the screws by tapping you'll have to scrape off the paint or varnish to reveal them. To reach the screws you can either drill down to their heads or cut out the filler or dowels with a small firmer chisel. Use penetrating oil to loosen the screws, then withdraw them using a screwdriver. Lift the surround clear of the wall and remove the battens.

Blocking the opening

Once you've removed the fire surround and hearth you can block off the opening in a number of ways. You may, for instance, be able to use the opening as a display recess for hi-fi or ornaments by fitting shelves inside. You'll have to fit a ventilated roof to the opening, strong enough to contain any soot or debris that might fall down the flue.

But if you want to block off the opening completly you can fit a panel of plasterboard over the cavity, or fill it with bricks or blocks. You can then plaster over the panel and decorate it with paint or wallpaper to match adjacent walls.

If, on the other hand, you want to fit a gas fire against the chimney breast, you can block off the opening with a sheet of asbestos-free insulating board with a hole cut in to take the vent outlet from the fire.

Whichever method you use, you'll have to fit a ventilator – which can be a simple metal, plastic or plaster grille – to prevent moisture collecting inside the flue and forming damp patches on the face of the chimney breast. The easiest place to put the grille is in your panel, although you could fit it in the wall higher up, or even at the side of the chimney breast. You'll also have to 'cap' the chimney pot with a proprietary ventilator or a half-round roof tile to provide a flow of air and to keep rain out (see *Ready Reference*).

Boarding up the opening

The advantage of fitting a panel to the opening is that you, or a future owner of the house, can easily restore the fireplace to working order.

To fit the panel first make up a simple frame of 50 x 50mm (2 x 2in) sawn timber to fit snugly inside the opening. On very wide openings you'll have to fit a central vertical batten to support the panel. If you want to plaster the surface flush with the wall you'll have to recess the frame about 12mm (1/2in) from the face of the chimney breast to allow for the thickness of the plasterboard and the skim of plaster.

If you simply want to decorate the panel with wallcovering you can recess the frame about 9mm (3/8in) from the finished suface of the chimney breast so that the plasterboard is at the same level.

Secure the frame – which can be simply butt-jointed – to the wall with masonry nails about 50mm (2in) apart; if the top of the

BOARDING UP THE OPENING

1 *If you want to cover the recess with plasterboard, you'll have to fix a stout frame of 50x50mm (2x2in) timber within the opening.*

2 *Where the sides of the opening are damaged you won't be able to fix the studs securely; nail short supporting battens at each corner of the frame.*

3 *Cut a panel of plasterboard to fit over the opening and nail it to the frame using 25mm (1in) plasterboard nails. Recess to allow for plastering.*

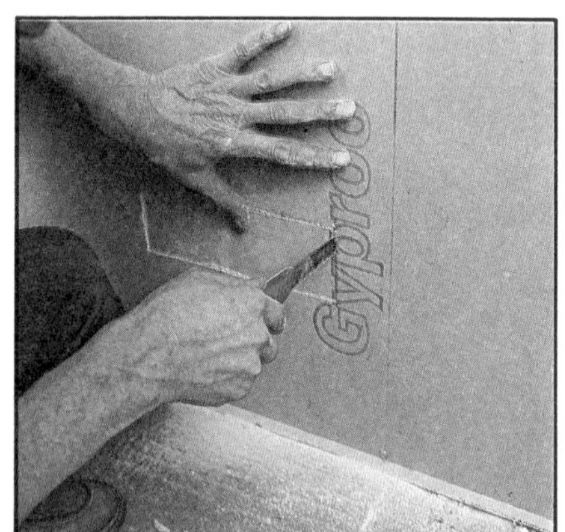

4 *If you're fitting a plastic ventilator, mark out the shape of the grille on the panel, just above skirting level, and cut it out using a pad saw.*

5 *Screw the first part of your plastic grille to the panel, locating the screw in a stud, and clip on the top piece. Remove the grille before plastering.*

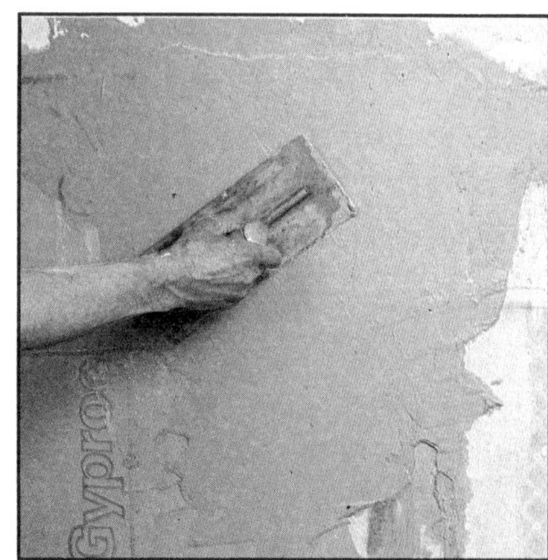

6 *To finish the panel to match the adjoining plastered walls, spread on a thin layer of Thistle Board finish; refix the ventilator when set.*

opening is curved you can wedge offcuts of timber between its lower edge and the top

Cut a piece of 9mm (3/8in) thick plasterboard to fit tightly inside the opening using a sharp trimming knife held against a steel straight edge. Nail the panel to the framework using 25mm (1in) galvanised plasterboard nails. Don't forget to fit it ivory side out if it's to be decorated directly with paint or wallpaper, and grey side out if it's to be plastered.

Measure the size of your ventilator and transfer its dimensions to the board. It's best to position the ventilator near the bottom of the opening. Drill a row of holes along the guideline to form a slot so you can insert a pad saw to cut out the waste.

If you're going to plaster the surface, fit a length of expanded metal lath over the joint between the brickwork and the panel to prevent the plaster surface from cracking along this line.

Plaster the surface or decorate it with wallcovering, then fit the grille.

Bricking up the opening

If you want to make a permanent job of blocking off your fireplace you should fill the opening with bricks or blocks and plaster the surface to match adjacent walls.

Your new brickwork should ideally be 'toothed' into the sides of the opening, by removing the half bricks around the perimeter at alternate courses and inserting new whole bricks to continue the bond. Install an air brick in the centre of the opening just above the skirting level. If you don't tooth in the new brickwork, you'll have to fix strips of expanded metal lath over the joint between old and new brickwork with dabs of plaster or masonry nails to prevent the plaster finish cracking.

When the mortar has set, rake back the joints to about 13mm (1/2in) deep to key the surface for plastering. Apply one coat of Carlite Browning plaster (see Working with plaster) to the brickwork – except for the airbrick – followed by a skim coat of Carlite Finish plaster to bring it flush with the wall. When the plaster has set, trowel over it (without any plaster) to polish the surface. Lubricate the trowel with water splashed on from a brush so that you don't score the smooth finish with the edge of the blade. Use light strokes of the trowel.

Fit a plastic or metal grille over the air brick. If you use concrete blocks to fill in the opening, you'll have to cut some to size to maintain the bond; you can fill in small gaps around the air brick with bricks. Don't forget to fix metal lath over the joint with the brickwork.

Fit a new length of skirting across the chimney breast.

BRICKING UP THE OPENING

A more durable method of blocking off your fireplace opening is to brick up the recess and then to plaster the surface to ensure that it matches the rest of the chimney breast and the adjacent walls of the room.

The first steps in bricking up your opening are to remove the hearth and surround, make good the floor and chop away the plaster from around the opening to reveal the old brickwork.

Chop out the half bricks at the sides of the opening so that you can 'tooth-in' the new brickwork to make a firm, bonded joint. Lay an air brick just above skirting level to allow for ventilation in the flue.

Continue bricking up the opening, toothing-in the bricks at alternate courses; then plaster the surface when complete to match the rest of the chimney breast. Don't plaster over the air brick. Fit a plaster grille on the finished surface.

USING BUILDING BLOCKS

If your fireplace opening is especially large you can use building blocks instead of bricks, which will enable you to span the distance much more easily. Where the size of the opening is less than a whole block you can fill in with half blocks or with bricks. Leave a gap for the ventilator when plastering the surface and fit a plaster grille on top. It's best to fix strips of expanded metal lath over the join between the brick- and blockwork.

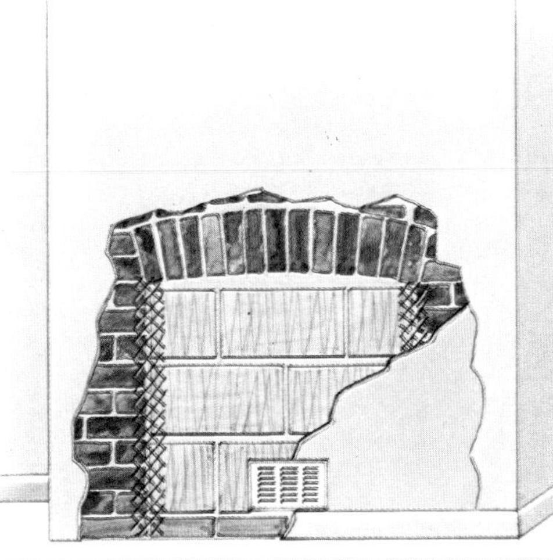

REPLACING AN OLD CEILING

If you've a drab, old ceiling that's badly cracked or sagging you can replace it with a completely new one made of plasterboard, which you can either decorate directly or plaster first.

A weak or faulty ceiling is not only unsightly but also, if it should partially collapse, dangerous. So it's wise to check it periodically for signs of deterioration and put right any defects immediately.

If the damage to your ceiling isn't widespread you can sometimes make a 'patch' of plasterboard, which won't be visible after you've redecorated (see *Ready Reference*). However, if the problem's more serious or if you think that the ceiling looks particularly shabby, it's usually better to take it down completely and start again from scratch.

How ceilings are made

The ceilings of your house are probably constructed in one of two ways, depending upon the age of the property and whether any renovation has already been carried out. Older houses usually have ceilings consisting of thin wooden battens called 'laths' nailed to the underside of the joists above, and clad with a thin coat of lime plaster reinforced with hair fibres to give a smooth, flat surface.

Modern ceilings, however, are made in a much simpler way: sheets of plasterboard are nailed to the joists and can be decorated directly or plastered first.

Causes of damage

Damage to your ceiling can be caused in a number of ways; not always due simply to failure of the original structure. If, for example, the joists in your loft are overloaded with your household storage, or you've hammered back loose floorboards in an upstairs room (it's best to screw them back), it's possible that you'll have jarred loose the plaster 'nibs' that secure the plaster to the laths. These nibs were formed when the original plaster was spread on, and consist of a ridge of plaster that's squeezed up between the laths to anchor the whole ceiling in place when set. If the nibs are broken, parts of your ceiling are likely to sag and may even fall away.

Damage might also have been caused by someone in the loft accidentally putting their foot – or dropping something – through the ceiling, or by water from leaking plumbing ruining the surface.

It's not likely that all or part of your ceiling will collapse without warning: usually you'll see signs such as bulges or large cracks appearing. If your loft is directly over the suspect ceiling, you can easily judge how well the plaster is keyed to the laths from above. Alternatively, you can lift a few floorboards in the room above the ceiling to inspect the nibs.

Unless a catastrophe such as a burst water tank occurs, it's rare for a complete plasterboard ceiling to need replacement. But it's quite likely you'll have to put right minor defects caused by accidents in the loft or by the fixing nails working loose.

Knocking down an old lath-and-plaster ceiling is a very messy job, so it's worth anticipating trouble should bulges or cracks appear, by covering the ceiling with sheets of plasterboard.

Making a new ceiling

Before you can put up your new ceiling you'll have to hack off the original lath and plaster surface to reveal the joists. You can then nail up sheets of plasterboard and finish off the surface with a skim of plaster or simply decorate with paper, paint or a textured finish.

There are a number of types and thicknesses of plasterboard and it's important to choose the right one for the job. The standard type is British Gypsum Gyproc wallboard and consists of an aerated gypsum plaster core encased in thick paper liners. One side has a smooth, ivory-coloured surface suitable for direct decoration, and the other face a grey surface, which you fix outermost when you're going to finish the ceiling with a skim coat of plaster.

There's also insulating wallboard, which has a veneer of aluminium foil on the grey face. It's useful for upstairs ceilings when used in conjunction with conventional loft insulation.

Wallboards are available with three different types of edges: squared, bevelled or tapered. The type with a tapered edge is the one to choose for your ceiling because it's designed for smooth, seamless jointing.

There are also two thicknesses of plasterboard: the 9.5mm (3/8in) thickness is used where the ceiling joists are a maximum of 450mm (18in) apart; the 12.7mm (1/2in) thickness is used for maximum joist spacings of 600mm (2ft).

There are many sizes of board but the most commonly available in builder's merchants are sheets 1800mm (6ft) and 2400mm (8ft) long by 1200mm (4ft) wide. The bigger the sheet you have the fewer the joints you'll have to fill, but you'll find the boards very heavy to lift to the ceiling, so it's probably best to go for the smaller size.

If the joists are widely spaced, you can use a board called Gyproc plank for the repair. It's available in one thickness (19mm/3/4in) and one width (600mm/2ft), but in several lengths, the most common being 2400mm (8ft) with an ivory face and tapered edges.

If you're replacing the ceiling of an upstairs room you should use a special board called Gyproc vapour-check wallboard, which has a water vapour-resistant, blue-tinted plastic film bonded to the grey side, leaving the ivory surface exposed for direct decoration. The plastic film stops water vapour from inside the

REMOVING THE OLD CEILING

1 *Working from a hop-up or a sturdy testle, start to hack away the old plaster ceiling using the claw of a large claw hammer.*

2 *When you've removed a section of plaster, prise away the laths. Hack away from you so you don't pull down the ceiling on top of you.*

3 *Work your way across the room, removing sections of plaster, then the laths. Bag up the plaster and bundle up the laths for disposal.*

4 *When you've cleared away the old ceiling, work your way along each joist to remove all the lath-fixing nails using a pair of pincers.*

building passing into the roof space above.

Where you want to repair and insulate your ceiling in one go, you should use Gyproc thermal board. This consists of standard plasterboard bonded to a backing of expanded polystyrene, with a vapour-check membrane between the two. It has an ivory surface and tapered edges for direct decoration. It comes in several thicknesses ranging from 22 to 65mm (7/8 to 2½in); it's 1200mm (4ft) wide and the usual length is 2400mm (8ft).

Handling plasterboard

Carrying sheets of plasterboard is a two-man job in most cases, and you'll probably find it easiest to carry on edge. Store them flat in a dry place, such as your garage or, better still, in the room where you're going to fix them. If this isn't possible, have the boards delivered on the day you want to fix them. If you're going to collect the boards, don't stack more than two or three sheets at a time on your car roof rack, and use a couple of long lengths of stout timber to support them as they're quite brittle.

If you find the full-size boards too difficult to handle, you can cut them in half, using a sharp trimming knife held against a steel straight edge. Cut through the ivory face with the knife and crack the board over a batten. Fold back the board to form a crease and run your knife blade along the fold.

Removing the old ceiling

There'll be years of accumulated dust and dirt above an old ceiling, so if your loft is directly above, it's worth hiring an industrial vacuum cleaner to clear away as much of the debris as you can.

It's best to wear overalls or old clothes, a dust mask and a pair of stout gloves. You'll

READY REFERENCE

SUPPORTING THE BOARDS

To avoid having to rig up complex working platforms for you and your helpers while you nail up the boards:

- make T-shaped supports from 50x25mm (2x1in) softwood with reinforcing battens (A)

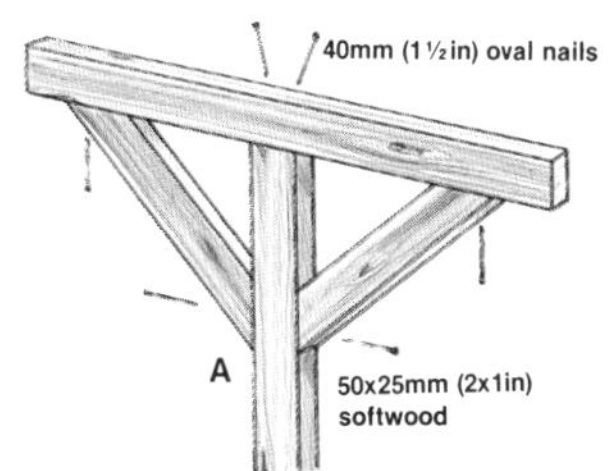

- get your helpers to use the supports at each end of the board to hold it against the joists while you nail it in place from a hop-up (B).

PLASTERING

If you want to plaster the surface of your new plasterboard ceiling:

- fix up the sheets grey-side-down, using plasterboard nails
- spread a thin layer of Thistle Board Finish plaster along the joints, about one arm's length at a time
- embed 75mm (3in) wide hessian scrim in the band of wet plaster using your trowel and leave to set
- spread a 3mm (1/8in) thick layer of plaster over the entire surface of the ceiling
- when the plaster has set, trowel over the ceiling without any plaster, to polish the surface. Lubricate your trowel blade with flicks of water from a brush.

FIXING THE NOGGINS

1 *Tidy up the perimeter of the walls; then mark the centre line of each joist on the wall as a guide to nailing up the plasterboard sheets.*

2 *Nail up long 50x50mm (2x2in) noggins parallel with the joists to support the boards; use a batten to set them level with the joists.*

3 *To support the long edges of the boards, fix noggins between the joists. Use a batten marked with the board width to mark their centres.*

4 *Skew-nail 50x50mm (2x2in) noggins between each joist, placed so that the plasterboard will cover half their width.*

need a helper (or helpers) to lift the boards into place and it's a good idea to wear a crash helmet, a hard hat like those worn on building sites, or some sort of padded cap so you can use your head to support the boards while you're driving in the nails.

Clear the room of all furniture, roll back and remove your carpets or other floor-covering and seal the gaps under doors with rolled-up sheets to stop dust blowing into the rest of the house. You'd also be wise to cover the floor with dust sheets or heavy gauge polythene to help you collect the rubble, and open the windows to ensure good ventilation.

Keep lots of old sacks or thick polythene bags handy to carry away the debris. You may also find it useful to keep a house plant spray close to hand to douse the dust.

After that it's just a matter of hacking away the plaster from above with a claw hammer or a club hammer and bolster chisel. Then you can prise away the laths and extract the nails that are left protruding from the joists.

If you've a central pendant light on your ceiling, or other ceiling-mounted electrical fittings, you'll have to remove these before you start to hack off the plaster. Switch off the electricity supply at the mains and disconnect your fitting leaving only the cable – suitably insulated – hanging down. If you can get into the loft, pull up the flex until you've installed your new ceiling. Otherwise make a hole in the plasterboard and draw the flex through it before fixing the board in place. Locate lights under joists, or nail a batten between adjacent joists, to provide a firm fixing for the rose.

Fixing the plasterboard

With the joists exposed and clear of nails, you should mark their centre points on the adjacent walls as a helpful guide when positioning the fixing nails.

The board should be fixed with the long paper-covered edges lightly butted together at right angles to the run of the joists, with the grey, foil or polystyrene surface against the joists. The ends of the boards must be located centrally over a joist so you'll probably have to cut them to the correct length.

If the joists are more than 450mm (18in) apart, you'll also have to support the edges of the boards. This means nailing 100x50mm (4x2in) timber battens called 'noggins' between the joists at these positions. Fix up the noggins by skew nailing (see Building a stud partition wall) before you start to nail up the boards. The noggins will ensure that your new ceiling is set perfectly flat and rigid. However, you can omit noggins if the joists are fairly closely spaced, or if you are using 19mm (¾in) thick Gyproc plank.

Secure the boards to the joists with galvanised plasterboard nails. Use nails 30mm (1¼in) long for 9.5mm (⅜in) thick boards; 40mm (1½in) long for 12.7mm (½in) boards. Thicker boards, such as thermal board, should be fixed with nails that are long enough to sink at least 25mm (1in) into the joists.

To enable you to reach the ceiling you'll need to rig up a platform such as a hop-up or planks between stepladders. Alternatively, you can make T-shaped timber supports (see *Ready Reference*), which will enable your helpers to hold up the boards at each end from floor level.

Drive home the nails firmly without the head fracturing the paper surface; the final hammer blow should leave a slight depression, which you can fill later. Nail each board to every joist and noggin at 150mm (6in) centres, starting at the centre of each board and working outwards. The nails shouldn't be closer than 13mm (½in) from the ends of the boards and 10mm (⅜in) from the edges.

Try to arrange the boards so that the cut edges fit into the internal angles at the sides of the room. When you're forced to have cut edges within the ceiling area, you should stagger the end joints and arrange the boards so that they fall mid-way over a joist, with a 3mm (⅛in) gap between each.

Sealing the joints

If you're going to plaster your new ceiling, you'll have to seal the joints with hessian scrim (see *Ready Reference* and Plastering plasterboard) and spread on a skim coat of special Board Finish plaster. But if you just want to decorate the surface with paint or paper you should seal the joints with a special tape.

You'll need Gyproc joint filler, joint finish

FITTING THE PLASTERBOARD

1 The corners of your walls won't be truly square, so scribe your first plasterboard sheet to fit. Set the angle on a profile gauge.

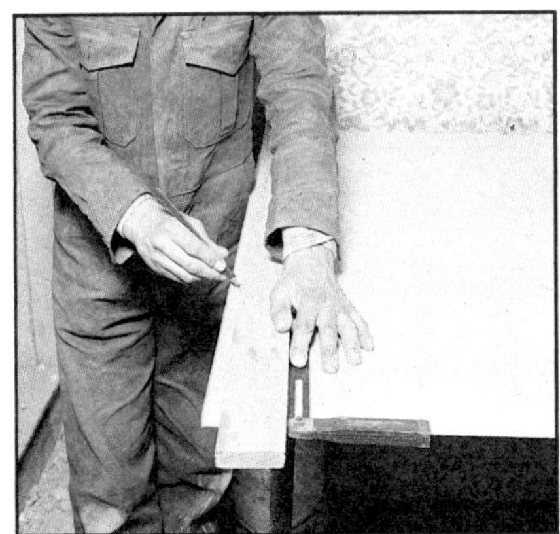

2 Transfer the angle to your first sheet of plasterboard and, using a straight-edge length of timber as a rule, mark the waste and cut it off.

3 Lift up your first plasterboard sheet and position it squarely across the joists with its long inner edge square on the intermediate noggins.

4 Nail the board at 150mm (6in) centres. Once you've made enough fixings to hold the board you and your helper can work simultaneously.

5 Line up a straight-edged length of timber with the centre of each joist, as indicated by the guidelines you drew on the wall, and make intermediate fixings at 150mm (6in) centres.

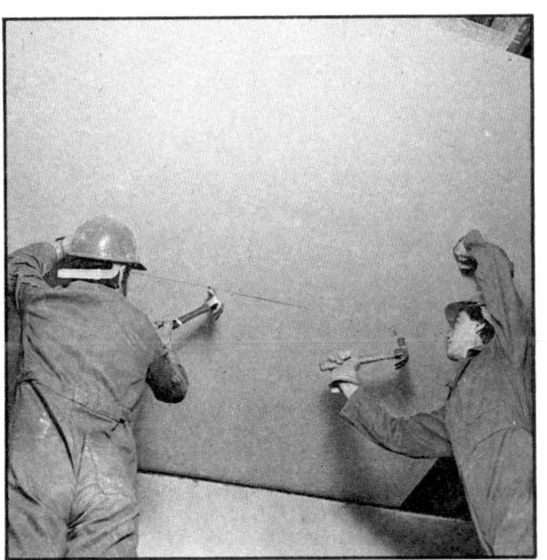

6 When you've nailed up the first board, lift up the second, butting it lightly against the first, and nail it to the joints and noggins. Continue in this way to complete the ceiling.

READY REFERENCE

MAKING A PATCH

If only a small section of your ceiling is damaged you can make a patch of plasterboard. To do this:

- tidy up the edges of the hole and remove any loose laths
- or, if the ceiling is plasterboard, cut out the damaged section or remove the whole board
- cut a new piece of board to fit just inside the hole
- fix the panel to the joists, grey-side-down, with plasterboard nails

- wet the edges of the board and spread on backing plaster to fill the gap between the board and the ceiling
- leave the plaster 3mm (1/8in) below the surface of the ceiling to allow for a skim coat of plaster
- apply a skim coat of Thistle Board Finish to the entire surface of the patch to bring it flush with the rest of the ceiling

- when the plaster has set, trowel it smooth using water to lubricate your trowel. You could then paper the ceiling or apply a textured paint as a decorative finish.

and joint tape and, to apply them, a 200mm (8in) jointing applicator, a 50mm (2in) taping knife, and a jointing sponge.

To seal the joints you have to use the applicator to spread an unbroken band of joint filler to fill the taper between the edges of the boards. Next you cut the jointing tape to the exact length of the joint and gradually press it into the filler using your taping knife.

Use the applicator straightaway to spread another band of filler over the tape. Here, you should aim to fill the taper level with the surface of the boards. Inevitably, there'll be some surplus material at the edges of the band of filler and you should wipe this away immediately, using your sponge moistened with water.

After about an hour the filler will have set, though it might not feel dry. Use your applicator to spread a thin film of joint finish over the joint. Dampen the sponge again and carefully 'feather out' the edges of the band of finish to smooth any ridges. Allow this first coat of finish to set then apply a second and feather out the edges.

While the joints are drying you can tackle the nail head depressions by applying a thin coat of joint filler followed by a thin coat of joint finish.

When the joints have dried, there'll be a slight difference in the surface texture between them and the board. You can even this up by spreading a slurry of joint finish over the whole ceiling, or you can apply, by brush or roller, a material called Gyproc drywall top coat. One coat is enough to prepare the ceiling for normal decorating; two coats provide a water vapour-barrier.

If your wall is solid and plastered, there'll be a fairly wide gap of the angle with your new ceiling, which you'll have to fill with plaster. But you can fill the joint between the new ceiling and wall using the basic technique as for flat joints. First, fill any gaps with joint filler, then cut the tape to length. Crease it firmly down the middle, apply a thin band of joint finish to each side of the angle between the ceiling and the wall and press the tape in place.

Run a thin layer of joint finish over the top and feather out the edges with your damp sponge. When this has dried, apply a band of finish to both sides of the angle and feather out the edges.

Finishing the ceiling

When you've filled the joints and have treated the entire surface of your ceiling with slurry of joint finish or drywall top coat you can decorate the ceiling with emulsion paint for a smooth finish, put up woodchip or a relief wallcovering (and again cover with emulsion paint), brush on a textured compound or simply hang a wallpaper with a pattern to match your wallcovering.

FINISHING THE CEILING

1 *To conceal the joints between the boards you'll have to seal them. Mix up some joint filler and spread a band along the tape.*

2 *Cut a length of jointing tape to fit the joint and press it into the filler using the special taping knife to bed it evenly.*

3 *Again using the applicator, spread another band of filler over the joint, covering the tape. Smooth out any air bubbles trapped underneath.*

4 *Remove any surplus filler from the edges of the joint with the sponge moistened with a little water, and leave for about one hour to set.*

5 *While the joint's drying, fill the nail head depressions with filler. Mix up some joint finish and spread it over the joint and then the nail heads.*

6 *Feather-out the edges of the joint with the sponge so that the surface is flush with the plasterboard. Apply top coat and decorative finish.*

BLOCK PARTITION WALLS

Building a partition is the easiest way to rearrange the rooms of your house – to improve storage or simply to divide a large area into two smaller parts. Lightweight building blocks make a particularly solid structure, and they're easy to work with.

Even with the most ingenious space-saving schemes you may find that one room is simply not enough to cope with all the different activities and storage needs of your household. Often your problems would be solved if you had two rooms where there was one, and a partition is the obvious solution. You could, for example divide an open-plan living and dining room, or make a second WC or a utility room. Or it may simply be that you are dissatisfied with the way your rooms are laid out, and wish to rearrange them. Pages 142-146 describe how you can build a timber-framed partition wall, which you can clad with plasterboard and decorate. But there are advantages to using concrete blocks instead: they give a stronger, more sturdy structure than timber, help to reduce the transmission of sound through the wall and provide a solid fixing for items such as book shelves and picture frames. Blocks also provide a greater degree of heat insulation. Because of their size you can build a fairly substantial floor-to-ceiling room-width wall relatively quickly which has the advantage that you can make it load-bearing.

Types of blocks

Building blocks are cast from concrete and come in a range of sizes with various densities (see *Ready Reference*).

Facing blocks have one patterned or decorative face, which is intended to be left bare as a finished surface.

Aerated blocks are the most widely available and the best for an internal partition. They're lightweight, can easily be grooved to take electric conduit or pipes and will take nails and screws. They also provide good heat and sound insulation.

Cellular blocks have cavities which don't go right through the block. They're laid closed end uppermost to give a continuous surface for spreading on the mortar bed for the next course. They have a rough surface which must be plastered.

Dense and **lightweight solid blocks** are rectangular, and can be used for most building work; some can even be used below dpc level. Load-bearing and non-load-bearing types are available; they'll take screws and nails, and can be plastered.

Jointed blocks have tongued-and-grooved ends for slotting together into a strong, load-bearing structure. They have good sound and heat insulation.

Hollow blocks have cavities, which can take pipes and conduits, or they can be filled with concrete or fitted with mild steel strengthening rods.

The best blocks for use inside are the aerated type, which are light in weight and so easier to handle than their heavyweight counterparts, and reduce the total load of the wall as well. The most commonly available size is 440 x 215 x 100mm (17 x 8½ x 4in). They're laid in a similar bonding pattern to bricks for strength, and have mortar joints between.

Planning a partition

You must plan the shape, size and position of your partition wall if you're not to end up with two dull rooms with unattractive proportions, but the most important factor to take into account is whether your floor can support a solid wall of blocks.

If you're building onto a solid concrete floor you can position the wall where you want. Don't worry if the surface isn't perfectly smooth and flat: you can accommodate any

READY REFERENCE

TYPES OF BLOCK

Typical sizes.
height: 140, 215mm
length: 440, 515, 610mm
thickness: 50, 60, 75, 90, 100, 140, 190, 215mm

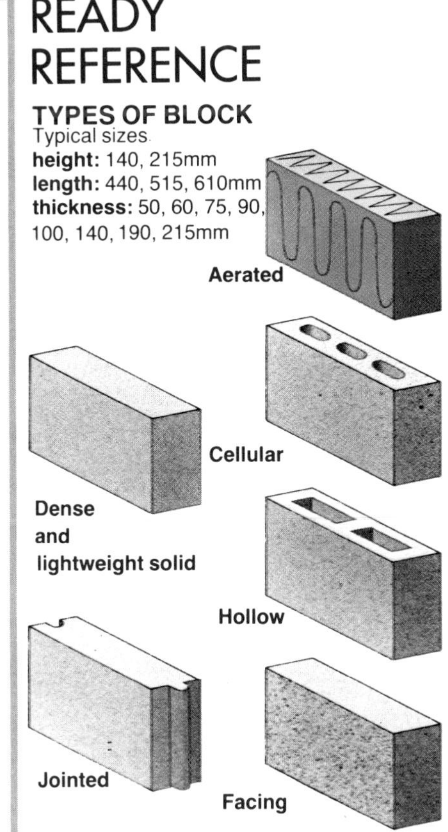

LAYING THE SCREED

1 *Spread a 150mm (6in) wide screed of mortar onto the floor where you want the partition and smooth it to about a 9mm (3/8in) thickness with your trowel.*

2 *Scribe the line on the screed with a trowel against a builder's square to mark the outside of the partition; extend it with a timber straight edge.*

3 *Measure along the mortar screed to the position where you want the wall to turn a corner and lay a second screed at right angles to the first.*

4 *Scribe the line around the corner of the screed using your trowel and builder's square and extend it if necessary, with a long timber straight edge.*

5 *Dry-lay the blocks on the screed against the scribed line, starting at the existing wall, leaving finger-thick joints between each of the blocks.*

6 *When you've positioned the first course of blocks you can erect the door frame. Check with a spirit level that its position is truly vertical.*

unevenness in the thickness of the mortar screed on which you lay the blocks.

A suspended timber floor causes more problems, although you can build onto it if the partition doesn't impose too great a strain on the structure. To ascertain this it's wise to consult your local council's Department of Building Inspectors.

Ground floor joists are usually spaced at 400mm (15in) intervals and supported on timber wall plates, which in turn are mounted on dwarf brick walls 1.3m (4ft 6in) apart over concrete foundations. Here you can build the partition on a stout timber floor plate, which will spread the load of the wall over the floor area. The floor plate should be of 100 x 50mm (4 x 2in) unplaned timber, nailed or screwed to the floor. It's a good idea to fix a 100mm (4in) wide strip of carpet underfelt underneath the plate to help reduce any sound transmission. You can nail strips of expanded metal lath on top of the plate to improve the mortar bond with the blocks.

Upper floors of timber construction aren't sturdy enough to support a blockwork partition and here you should build one with a timber frame. Similarly, upper floors of reinforced concrete construction aren't designed to support the weight of a full-width, room-height block wall, but you can safely build short-length dividers and half-height partitions, the top section of which can be glazed.

Structural requirements

The type of wall you wish to build is entirely up to you. It can be floor-to-ceiling and full-room-width: it can incorporate a window or door (see pages 95-99) or it may be a half-partition. But if you choose a full-width wall there are some important structural rules to keep in mind.

On very long walls you'll have to install a 'vertical movement joint' of flexible mastic every 6 to 9 square metres (20 to 30ft) – usually against a door or window frame – to allow for flexing of the structure. A wall of this size should also be 'toothed' into the existing walls (see *Ready Reference)*.

When it comes to finishing off the wall, the choice of surface is again a matter of choice. Concrete blocks can be left bare or simply painted with emulsion, but they're not attractive to look at, and aren't really acceptable anywhere except in a garage or workshop. Other options are to match the partition to adjacent walls, or clad it with plasterboard over wooden battens and decorate it with paint or wallcovering.

Another thing to bear in mind when you're still at the planning stage is the lighting of a newly-partitioned room. Only the smallest of rooms such as a cupboard or WC can cope without natural light, although by incorporating glazing panels in the partition

LAYING THE BLOCKS

1 Remove the blocks and apply a bed of mortar to the screed, against the scribed line. 'Furrow' the mortar with your trowel so you can bed the block accurately.

2 'Butter' one end of the first block with mortar and furrow it with your trowel. The mortared end of the block is butted up to the existing wall.

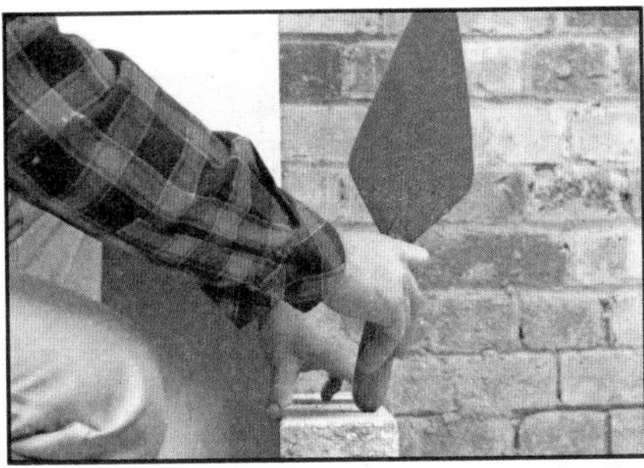
3 Position the block on the mortar bed, parallel with the scribed line on the screed. Tap it gently into place using the handle of your trowel.

4 Move to the other end of the partition and lay the two corner blocks on a bed of mortar with finger-thick mortar joints between them.

5 Place your builder's level across the tops of the first and corner blocks and tap them level, then fill in the intermediate blocks of the first course.

TIP

6 As you tap the blocks into place, some mortar will be squeezed out of the joints. Scoop this off with your trowel and re-use it for other joints.

you can make use of light from windows in adjoining rooms. It's illegal, though, to position a partition wall so that it divides a window, and it's not, in any case, an attractive solution.

Another important factor in positioning your partition is whether you'll block access to underfloor cables and pipes. You must also consider whether electricity, gas and water supplies, and even the central heating system, can be extended into your new room. Also allow for adequate ventilation such as an extractor fan or a window with an opening fanlight.

The easiest way to provide access to the new room is to build a doorway into the partition. otherwise you'll have the extra job of cutting one in an existing wall. Whichever method you choose you'll have to fit a lintel over the door as support for the blockwork or brickwork above (see pages 95 -103 for details). Alternatively you can make a room-height frame (see *Ready Reference*). in effect dividing the partition into two parts and providing the structure with greater strength. The upper part of the frame can be glazed with wired safety glass or panelled with plasterboard.

Before you decide upon a final plan for your partition it's wise to check with your

READY REFERENCE

FITTING A DOORWAY

If your partition is to have a door, use a room-height frame to increase the stability of the wall.

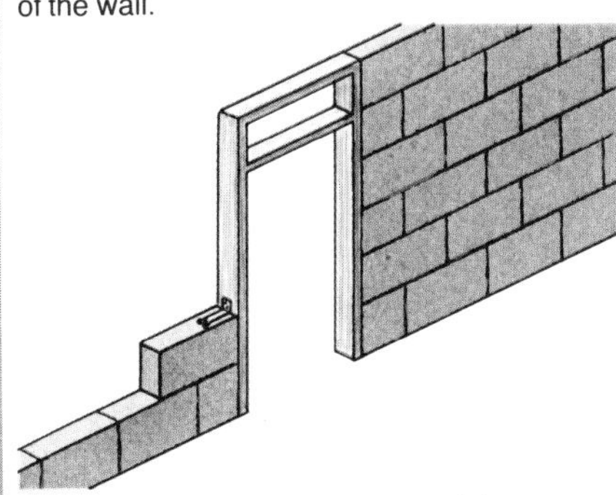

- screw it at the top to the ceiling joists
- secure it at the sides with galvanised metal frame cramps at every second course
- fill the gap between the top of the doorway and the ceiling with wired safety glass, plasterboard or wood panelling.

If you use a normal-size door frame:

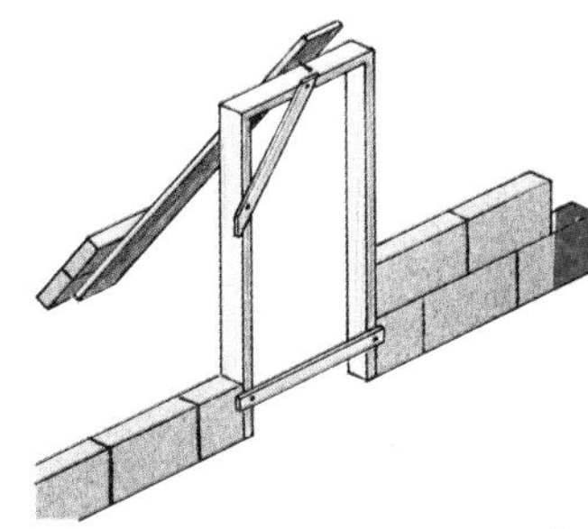

- fit it with temporary timber battens called 'strainers' across one corner and between the uprights at the bottom to prevent the frame from warping under the weight of the blocks
- erect the frame exactly where you want it and prop it in place with a 3m (9ft) long plank with a nail driven into the top.

FORMING A CORNER

You'll have to cut some blocks to size at the corners and the ends to maintain the stretcher bond throughout the rest of the wall.

- form the corner at ground level with full blocks then cut three-quarter-size blocks (A) for the second and then for the alternate courses
- use bricks or cut 100mm (4in) pieces of blocks (B) to use at each course on alternate sides of the corner.

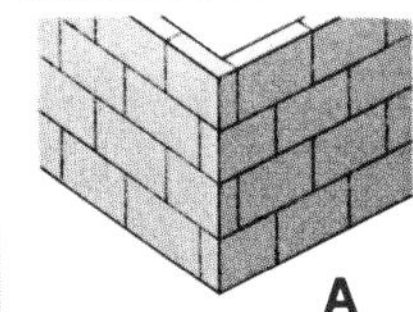
A

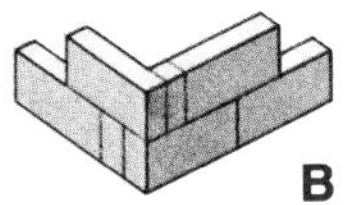
B

BUILDING THE PARTITION

1 Build up the corner and end of the partition to the fourth course, 'stepping back' at each course. Check the angle with a spirit level.

2 Fix a stringline as a guide to laying the intermediate blocks. Push the pin into the vertical joint, with the string wound over the pin.

3 At the other end, take the string over the last block, around the corner and push in the pin diagonally opposite. Use a brick to hold the string in place.

4 Infill with blocks to the stringline. Butter one end of each block and lay it on a bed of mortar applied to the laid blocks to make fitting easier.

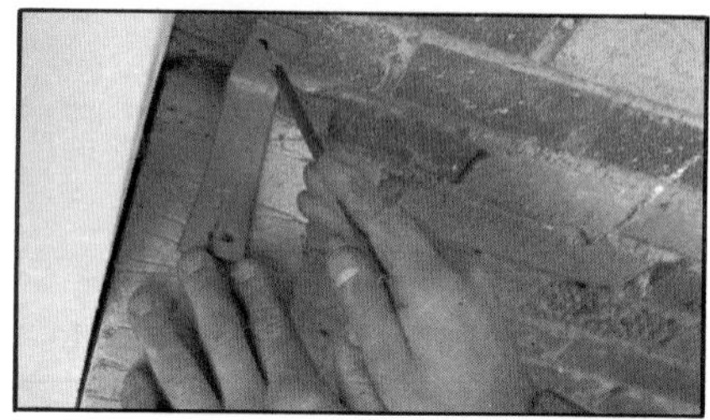

5 Secure the partition to the wall and door frame with metal frame cramps at every second course. Mark the screw holes on the wall with a pencil.

6 Drill holes at these marks to take the fixing screws for the frame cramp and insert plastic or wooden wallplugs into the holes.

7 Screw the frame cramp to the wall. There's no need to cut a recess in any of the blocks; the cramp is buried neatly in the mortar joint.

8 A mortar joint is all you'll need at ceiling level. Apply mortar to the top edge of the block and to the blocks already laid to make fitting easier.

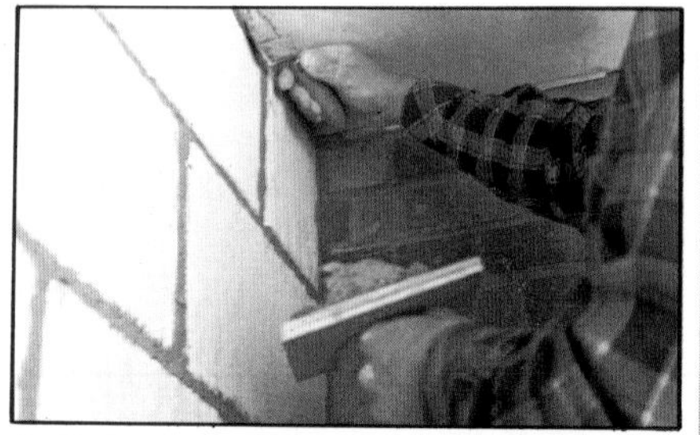

9 Point the ceiling joint with a small pointing trowel. Use a small hawk to carry the mortar. Allow the structure to set before finishing.

local council's Building Control Department that you aren't infringing any Building Regulations. If you don't, and are found out, you could be required to demolish the wall.

Marking out the partition

Before you start to build the wall on a solid floor you must remove any loose floorcovering such as linoleum, sheet vinyl, carpet or tiles, so that you have a firm, flat base for the blockwork. You can lay the blocks directly onto thermoplastic, vinyl or ceramic tiles as long as they're firmly stuck down. You should also remove any wallcovering from the existing wall where the partition is to be fixed.

Mark on the floor in pencil or chalk parallel lines that indicate the exact position of the partition, and carry them up the adjacent wall or walls to ceiling height. Use a spirit level and a long timber straight edge to ensure these guidelines are evenly spaced and straight. Next, chisel out a section of the skirting board and ceiling moulding, if any, between the guidelines so that the blocks can be laid flush with the wall; this makes for a sturdier structure and means that you don't have to cut any blocks to shape. When the partition's built you can fit new skirting and moulding to match that on the adjacent wall. You also need to chop out 'bonding pockets' in adjacent walls so that the new structure is properly tied in (see *Ready Reference)*.

To support the blockwork during construction you'll need a temporary timber batten called a 'wall profile', which you should nail lightly to the existing wall against the guideline marking the outside face of the partition. You can make use of a chimney breast or other feature that extends into the room as a wall profile (see photographs).

Space out the blocks dry for the first two courses, without using any mortar, so that you can plan the best bonding pattern with as few cut blocks as possible. You'll probably find that stretcherbond, as used in brickwork, is the best and simplest arrangement. When placing the blocks, leave finger-width joints between them (see photograph 5, page 138). You'll have to cut some blocks to size for corners and ends to maintain the bond throughout the

CUTTING A SOLID BLOCK

1 *To cut a block into two pieces, measure across one end the distance required and scribe a line at this point using a bolster chisel.*

2 *Tap the bolster chisel along the scribed line with a club hammer to make it deeper but don't hit it too hard or you may shatter the block.*

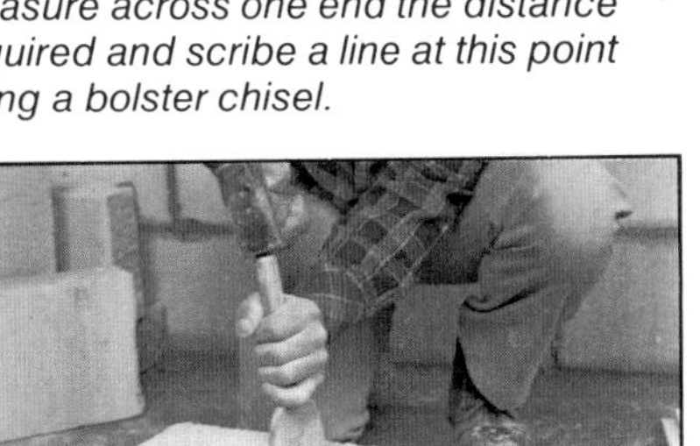

3 *Turn the block onto one face and scribe a line across, then around and onto the other face. Tap lightly along the line with the bolster chisel.*

4 *Place the bolster chisel on the centre of the face side of the block and hit it sharply with the club hammer to break it cleanly along the lines.*

rest of the wall. Alternatively, you could use bricks to fill in some spaces. When you've dry-laid the first two courses of blocks you can fix the door frame, if any, in place. Hold it erect with a temporary timber strut (see *Ready Reference*).

Laying the blocks

Once you're satisfied with the position and bond of the dry-laid blocks you can remove them and start to relay them, this time bedding each in mortar.

To do this, spread a 150mm (6in) wide screed of mortar about 9mm (3/8in) thick onto the floor over your guidelines and trowel it smooth. Then scribe a guideline marking the outside face of the wall onto the screed using your trowel and a timber straight edge. Bed the first course of blocks in mortar up to the line, starting at the wall profile. Make a mortar joint with the existing wall.

The most accurate way of working is to build up the corners or ends of the wall first, to about four courses, so that a string line can be stretched between them at each course as a guide to laying the intermediate blocks. Check constantly with a long spirit level as you work that the partition is level and upright.

Unlike brickwork, laying blocks that are going to be plastered or boarded doesn't call for perfectly regular size joints between the blocks. So long as there's sufficient mortar between each one, and they're laid so that the face is vertical, minor irregularities in thickness don't affect the strength of the structure. If you're building a large, room-width wall it's wise to carry out the work over two days to give the mortar in each part time to set. To tie the partition to the existing wall of the house you can either recess alternate courses of blocks into bonding pockets cut into the wall or insert galvanised metal frame cramps at these positions (see photographs 5-7, page 140).

Finishing the partition

If your partition is to be in a garage or utility room you might not feel a perfectly smooth, plastered finish is necessary, in which case you can simply form 'struck' or bevelled mortar joints with a bricklayer's trowel as you work and then give the wall a coat of emulsion paint when the mortar has set.

On the other hand, if the partition is to be in a habitable room within the house you won't want to see the outlines of the individual blocks. Here you should spread on two coats of plaster (see Working with plaster). You can then either paint or wallpaper the partition to match adjacent walls.

Alternatively, you can screw or nail horizontal timber battens directly to the blockwork and clad the surface.

READY REFERENCE

EXPANSION JOINT

If you're building a very long wall it's advisable to include a vertical 'movement' joint every 6 to 9m (20 to 30ft). Using a flexible mastic compound (available from builders' merchants) instead of mortar, the joint allows for a little movement in the solid structure and prevents cracks appearing in the mortar joints. If you're including a door or window, alongside the frame is an ideal place for the joint.

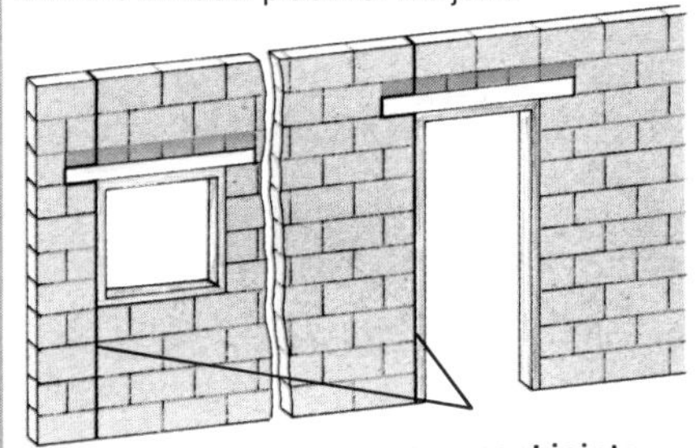

BONDING POCKETS

To form a strong bond with an existing brick or block wall and to prevent sideways movement of the partition, 'tie-in' the blocks at alternate courses.

- cut bonding pockets 50mm (2in) deep x 130mm (5in) wide x 245mm (9½in) high (to allow for mortar joints) in the existing wall with a club hammer and cold chisel
- slot the last whole block of alternate courses into the bonding pockets, bedding them in mortar.

LINTEL OVER DOOR

You'll need to install a lintel over the doorway to support the blocks above. Because of the lintel's size you'll have to fill the gap above with bricks or cut blocks.

For more information about using lintels see BUILDING TECHNIQUES 13.

BUILDING A STUD PARTITION WALL

Building a partition wall gives you two rooms where you only had one before. Surprisingly, you don't have to be a skilled craftsman. Here's how to build a simple framework.

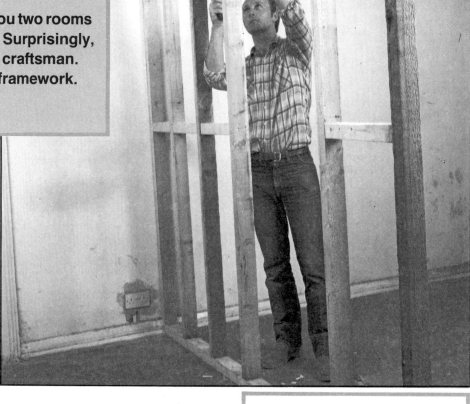

Sometimes, even after the most careful planning and the cleverest space-saving schemes, one room just won't do all the jobs you want it to do. Perhaps you've got a combined kitchen and dining room, but you could really do with one of each. Maybe the house needs a second toilet. Or, try as you may, you can't squeeze everyone into the available bedrooms.

In any of these situations the answer could be to build a timber-framed partition wall. That may sound daunting, but it's not. An ordinary partition — even one that stretches from floor to ceiling and right across the room — needs only simple carpentry and easily obtainable materials. You can even incorporate a door, overhead glazing, or a serving hatch without much extra trouble.

Putting together a partition is simplicity itself. One long piece of wood (the 'head' or 'top plate') is fixed to the ceiling. A second piece (the 'sole plate') is fixed to the floor. Uprights run between them; these are the 'studs', which is why the structure is usually called a 'stud partition'. Between the studs run short horizontal spacers called 'noggins'. That's the framework, and all you do after building it is to nail sheets of cladding, which are usually plasterboard, to it.

The planning stage

A partition wall will make quite a difference to your house, and it needs to be made properly. Here, as so often, thoughtful planning is the key to success.

Be careful, for example, that you do not accidentally create two narrow, gloomy cupboards. Think about lighting in particular. Only in very small rooms such as toilets can you rely solely on artificial light. Elsewhere, you may be able to 'borrow' light through windows in the partition itself. Existing outside windows may take care of the situation — but you should avoid, at all costs, the temptation to site the partition so that one window sheds half its light on each side. It will look terrible, and it's against the law.

Ventilation needs similar attention. A habitable room must either have a mechanical ventilator, or one or more ventilation openings so constructed that 'their total area is not less than one twentieth of the floor area of the room and some part of the area is not less than 1.75m above the floor'. In other words, a room 3 × 4.5m (10 × 15ft) needs a window about 840mm (33in) square; and the top of the window must always be above head height.

Another point to consider is access. You'll do well to plan the partition so that you don't have to put a new doorway in an existing structural wall. It's far less work and just as effective to include one in the partition.

You must also consider how the ceiling joists run. This is important because you'll have to fix your partition to them, not just into the ceiling plaster. They're probably spaced regularly, but you'll have to find their exact positions by tapping and making small holes, or by removing the floorboards above. If they lie at right angles (or nearly) to the intended line of your partition, there's no problem. If you want the partition to run in the same direction as the joists, think carefully. You'll have to position it directly underneath a joist, fit a new joist and fix it to that, or fit 50 × 50mm (2 × 2in) bridging pieces between existing joists at regular intervals and fix the top plate to them. Moreover, an especially long and/or tall partition may be too heavy for the floorboards alone to support — so you'll have to make similar decisions about the floor joists.

Have a look at the electricity, gas and water supplies, and see that any necessary modifications to these won't be too difficult to make.

READY REFERENCE

USING JOINTS

Although skew-nailing will make quite a strong framework, there is a slightly more complex and craftsmanlike alternative. This is to use a wide housing joint instead. Mark out the housings across both plates at the same time with a try-square.

The four corners of the frame will have rebate joints; so will the bottom corners of the door opening.

And lastly ring the local council. Unless you are converting a house into flats, you don't need planning permission. But you can't be too careful where the Building Regulations are concerned, because they deal with things like fire hazards and proper ventilation. The council should be able to tell you whether your plans conform.

When you've thought about all this and worked out a likely scheme, it's a good idea to sketch it out on paper. If there's a hidden snag, you'll find it staring at you in black and white, and you can deal with it before it causes any trouble.

Constructional details

Something you'll need to decide is how far apart the studs should be. Studs set at '600mm (2ft) centres' (i.e., with their centres that distance apart) give what is really the maximum spacing, and 450mm (18in) will make an even more rigid structure.

You should also measure and take into account the sizes of whatever cladding material you'll be fixing to the wooden framework. Plasterboard, for example, is standarised at 2440 × 1220mm (8 × 4ft) and 3050 × 1220mm (10 × 4ft). You might therefore want to arrange the studs so that there's one every 1220mm (4ft). Putting them at either 600mm (2ft) or 400mm (16in) centres would ensure this.

The door opening, of course, needs to be wider. Take its size from that of the door you plan to use, plus 3mm (1/8in) clearance either side and thickness of extra 'lining' pieces of, say, 100 × 25mm (4 × 1in) wood, fixed round its inside at top and sides. These should be wide enough to cover the edges of the cladding on both sides of the partition. A window opening should be lined in the same way (see *Ready Reference*, page 146).

It's unlikely, of course, that you'll be able to fit an exact number of whole sheets of cladding from wall to wall or floor to ceiling. So you'll need to cut some to fit. Besides, the walls and ceiling may not be dead straight or true, so you'll need to mark and cut the edges of the sheets which adjoin them, to make them fit snugly. Luckily, plasterboard is extremely easy to cut.

Noggins need only be placed 1220mm (4ft) above the floor, and again at 2400mm (8ft) if the ceiling is higher — assuming you'll be using 2400 × 1200mm (8 × 4ft) sheets.

Starting work

First, of course, you'll need to buy your timber. This is made easy by the fact that all the pieces (except for door and window linings, which are added later anyway) are the same cross-sectional size. This can be as massive as 100 × 50mm (4 × 2in), but 75 × 50mm (3 × 2in) is quite big enough for most purposes, and 75 × 38mm (3 × 1½in) will

FIXING TOP AND SOLE PLATES

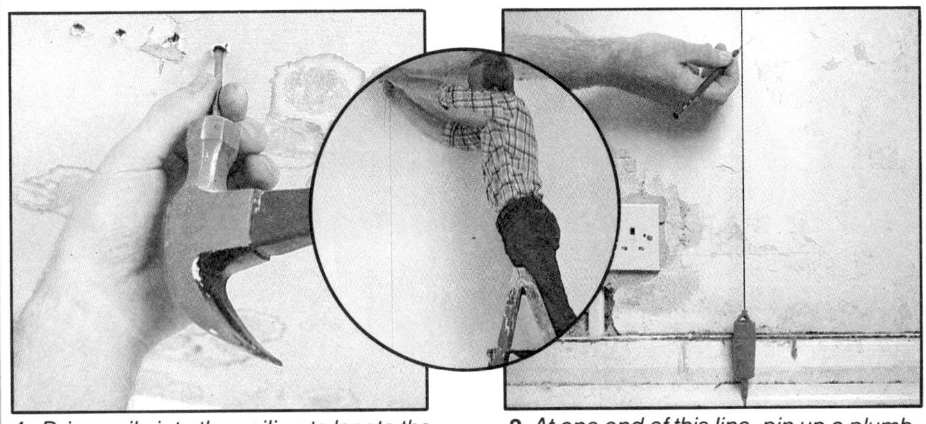

1 *Drive nails into the ceiling to locate the exact centre of the joist (or joists) to which the top plate will be fixed. Mark the new wall's position on the ceiling.*

2 *At one end of this line, pin up a plumb-line to mark the exact centre of the top plate (inset). Mark a true vertical line down the side wall to floor level.*

3 *Cut the sole plate to length. Suspend the plumbline further along the ceiling line and position the sole plate using plumbline and wall marks as guides.*

4 *Nail the sole plate in place (inset – use screws and plugs on solid floors). Then cut the top plate to length and drill screw clearance holes through it.*

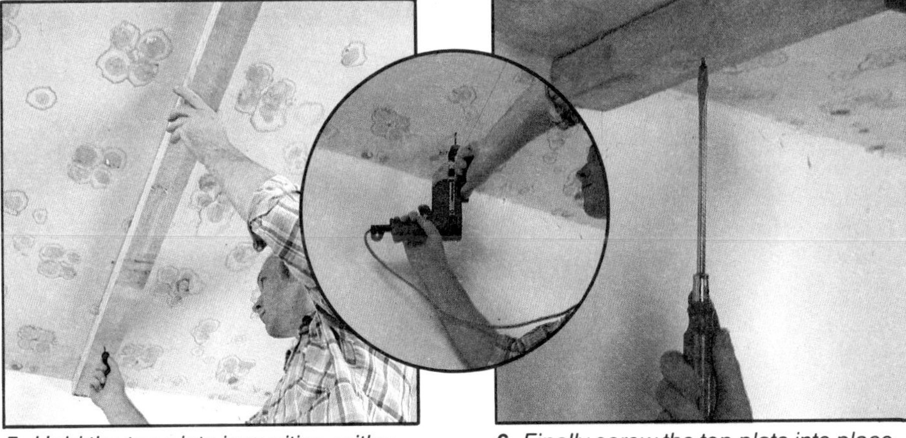

5 *Hold the top plate in position, with a helper or a stud to support it, and mark the screw positions. Drill pilot holes into the joist or joists (inset).*

6 *Finally screw the top plate into place using 90 mm (3½in) screws. Check that it is precisely aligned with the sole plate by suspending the plumbline at each end.*

PUTTING IN THE STUDS

1 *Measure the distance between the two plates, add 3 mm (1/8in) and cut the end stud to this length. Drill clearance holes, then mark the wall through them.*

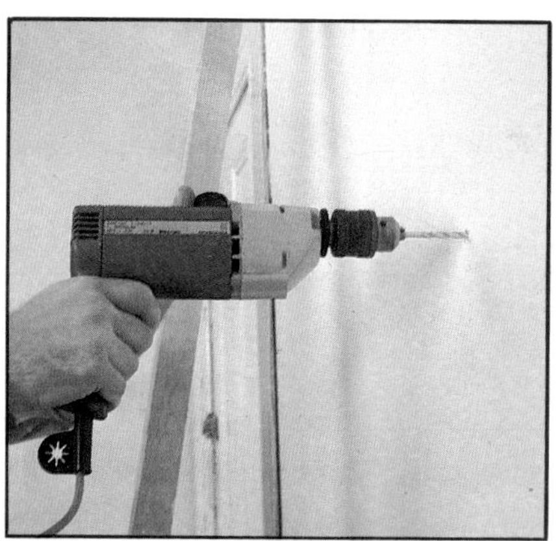

2 *Drill holes for wall plugs at each point, using a masonry drill. Be sure to drill deep enough to penetrate the masonry beneath. Insert wall plugs.*

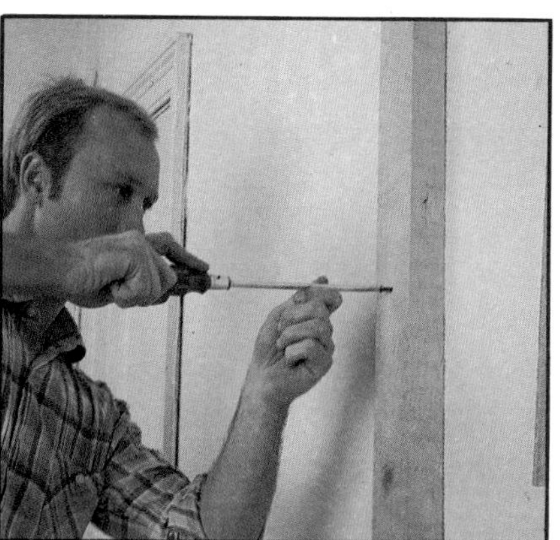

3 *Tap the stud into place – it should be a tight fit – and drive in the fixing screws. The same procedure is used to complete the other end of the framework.*

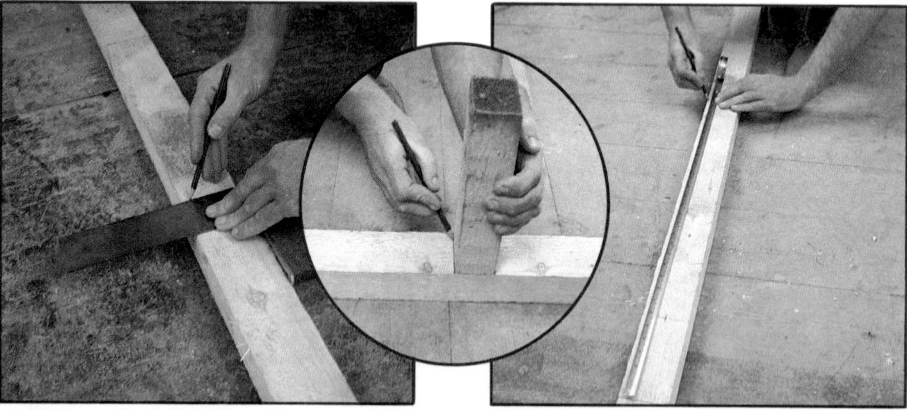

4 *Mark the stud position on the sole plate – usually at 610mm (24in) centres. Square a line across the sole plate with a try square at each mark.*

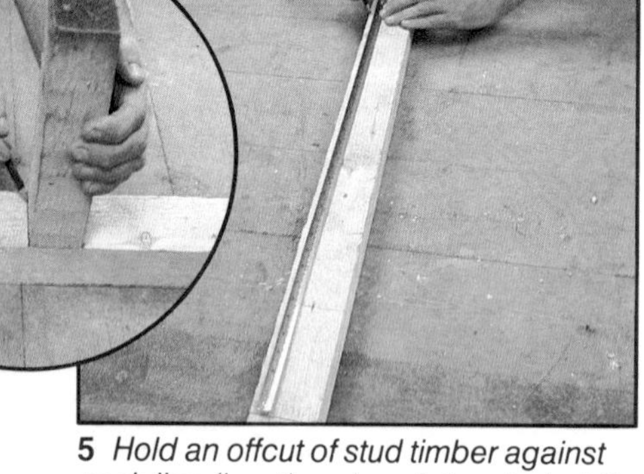

5 *Hold an offcut of stud timber against each line (inset) and mark the stud width. Leave a wider gap between studs at door openings, allowing for door linings too.*

6 *Measure and cut each stud as in* **1**, *tap it into place with its foot on the mark made on the sole plate, and check that it is precisely vertical.*

7 *Temporarily nail an offcut to the sole plate beside the base of each stud to stop it from moving out of position as you nail it into place.*

8 *Drive two nails at an angle through one side of each stud into the sole plate. Check that the stud hasn't moved, and then remove the offcut.*

9 *Drive two more nails down into the other side of each stud, and then repeat the skew-nailing process to nail the top of each stud to the top plate.*

PUTTING IN NOGGINS

1 *Mark across one edge of the studs, at the desired height, where the tops of the noggins should go. Cut the noggins exactly to length.*

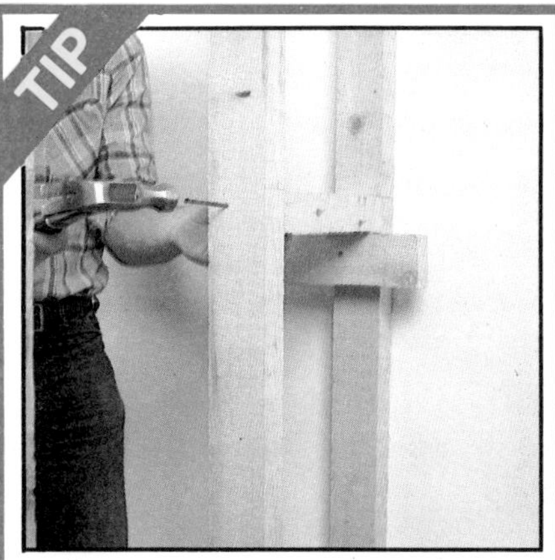

2 *Nail a steadying block, as for studs, under one end of the noggin; hammer a nail horizontally through the stud into the other end of the noggin.*

3 *Next skew-nail through the end of the noggin above the steadying block, and down into the side of the stud. Repeat the procedure for each noggin in turn.*

4 *In some cases you'll also need an upper row of noggins so that you can fix the top edge of the cladding to them. The procedure is exactly the same as before.*

5 *Each upper noggin has its centre (not top or bottom) aligned where the top edge of the cladding will come, so further cladding can be nailed to its upper half.*

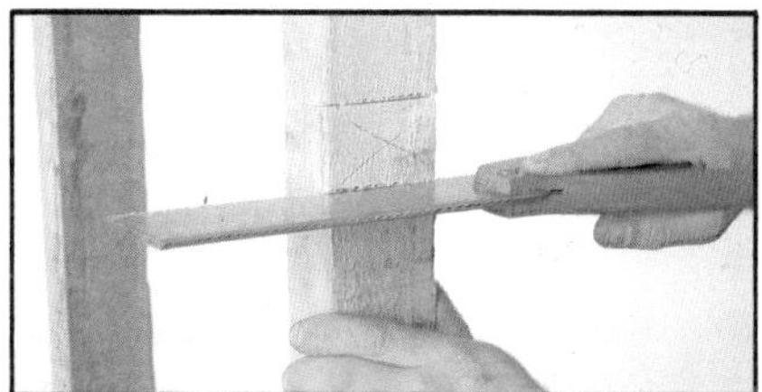

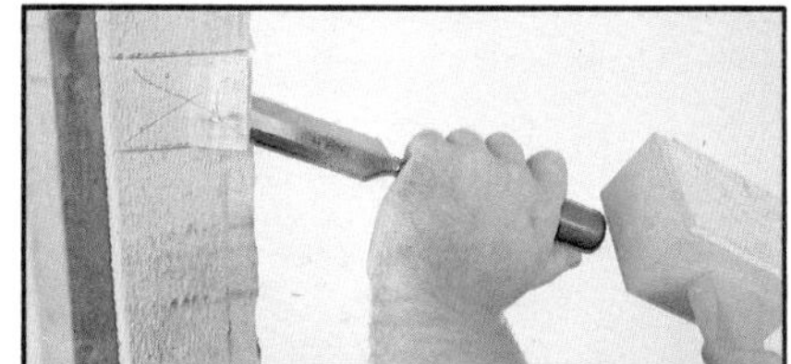

6 *The noggin above the door opening needs a firmer fixing. Use a tenon saw and chisel to cut housings (marked with the square and an offcut) across the studs.*

sometimes do for the top and sole plates. Buy ordinary softwood: it needn't even be planed smooth — just sawn.

Next, you should cut away the existing skirting board and cut or chip away the ceiling moulding, if any, so that the corners of the framework will fit closely into the angles between wall and floor and between wall and ceiling. Doing this will help to make the structure rigid and secure. However, for a light partition it's often omitted. (You need only cut away to fit the partition round skirting and moulding.)

Then cut the sole and top plates to length. (Keep each as a single piece of timber if at all possible.) Screw the top plate to the ceiling joist or joists, and use a plumbline to position the sole plate directly underneath it. Then nail or screw the sole plate through the floorboards and into the floor joist(s), or screw into a solid floor with the aid of fibre or plastic plugs.

Adding the studs

Now you can start on the studs. You'll have to measure separately for the length of each one, in case floor and ceiling aren't quite parallel. Skew-nailing using long round wire nails is a perfectly adequate way of fixing them for most purposes. You can also buy specially shaped metal connectors which you just nail into place. For an exceptionally sturdy job, cut housings at each stud position across the plates with a tenon saw and chisel, and simply fit the ends of the studs into them.

The last stage in building the framework is to cut and fix the noggins. Skew-nailing is, once again, the usual way of attaching them. They make better braces if you stagger them slightly, positioning them alternately higher and lower. But if you're going to fix the edges of the cladding to them, they'll have to be in a straight line. Either way, be careful not to make them too long. If you do, you'll probably still be able to squeeze them into position, but they'll bend the studs out of true.

The lintel (the noggin above the door opening) should be housed in the studs at each side for stability. If you are mounting cupboards on the wall, you may find it helpful to fix bearers for them in the same way.

Next, screw the door lining to its frame; the top piece is fitted to the side pieces with either rebate or barefaced housing joints already prepared by the manufacturers.

Pipes and cables

The final job before putting on the cladding (though you can do it after cladding one side) is to bore holes in studs and perhaps noggins, and run any essential pipes and cables through them. At the same time remember to nail or screw on blocks on which to mount light switches, power points etc.

See pages 162 -166 for details of plastering partition walls.

COMPLETING THE FRAME

1 *Measure between the housings, cut the noggin exactly to length, and nail it horizontally through the sides of the studs like the other noggins.*

2 *Now that the partition's framework is held straight and rigidly in position, you can cut away part of the sole plate to complete the door opening.*

3 *Finish off the door frame by screwing on lining pieces of planed timber. It's best to cut a rebate across each end of the top piece, using a tenon saw and chisel.*

4 *Then you can fit the sides of the frame into the rebates and screw them to the studs. Any windows (eg, over the door) are treated in just the same way.*

5 *Next nail the first piece of cladding to the studs and noggins, after cutting the edge to fit against skirtings, mouldings and uneven walls.*

TIP

6 *Holes for pipes and electricity cables should be drilled in studs and noggins before you complete the cladding.*

READY REFERENCE

TIPS FOR BETTER PARTITIONS

- To help you align the sole plate, you can nail the end in position before moving the other end round to centre it under the plumbline.
- You'll need at least No 10 screws for fixing to walls and ceiling; that means a 2mm (5/64in) diameter pilot hole, and a 5mm (3/16in) clearance hole through the timber.

- Studs should always be cut slightly too long, for a really tight fit; you should have to knock them in. But noggins should be just right — any less and they'll be loose; any more and they'll push the studs out of true.
- Door and window linings are made of planed timber, fitted together with rebate or barefaced housing joints, and just wide enough to cover the edges of the cladding either side. The lining is screwed to the framework. You can subsequently fit doorstop or glazing beading to its faces, and architrave mouldings over its edges to cover the join with the cladding.

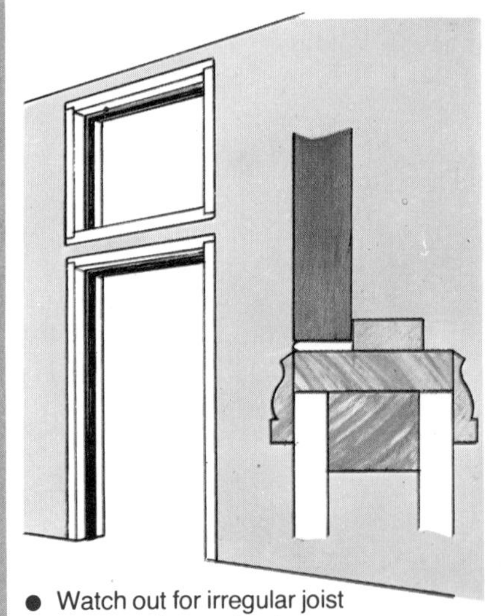

- Watch out for irregular joist arrangements, e.g. in alcoves.

CHAPTER 6

PLASTERING WALLS AND CEILINGS

Of all the skills the home builder needs, apparently the most difficult to master is the art of plastering. Watching a plasterer at work seems like observing a conjuror, yet speed is not the essence of mastering the technique. Once the various operations involved in getting the plaster from hawk to wall are understood, it's simply a matter of practice and gaining confidence. It's a skill well worth getting to grips with, since it is probably the most labour-intensive of all building operations, and once you've mastered it you will be able to tackle solid walls, stud partitions, even ceilings with ease.

BASIC PLASTERING TECHNIQUES

Patching small areas of plasterwork is a fairly straightforward job, but sometimes you'll need to replaster a whole wall. Before you can start, you have to learn the basic techniques.

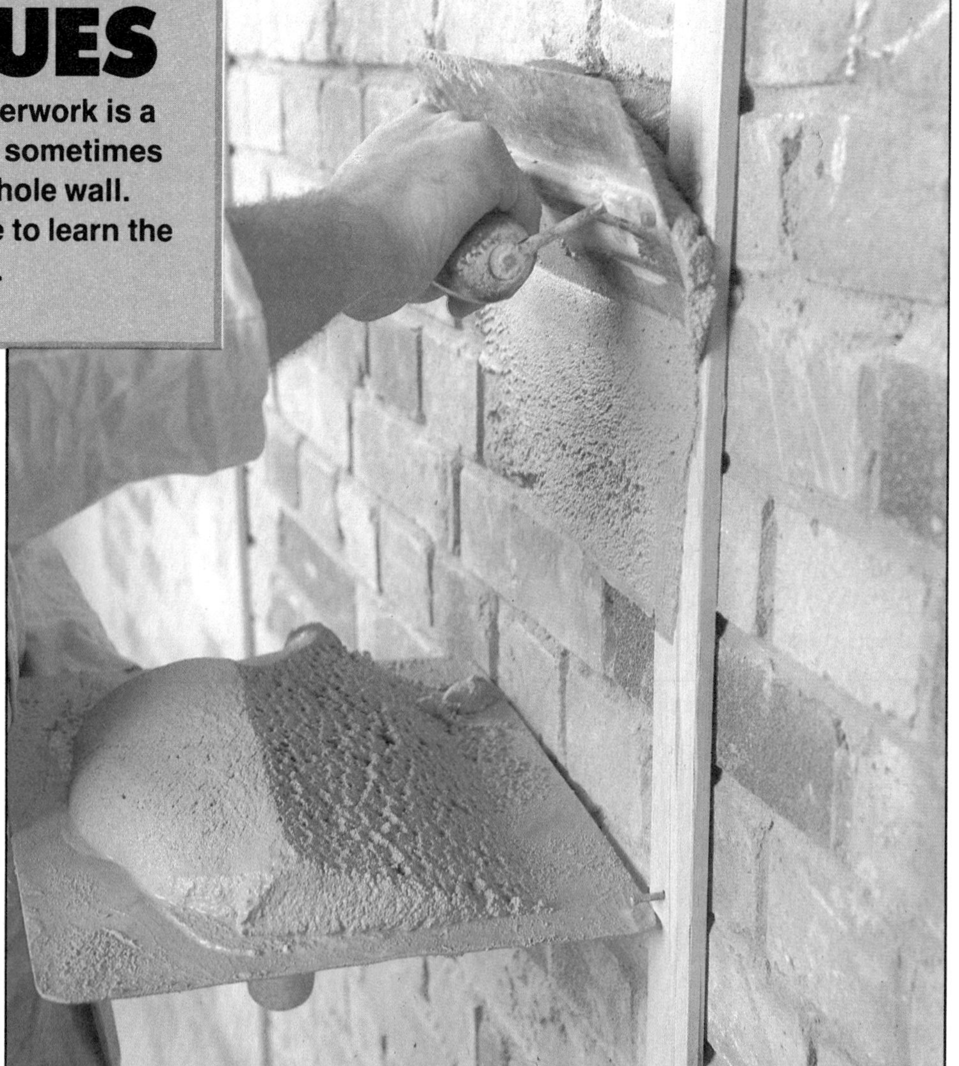

Plaster is used on internal walls to give a smooth, flat surface that you can decorate with paint, wallpaper or tiles. There are two basic types of plaster in common use. One is a mix based on a mineral called gypsum. The other, cement-based plaster, is used mostly as 'rendering' to weatherproof the exterior walls of a house. But it is also employed indoors, especially as part of the treatment of damp walls, or as an 'undercoat' for other plasters. Its disadvantages are a slow drying time and the possibility that mistakes in proportioning of the constituents could result in a weak mix.

Types of plaster

Gypsum-based plasters have largely superseded cement-based plasters. They are quicker-setting and usually available in pre-mixed form, which requires only the addition of clean water to make them workable. Another point in their favour is that they contain lightweight aggregates such as perlite and vermiculite instead of sand, so they're easier to use.

Ready-mixed plaster is usually spread onto the wall in two parts. The first is a backing or 'floating' coat, which is applied fairly thickly – up to 10mm (3/8in) – to take up any unevenness in the wall. The second is a finishing coat, which is spread on thinly – up to 3mm (1/8in) – and finished to give a smooth, matt surface.

Carlite is the most widely-used lightweight pre-mixed gypsum plaster and it's available in various grades for use on different wall surfaces, depending upon how absorbent they are: the plaster will crack if the wall to which it's applied draws moisture from it too quickly. Common brickwork and most types of lightweight building blocks, for instance, are described as having 'high suction', which means that their absorption rate is rapid. Concrete, engineering bricks, dense building blocks and plasterboard, on the other hand, have 'low suction'.

You can recognise which walls are high- or low-suction by splashing on a little clean water. If it's absorbed immediately, the wall is high-suction, but if it runs off the surface the wall is low-suction. If after this test you're still unsure, you can treat the wall with a coat of PVA bonding agent or adhesive, which, when brushed on, turns all backgrounds into low-suction, and both seals and stabilizes the surface.

For a high-suction background you'll need Carlite Browning plaster for the base coat; for low suction choose Carlite Bonding plaster. Use Bonding where the wall is of a composite nature (containing both high- and low-suction materials). Carlite Finish plaster is used as the final coat on Bonding and Browning plaster.

Preparing the surface

You'll achieve a smooth, flat and long-lasting plastered finish only if you've prepared the background properly If you're replastering an old wall of bricks or blocks, hack off all the old plaster using a club hammer and bolster chisel and examine the mortar joints. If they're soft and crumbly, rake them out and repoint them.

Lightly dampen the wall using an old paintbrush – this is essential if the new finish is to stick properly, and prevents the wall absorbing too much moisture from the plaster. New brick or block walls probably won't need any preparation before you plaster other than light wetting.

Smooth surfaces such as concrete and timber (used as lintels over doors and windows, for example), must be keyed to accept the plaster. You can do this either by nailing expanded metal laths (see *Ready Reference)* or plasterboard to them, applying PVA bonding agent, or by hacking a series of shallow criss-cross lines on the surface with a cold chisel.

Applying the plaster

Plaster is applied to the wall in a series of sections called 'bays'. These need to be marked out. One method is to use timber battens called 'grounds' lightly nailed vertically to the wall. Another method employs 'screeds', which are narrow strips of plaster. These are spread onto the wall

MIXING PLASTER

1 *Sprinkle handfuls of dry plaster onto clean water, breaking up any lumps between your fingers. Mix up equal volumes of plaster and water.*

2 *When the water has soaked into the dry plaster, stir thoroughly using a stout stick until the mix reaches a uniform consistency without any lumps.*

3 *Test the consistency of the plaster mix: Browning and Bonding plaster should resemble porridge and should be fairly stiff in texture.*

4 *When the plaster is mixed, tip it from the bucket onto your spot board, which should be positioned close to the wall you're about to work on.*

5 *Use the trowel to knead the mixed plaster on the spot board; if the mix is too sloppy sprinkle on more plaster.*

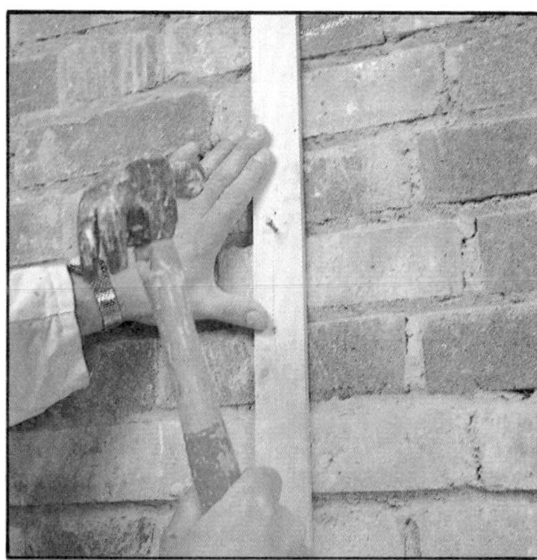

6 *Temporarily nail 10mm (3/8in) thick softwood battens vertically to the wall at 1m (3ft) spacings to act as grounds (thickness guides) during plastering.*

READY REFERENCE

TOOLS FOR PLASTERING

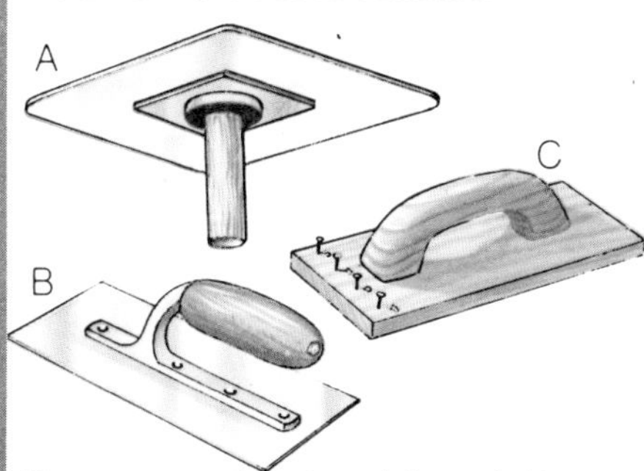

There are a number of specialist tools for plastering but the following are the basic requirements:

- hawk (A) – a 300mm (1ft) wood, aluminium or plastic square with a handle, used to carry plaster to your working area
- plasterer's trowel (B) – the basic tool for applying plaster to the wall; it has a thin rectangular steel blade measuring about 250 x 115mm (10 x 4½in) and a shaped wooden handle
- wood float (C) – generally used to give a flatter finish coat to plaster, it is rectangular in shape. It can be converted to a devilling float for keying surfaces by driving in two or three nails at one end so that their points just protrude
- rule – a planed softwood batten measuring about 75 x 25mm (3 x 1in) and about 1.5m (5ft) long, used to level off the floating coat when applied between screeds, grounds or beads
- water brush – used to dampen the wall and to sprinkle water on the trowel when finishing
- spirit level – for positioning the timber grounds accurately.

MAKING A SPOT BOARD

The spot board is used to hold the mixed plaster near the work area. Make one from a 1m (3ft) sq panel of exterior grade plywood mounted on an old table or tea chest so it's at a convenient height. Make sure it projects over the edge of the stand so you can place the hawk underneath when loading with plaster.

BUYING PLASTER

Large quantities of plaster are sold in 50kg (110lb) bags; smaller amounts for patching are sold in 2.5 to 10kg (5½ to 22lb) bags.

HOW FAR WILL IT GO?

- 10kg (22lb) of Carlite Browning laid 10mm (3/8in) thick will cover about 1.5sq m (1.8sq yd).
- 10kg (22lb) of Carlite Bonding laid 10mm (3/8in) thick will cover about 1.6sq m (1.9sq yd).
- 10kg (22lb) of Carlite Finish will cover about 5sq m (6sq yd).

USING A HAWK AND TROWEL

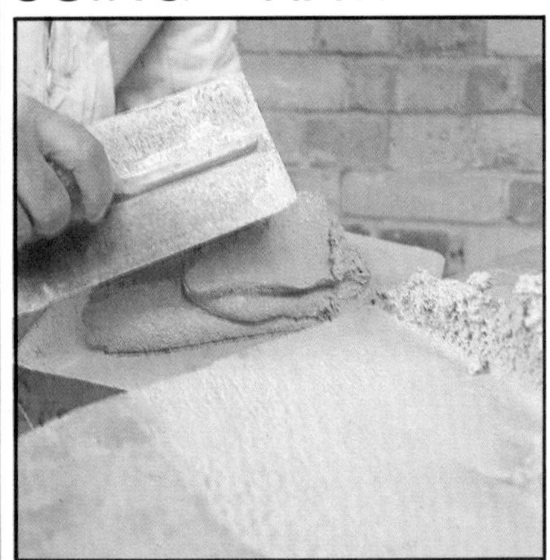

1 *Hold the hawk under the edge of the spot board and scoop a trowel-load of plaster onto it. Use the trowel to push the plaster into a neat mound.*

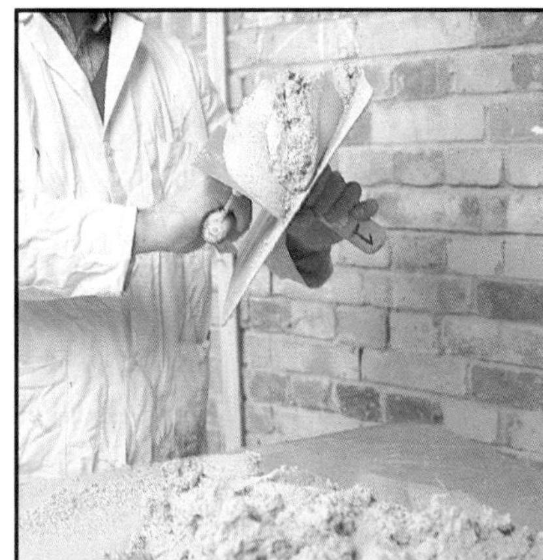

2 *With the hawk level, hold the edge of the trowel blade on it at right angles to the face. Push the trowel forward while tilting the hawk towards you.*

3 *When the hawk is vertical push up with the trowel, which should still be at right angles to the hawk face, and scoop off the plaster.*

4 *Return the hawk to the horizontal position and keep the trowel upright with the plaster on top. This whole sequence takes only a few seconds.*

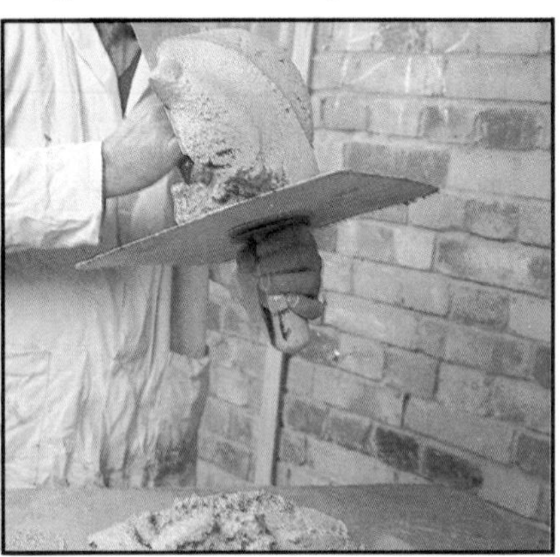

5 *Without hesitating, tip the trowel forwards to return the plaster to the centre of the hawk. Don't drop it from too great a height or it will splash.*

6 *The mound of plaster should keep a roughly rounded shape, if it's of the right consistency. Practise this loading technique several times.*

from floor to ceiling, generally using wood blocks called 'dots' at the top and bottom as thickness guides.

The distance between these markers can vary according to your skill in applying the plaster, but 1m (3ft) is an easily manageable width for the beginner. Screeds and grounds are essentially guides that enable you to apply the backing coat to the correct thickness, and when the plaster's been applied to one bay it's smoothed off level with them using a timber straight edge called a 'rule'

Expanded metal screed beads for flat surfaces and angle beads for external corners (see *Ready Reference)* serve the same purpose as timber grounds.

Applying a plaster screed to the correct thickness takes some practice and it's much easier to use timber grounds or metal beads. When you've plastered one bay using timber grounds as guides, leave the plaster until it is partially set; then remove one of the battens, move it along the wall about another 1m (3ft) and refix it.

Plaster this second bay using the edge of the first one as a thickness guide, and rule off the surface carefully. Carry on in this way until you've covered the whole wall.

To ensure the finishing plaster will adhere to the backing coat the latter must be 'keyed' using a tool called a devilling float. This is a wooden or plastic block with nails driven in from the top so that their points just protrude through the base, and it's used to scratch the surface of the backing coat lightly.

Two thin coats of finishing plaster will give a smooth and flat surface. The first coat is applied from bottom to top, working left to right if you're right-handed, right to left otherwise, and is then ruled off. The second coat is applied straight away and then flattened off to produce a matt finish. When this has been done you return to the starting point and, with the addition of a little water splashed onto the wall, you trowel over the entire surface. When the plaster has hardened, trowel the surface again several times, applying water to the surface as a lubricant to create a smooth, flat finish.

Mixing the plaster

Cleanliness in mixing plaster is of prime importance because any dirt or debris that gets into the mix could affect the setting time and mar the finish. Keep a bucket of water nearby for cleaning the tools and don't use this water for mixing the plaster – use clean, fresh tap water. See the photographs on page 149 for how to mix the plaster.

PRACTISING ON THE WALL

1 *Place some plaster on your hawk and move over to the wall you're going to plaster. Repeat the process you've just practised (left, 2 to 6).*

2 *Take about half the plaster from the hawk. Keeping the trowel horizontal, position yourself near the right-hand timber ground.*

3 *With the right-hand edge of the trowel resting on the timber ground, tilt the blade up until its face is at about 30° to the wall surface.*

4 *Push the trowel upwards, keeping an even pressure on its heel, which rests on the timber ground. Decrease the angle of the blade as the plaster is spread.*

5 *When most of the plaster has been spread, the trowel blade should be parallel with the wall. Press in its lower edge to pinch in the plaster.*

6 *Spread a second trowel-full of plaster immediately to the left of the first one, taking care not to make the layer thinner by pressing in too hard.*

READY REFERENCE

MIXING PLASTER

When mixing plaster you'll need the following equipment:

- a trough about the size of a galvanized bath for large amounts, or a tea chest lined with polythene
- a 5 gallon (22 litre) bucket for small amounts
- a clean bucket containing clean water for adding to the mix
- a bucket to transfer mixed plaster to the spot board
- a clean shovel for stirring large mixes.

TIP: FIXING GROUNDS

It's important that the timber grounds are fixed truly vertical as they're guides to the thickness of the floating coat. On uneven walls, pad out gaps between the grounds and the wall with wood offcuts to make their faces vertical.

USING METAL BEADS

Expanded metal lath and plaster beads act as a thickness guide for applying plaster to the wall and levelling it off. They're bedded on plaster dabs and remain in the wall when plastering is complete.

- Screed beads are used instead of timber grounds or plaster screeds.

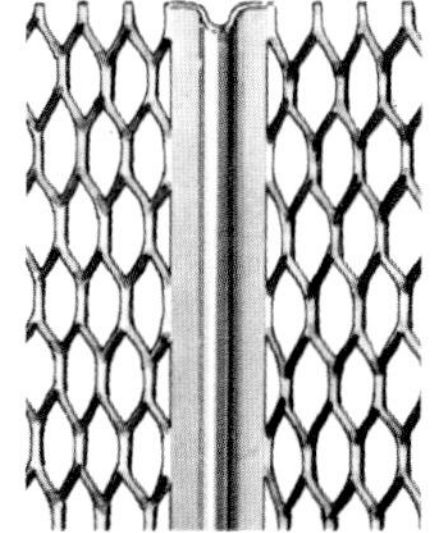

- Stop beads are used for plastering up to doorways or abutting skirtings

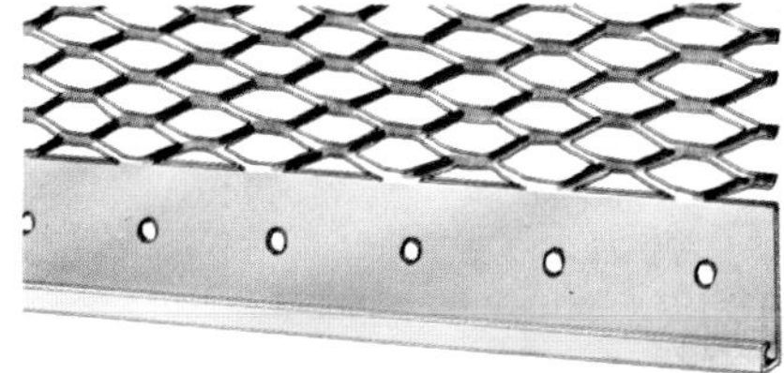

- Angle beads fit over external corners and protect the plaster from chipping.

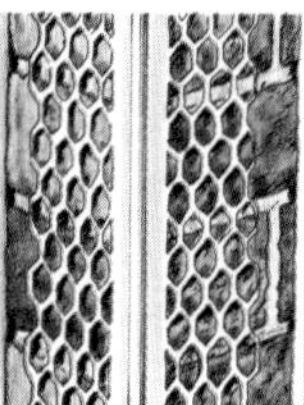

For plastering plasterboard see pages 162 -166

APPLYING THE FLOATING COAT

1 After your trial run, scrape the plaster from the wall, clean and dampen the surface, then start plastering at the bottom right of the bay.

2 Work your way up the bay, spreading on the plaster in rows. You'll have to rig up a trestle to reach the top of the bay.

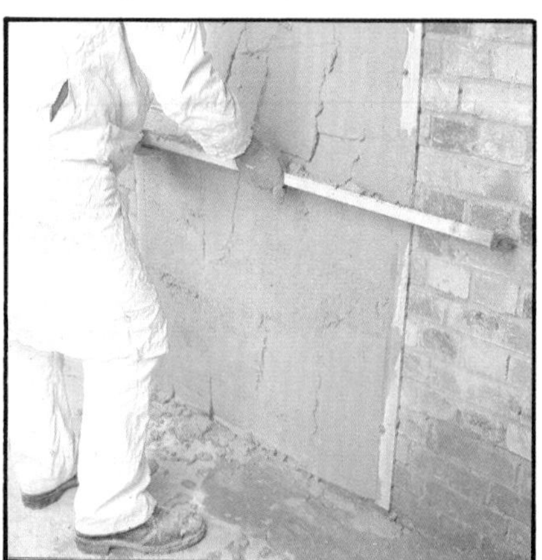

3 Use a timber batten to rule off the plaster level with the face of the timber grounds. Draw the rule upwards with a side-to-side sawing action.

TIP

4 Look for any hollows in the surface and fill them in with more plaster. Then rule off again as before.

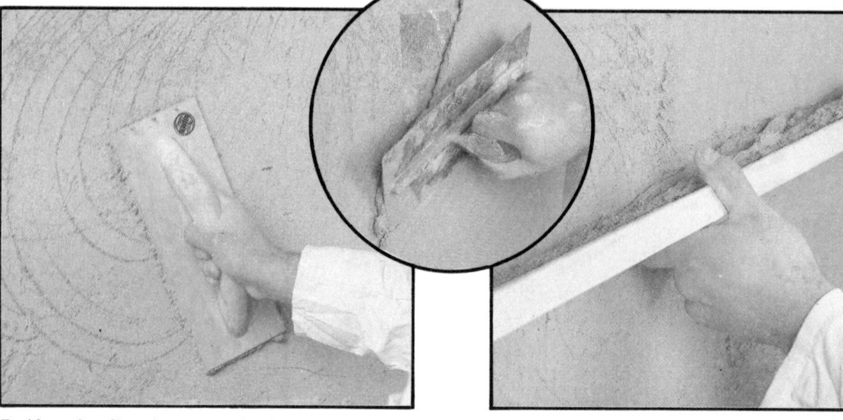

5 Key the floating coat before it sets with a devilling float. Wet its base and keep it flat to the wall so the nails score shallow marks in the plaster.

6 When the first bay has set, refix the right-hand ground 1m (3ft) away. Plaster between this and the raw edge of the first bay (inset), then rule off carefully.

APPLYING THE

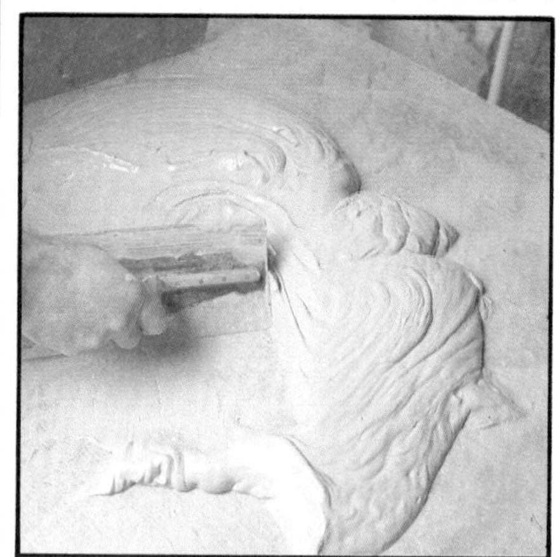

1 Mix the finishing plaster in a bucket, then transfer it to the spot board and knead as before. It should have the consistency of melting ice-cream.

5 Once you've applied the finishing coat of plaster to this area, return to the starting point and apply a second even thinner coat immediately.

Basic techniques: Floating
Patching small areas of damaged plasterwork is fairly straightforward but plastering a whole wall calls for a degree of skill in using the various tools that can only be achieved by practice.

When you've mixed the plaster place it on the spot board. If you're right-handed, hold the hawk in your left hand and the trowel in your right (vice versa if left-handed). Grip the trowel so that your index finger is against the front shank, the toe of the trowel pointing left. The knuckles of your right hand should be uppermost. The hawk should rest in the left hand on your thumb and index finger.

To load the hawk, place it under the edge of the spot board, scoop a small amount of plaster onto the hawk and move it away.

Hold the hawk level and place the bottom

INISHING COAT

2 When the floating coat has hardened – after about two hours – scoop a trowel-load of finishing plaster onto your hawk and move over to the wall.

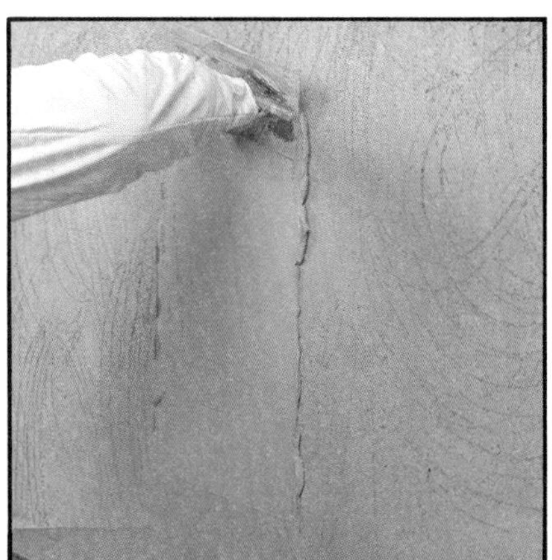
3 Spread half of the amount onto the floating coat, working from bottom to top. Keep the coat very thin. Apply the other half, blending the two.

4 Work from bottom left to top right over an area about 2m (6ft 6in) wide, with broad, sweeping arm movements.

6 Trowel off any ridges or splashes on the finishing coat with light downward strokes; hold the wetted trowel blade at about 30° to the surface of the wall.

7 When the finishing coat has hardened, trowel again using water to lubricate the blade. This polishes the surface and gives a smooth, flat finish.

8 You probably won't be able to complete a whole wall in one go. If you do break off, scribe down the plaster and scrape off to form a neat edge.

edge of the trowel on it, blade at right-angles to the hawk face. As you push the trowel forward against the plaster, tilt the hawk until it's almost vertical, keeping the trowel at right angles to the hawk face. Push the plaster off the hawk and gently slide it all back. Don't drop it from too great a height or too fast as it will splash. Repeat this several times before attempting to spread the plaster onto the wall.

When you're fairly confident, move to the wall and repeat the operation, but only remove half the plaster from the hawk. Keeping the trowel horizontal, place the lower edge hard against the wall at chest height. Open the gap between trowel and wall to about 5mm (1/4in), tilt the trowel up until its face is at about 30° to the wall surface and then move the trowel upwards. The gap is similar to a valve and controls the thickness of plaster applied to the wall. As the material is spread evenly and disappears from under the trowel, decrease the angle between trowel and wall so that you apply the last of the plaster with a pinching movement between the trowel edge and the wall. This prevents the plaster from sliding down. Repeat this until you get the plaster to stay on the wall.

After your 'practice run', scrape the plaster from the wall, and apply a 'floating' coat of backing plaster between the grounds; don't worry about any ridges or hollows at this stage but aim to get coverage of an even thickness all over.

Rule over the plaster and fill in any hollow areas, then rule again. Before the plaster has set, lightly key it with a devilling float.

Basic techniques: Finishing

Carlite plaster sets in less than two hours, so you should apply the finish coat as soon as possible after the floating coat has hardened. Use the hawk and trowel as if applying the floating coat, but take less plaster onto the hawk and apply a very thin coat to the floating coat, working left to right and from bottom to top. Smooth out all ridges to leave the surface as flat as possible. Once you've covered the undercoat, repeat the operation. Lightly sprinkle water onto the face of the trowel using a brush. With the trowel blade at an angle of 25 to 30°, trowel over the finish coat with long straight sweeps to achieve a smooth, flat finish.

Leave the plaster until set, then trowel once more, aided by water and harder pressure, to polish the surface.

PLASTERING ANGLES AND REVEALS

When you're plastering large areas, you'll have to cope with corners sooner or later. Metal angle beads make it easy to get perfect corners every time.

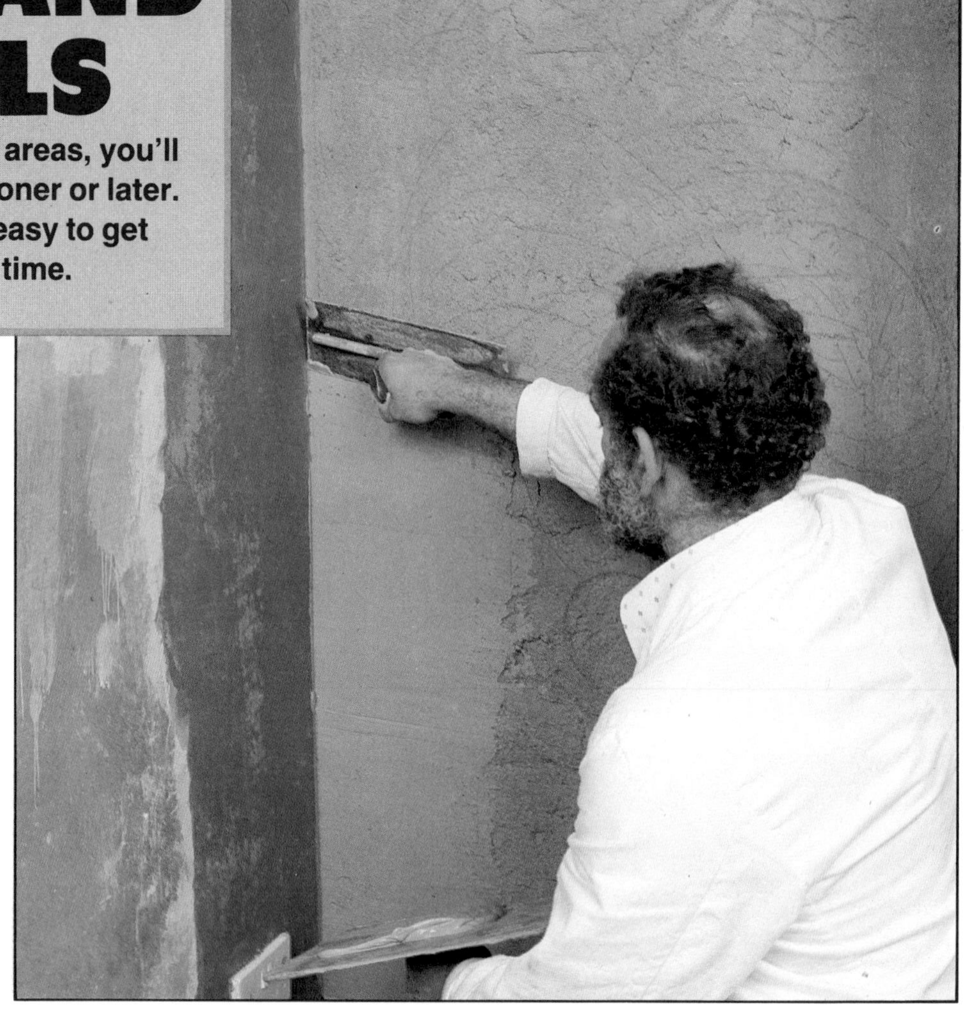

Applying plaster to walls may seem like a daunting task, but you will have seen from other sections on pages 148 -153 and 162 -166 that, providing the right techniques are practised and used, professional-quality finishes can be quite easily obtained on both exposed brick and plasterboard surfaces.

In these earlier articles, plastering was confined to flat, uninterrupted surfaces, but in practice there will usually be a certain amount of finishing off needed at internal and external corners, and around door and window openings. You will need to learn a few more techniques to deal with these, although the method of applying the plaster, and the tools for doing so, are basically the same as those detailed previously with a few exceptions.

Internal angles

You are likely to meet two types of internal angle when plastering. The first is where your newly-plastered wall meets an existing hard plaster surface on the adjacent wall, and the second is where both adjacent walls are being plastered simultaneously.

Where you have a hard surface to work to, apply your floating coat to the wall in the normal manner. Then rule the plaster outwards from the corner, using the wooden rules vertically instead of horizontally. Key the plaster well with a devilling float and then cut out the internal angle. This is done by laying the trowel flat against the finished surface so that it is at an angle of 30 to 40° to the vertical and then moving it into the corner until the tip of the toe cuts into the fresh plaster. Move the trowel up and down the angle and then repeat the procedure with the trowel flat against the floated surface and its tip against the hard plaster. This will cut out the corner cleanly. Leave it to harden.

Next, apply the finishing plaster, ruling vertically away from the angle with a feather-edge rule and cut into the corner as before. The second coat of finishing plaster should be trowelled in to form a flat surface. Just before the plaster hardens fully, pass a wooden float up and down the angle.

When you are satisfied that the angle is straight, you can finish it off. To do this, hold your trowel so that its toe is flat against the finished wall with one corner just touching the new plaster at the angle. By moving the trowel down the entire length of the corner you should be able to produce a clean and sharp internal angle.

When plastering two adjacent walls at the same time, the procedure for dealing with the internal angle is basically the same, but extra care is required because there is no hard surface to work from. You can use a special internal angle trowel for finishing off the angle smoothly, but it is probably not worth buying one unless you will be doing a lot of plastering.

External angles

Although it is possible to finish off external corners freehand, considerable skill would be needed; for the do-it-yourselfer there are two simple methods which will produce successful results without too much trouble. Probably the easiest of these is to use a metal angle bead, which has the added advantages of allowing simultaneous plastering of both walls and providing an extremely durable corner. The other method is to use a timber rule to form first one side of the angle and then the other.

The metal angle bead will provide a true, straight arris (corner) that will not chip. It comprises a hollow bead, flanked by two bands of perforated or expanded metal lath. Two versions are available (see *Ready Reference*): one that will take the full thickness of a floating and finishing coat of plaster, and another that is shallower for use with plasterboard. The latter is called a 'thin coat' bead.

The first type of beading is fixed by means of 30mm (1¼in) thick dabs of backing plaster applied at 600mm (2ft) intervals to both sides of the angles. After pressing the bead into place on the dabs, it is trued up and straightened with the aid of a straightedge and plumb line. Then the plaster dabs are allowed to harden before the floating coat is applied. Alternatively, the bead can be bedded in a thin strip of plaster running the full height of the wall as this makes truing up easier. It can even be pinned in place with galvanised nails, trued up and then secured with plaster pressed through the lath on the

FITTING ANGLE BEAD

1 Cut a length of angle bead to fit the height of the corner to be plastered, then spread a screed of Carlite Browning plaster down the angle.

2 Position the length of angle bead on the corner and bed its mesh wings gently in the screed. Don't press hard until you've set the bead correctly.

3 Using a long, straight-edged length of timber and a spirit level, check the plumb of the angle bead, and adjust it if necessary before the plaster sets.

4 When you've accurately positioned the angle bead on the corner, spread plaster over the mesh wings to secure it, then check its alignment again.

READY REFERENCE

TOOLS AND EQUIPMENT

You'll need the following tools and equipment for plastering angles and reveals:

- metal angle bead of the appropriate type, plus tinsnips and hacksaw for cutting it to length
- a hawk to hold the plaster close to the work
- an angle trowel (A)
- a steel trowel (B) for spreading and smoothing the plaster
- a devilling float (C) for scoring the floating coat
- a floating rule – a planed softwood batten about 75 x 25mm (3 x 1in) and 1.5m (5ft) long, to level the floating coat up to the angle.
- a reveal gauge for gauging the plaster thickness within the reveal, plus a try-square
- a spirit level for positioning angle beads.

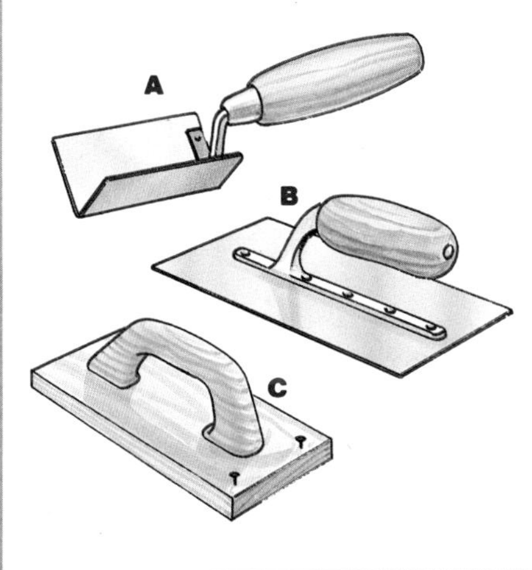

wall surface beneath. Thin coat bead is usually fitted by nailing (with galvanised nails) through the side wings into the wooden batten behind the plasterboard.

Whatever type is being used, the bead can easily be cut to length, using tinsnips to cut through the wings and a hacksaw to cut the nosing.

Once the bead is secure, it may be used as a screed for the floating coat. When this has become sufficiently hard, it should be cut back with a steel trowel to just below the level of the bead nose to allow room for the finishing coat. This coat is applied in the normal way, using the bead as a guide. When the finishing coat has been trowelled off to the angle bead, a sharp, clean and hard arris should be left exposed. The same method is used when applying the finishing coat to thin coat bead.

If you are using a wooden rule as a guide for plastering an external corner, you should nail it first to one of the walls so that it projects beyond the corner ready to act as a ground for floating the other wall. Once the floating coat has hardened on the first wall, the rule should be removed and nailed to the wall just floated to enable the second wall to be treated in the same way. Once again, you should wait until the floating coat has hardened and then remove the rule.

The next stage is to reposition the rule so that the wall floated in the first instance may be finish-coated. Wait until this coat has hardened, remove the rule once more and refix it to the second wall so you can complete the finishing coat. When this has hardened the rule may be removed.

When you remove the rule after applying the last finishing coat, you will probably find that a slight selvedge will have formed behind the rule. This should be removed by using the back edge of the laying trowel in a scything action, working away from the angle.

Sometimes you may find this tendency for a selvedge to form will allow you to finish the second wall without the aid of the rule. When you finish up to the rule on the first wall, a slight selvedge will form between the rule and the floating coat on the second wall

To finish off the plaster angle, a Surform block plane may be used lightly to produce a perfect slightly rounded corner. Alternatively, use a piece of fine abrasive paper. As with internal angles, there is a special trowel for finishing off external angles. Using angle beads or rules means you simply will not need one, however.

Door and window reveals

The narrow strips of wall at door and window openings, which are normally at right angles to the main wall surfaces, are known as

APPLYING A FLOATING COAT

1 *When the plaster retaining the bead has set, apply a floating coat between the bead nosing and the original, hard plaster at both sides of the angle.*

2 *Use a short wooden floating rule to level the backing plaster at each side of the corner. Draw the rule up the angle from bottom to top.*

3 *Place the trowel flat on the plaster, with one corner against the nosing; draw the trowel down the wall to cut in a margin for the finishing coat.*

TIP

4 *To plaster an area of wall between two corners, set a 'dot' of plaster at the base of the wall; press a small strip of wood in its centre as a thickness guide.*

5 *Lay one end of your floating rule and spirit level on the dot and the other end on the original plaster surface so that they're truly vertical.*

6 *Spread on a screed of backing plaster between the dot and the original hard plaster and level it off with your long floating rule to an even thickness.*

7 *Make another dot and screed at the other side of the area to be plastered. Spread on a floating coat between the two screeds and then rule off the surface.*

8 *Score the hardening plaster lightly with a 'devilling float', then draw the float down the angle between the original plaster and the freshly-applied coat.*

TIP

9 *Trim off the excess plaster from the angle by running the trowel down the original, hard plaster, with one corner cutting into the fresh plaster.*

FINISHING

1 *When the backing coat of plaster has set sufficiently, apply the first coat of finishing plaster, spreading it between the nosing and the hard plaster.*

2 *Apply a second coat of finishing plaster immediately after the first and smooth the surface. Use an angle trowel with a little plaster to smooth internal corners.*

3 *Trim off the excess ridge of plaster left by the sides of the angle trowel using your steel trowel. Be careful not to disturb the corner itself.*

reveals. They may also be found at the sides of a chimney breast or on a plain pier.

Actually forming the corner in this is straightforward, using the methods described previously. However, there are two points which require special attention. These are the depth of the reveals (ie, the distance from the face of the main wall to the window or door frame, or back wall) and the thickness of plaster, or 'margin', at the frame or back wall. It is essential that the depth of the reveal is the same all the way round the opening and that the plaster is the same thickness across the reveal. This will ensure that a uniform amount of frame remains visible.

Setting the depth of the reveal is simple providing the metal angle bead or fixed wooden rule for the arris is the same distance from the top of the door or window frame as it is from the bottom. Make sure, too, that it is vertical when viewed from the front.

When applying the plaster you will need a reveal gauge (see *Ready Reference*) to make sure the plaster is the same thickness all the way round. This may be simply constructed from a piece of wood which should be at least 50mm (2in) longer than the width of the reveal. After installing the metal angle bead (or fixed rule) as for external angles, lay a wooden or plastic set square (which should be long enough to reach from the back of the opening to beyond the main wall) on to the sill or floor. Push one side of its right angle against the horizontal window frame member, or bottom of the door frame, with the adjacent side against the front edge of the metal angle bead or rule. Now lay your reveal gauge on top of the square so that its long edge is in line with the set square edge that runs from the frame to the rule or bead. Using a pencil, mark the window frame end of the gauge opposite the inner edge of the window frame. Drive a nail into the end of the gauge at the pencil mark, leaving approximately 25mm (1in) protruding.

The nail acts as a shoulder when using the gauge as a horizontal rule for the plaster, supporting the inner end of the wood as you run it around the inside of the frame and maintaining an equal distance from the wall. Thus the thickness of the plaster where it meets the frame will be exactly the same all round. Or cut a right-angled notch out of the gauge to form a shoulder.

The gauge is also used to complete the underside of the reveal, although you may need to adjust the nail position if the frame is deeper at the top than down the sides. This is plastered last, after the reveal sides have been floated, and the same technique is used (with the obvious difference that you are working on a horizontal 'ceiling'). Lay on the floating coat firmly, working to the angle bead or batten, and rule off with the reveal gauge before adding the finish coat.

READY REFERENCE

TYPES OF ANGLE BEAD

There are two basic types of metal angle bead. These are:

- ordinary bead, fixed to masonry walls with dabs of plaster, which takes a floating and finishing coat of plaster (A)
- thin coat bead, nailed over the angle, which takes only a thin plaster finish on plasterboard (B).

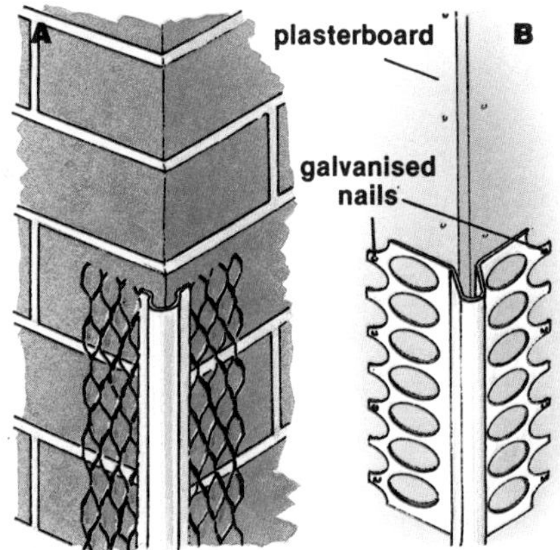

TIP: JOINING BEADS

Where you have to join two lengths of angle bead, fix the first length to the angle and insert a dowel – a short length of stout wire or a headless galvanised nail – into the nose of the bead. Then position the second length over the dowel and bed it into place.

MAKING A REVEAL GAUGE

To ensure that reveals are plastered squarely and with a uniform thickness of plaster, use a piece of softwood about 75mm (3in) larger than the reveal depth. Lay it in the reveal, against the angle rule or bead, and parallel to the blade of the square, and mark one end to match the margin on the frame. Cut out a notch and use the gauge as a short rule to level the plaster within the reveal.

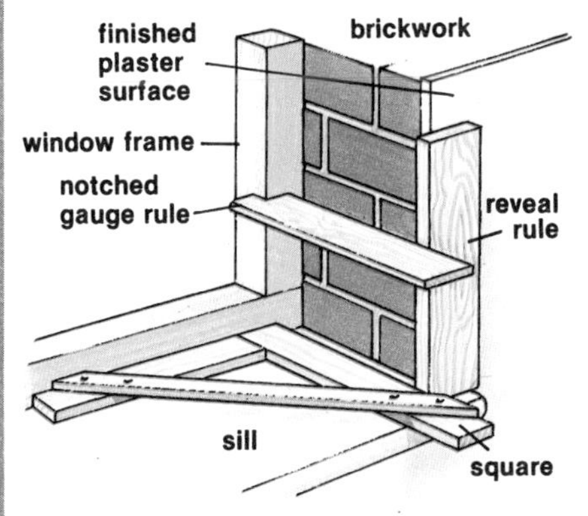

PATCHING PLASTER CEILINGS

Accidents in the loft, plumbing leaks, or simply old age can be responsible for cracks, sagging areas – or even holes – in a lath-and-plaster ceiling. But defects can often be repaired with a simple patch.

Ceilings may well be out of reach but they are, nevertheless, susceptible to damage. Accidents in the loft – a foot slipping off a joist, for instance – are a notorious cause of holes or cracks appearing. Plumbing leaks are another common source of damage, but the reason for the failure of your ceiling may simply be old age.

Types of ceiling

Modern ceilings are usually made from sheets of plasterboard nailed directly to the joists and skimmed with plaster. Accidents apart, it's unlikely that small areas of this type of ceiling will become damaged. If they do, however, you'll probably be able to remove part or all of the damaged sheet and replace it.

The ceilings of older houses, however, were usually constructed by a far more complicated method – narrow strips of timber called laths, usually measuring about 25 x 6mm (1 x 1/4in), were nailed to the joists with 10mm (3/8in) spaces between them, and the entire surface was coated with lime plaster, often mixed with animal hair fibres. The wet plaster, forced up between the laths, set to form ridges called 'nibs'. These anchored the ceiling surface in place.

Lath-and-plaster ceilings are prone to deterioration with age, and as a result of slight movement in the structure of the house. The usual signs of failure are cracks in the surface, sagging, or localised areas where the plaster has fallen away.

Whereas overall failure of your lath-and-plaster ceiling can really only be remedied by complete removal and replacement with a new one of plasterboard (see pages 132 - 136), minor damage can often simply be made good with an inconspicuous patch, using only basic plastering skills and tools (see pages 132 -136 and 148 -153). First of all, examine the entire ceiling to identify the type of damage and its extent; if the defects are widespread you may decide that replacement is the best bet in the long run.

Reaching the ceiling

So that you can work safely and in comfort close to the patch, you may need to hire slot-together platform tower sections or trestle stands fitted with sturdy scaffold boards – a ladder won't provide stable enough support. Your head should be about 150mm (6in) from the ceiling to enable you to work smoothly.

Covering a drab ceiling

If, on examination, you're sure that the ceiling isn't dangerously weak, and there are only minor damages to be made good, you could even conceal the poor finish by nailing up sheets of plasterboard to the joists, seal the joins, then plaster the surface.

Curing a sagging ceiling

The type of repair you make depends largely upon the extent of the damage to your ceiling. If the problem is sagging of the plaster in localised areas, for instance, this could be due to some of the nibs working loose, or even breaking off, separating the finished surface from the framework of laths.

You may be able to refix the plaster by forming new nibs. To do this you'll need to gain access to the top of the ceiling where the nibs are located – either in the loft or beneath the first floor floorboards. You will also need some assistance.

First of all, clean up the area where the nibs have come loose using a vacuum cleaner with a nozzle attachment. Then dampen the surface with water to prepare it for the new plaster: this prevents the plaster from drying out too rapidly, when it could crack.

Get your helper to push up the sagged area of ceiling gently against the underside of the joists using a home-made timber support (see *Ready Reference*). When the sagged area is back in place you can pour quick-setting plaster over the un-fixed area. Keep the ceiling propped until the new plaster has set fully, then remove the support carefully. Unless the ceiling is especially weak, the plaster should be sufficient to make a firm anchor.

There are basically two methods of repairing small sections of a damaged ceiling: the first applies where only the plaster has fallen

REPLASTERING SOUND LATHS

1 *Where the laths are sound, prepare the hole for replastering by scoring around the rough-edged perimeter with a trimming knife and metal straightedge.*

2 *Lever out the lumps of plaster within the scored lines using a trowel. Make the hole as regular in shape as you can and undercut the edges of the plaster.*

3 *Use the trowel to clear out the remains of the original nibs from between the laths, then brush away any dust. Dampen the laths with water.*

4 *Spread on a thin 'scratch' coat of Bonding plaster across the laths. Work away from obstructions such as electric cables, which should be insulated.*

5 *Use a tool called a 'comb scratcher' to key the surface. Scratch across the laths – not parallel to them – to avoid knocking out any plaster. Leave to set.*

6 *Apply a, thicker, floating coat of Carlite Bonding plaster to the patch, working onto the hard edges of the original plaster to make a firm bond.*

7 *Use an aluminium darby to rule off the floating coat. This will show up any hollows in the surface and will flatten any high spots.*

8 *Key and flatten the plaster with a devilling float, then spread on a coat of Finish plaster. Leave to set for 20 minutes, then apply a second coat.*

9 *When the finishing coat of plaster has begun to set, polish the surface with a dry trowel, paying particular attention to the join with the original plasterwork.*

REPAIRING A HOLE

1 *To repair a hole in the ceiling pull out the broken laths but leave any whole ones that aren't too bowed. Cut back the laths to the nearest joists.*

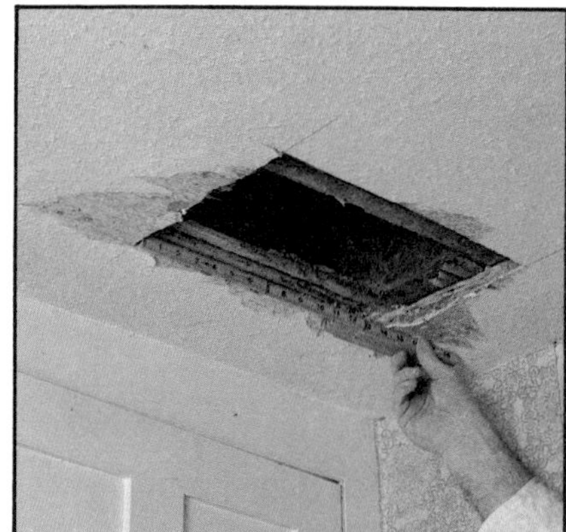

2 *Cut back the plaster to the centres of the joists, then measure the hole and cut out a piece of plasterboard to fit snugly inside.*

3 *Nail the plasterboard panel to the joists, grey side down, using 30mm (1¼in) long galvanised plasterboard nails at about 75mm (3in) centres.*

6 *Spread a coat of Carlite Bonding plaster over the entire patch, making sure it's pressed well into the edges of the hole for a better bond.*

7 *Flatten off the surface of the fresh plaster using a short wooden rule or a darby, then allow the plaster to stiffen slightly for about half an hour.*

8 *Use a devilling float to flatten the plaster to just below the finished level of the ceiling and to provide a key for the second coat of finish plaster.*

away from otherwise soundly fixed laths; the second where the laths themselves are loose, broken or even completely missing.

Plastering exposed laths

If the laths are intact and only areas of plaster have crumbled and broken away you may be able to replaster the framework as it was originally. One of the most important points to note is that the edges of the hole are sound. For this reason the first step is to cut back any loose plaster to firm fixings, making a regular-shaped hole.

Use a sharp trimming knife held against a metal straightedge to cut back the plaster It's also a wise precaution to undercut the edges of the hole to form a better bond for the new plaster.

There'll undoubtedly be some hard plaster wedged between the laths – the remains of the original nibs – and you can remove this by running the blade of your plasterer's trowel along the spaces. Brush away all dirt and dust from the laths, then dampen them with clean water to aid adhesion of the new plaster. When you apply the new plaster, the water in the mix will tend to be absorbed by the old plasterwork. This can cause serious staining to the original surface, so you can also treat the laths and the edges of the hole with a coat of PVA adhesive to improve adhesion of the plaster and minimise the suction of the background.

Spread on a fairly thin 'scratch' coat of Carlite Bonding plaster with a plasterer's trowel, working across the laths to force the plaster up between them. Use a tool called a 'comb scratcher' (see *Ready Reference*) to key the surface to accept the floating coat of plaster. Work across the laths – not along them – to prevent knocking out any of the fresh plaster, and don't criss-cross the scratches or you'll weaken the plaster key.

When the scratch coat has set, spread on the floating coat of plaster and rule it off using a darby (see *Ready Reference*). Again key the surface, this time using a devilling float and apply a coat of Carlite Finish plaster. After about half an hour you can spread on a final coat of finish plaster. When this has set, trowel the surface to a smooth, polished finish.

Always spread on the plaster away from any obstructions and towards the hard plaster edge of the hole to prevent an ugly ridge forming. Pay particular attention to this join between the patch and the original ceiling, cutting back the edges with your trowel when the plaster has started to set.

Using 'haired' plaster

When you're plastering an area of lathwork you can use a mix of sand and plaster with some fibrous material added – chopped sisal

4 Dampen the surrounding paper and the edges of the original plasterwork with an old paintbrush, to ease stripping the paper and to prepare the plaster.

5 When the paper is soft scrape back a 50mm (2in) margin around the hole so that the new plaster can be spread on to the correct level.

9 Trim up the patch using your steel laying trowel. Make sure the new plaster is recessed a little to allow for the thickness of the finishing coat.

10 Spread two coats of Carlite Finish plaster onto the patch and then polish the surface when it has begun to set, using a clean trowel lubricated with water.

is commonly used – to help bind it to the lathwork, only for the first coat. Or you can obtain special 'haired' Browning plaster, which has a fibrous content, for the first coat. For most small areas, however, this may not be an economical proposition.

Patching broken laths

Where any of the laths have worked loose you may just be able to refix them to the joists with nails, and then replaster as previously described. But if any of the laths are broken or bowed you must remove them as far as sound fixings at the joists. Cut back the plaster at the perimeter of the hole to the nearest joists, using a trimming knife. Then measure up for and cut a panel of plasterboard to fit snugly within the resultant hole.

Nail the plasterboard to the joists using 30mm (1¼in) long galvanised plasterboard nails at 150mm (6in) centres; then dampen the surrounding ceiling paper and existing plaster edges with clean water, both to soften the paper for removal and to prepare the plasterwork for the new coat. Scrape off the paper so you've a margin of bare plaster to blend the new finish into at the correct level.

Apply a coat of Carlite Bonding plaster to the patch and rule off the surface using a wooden feather-edged rule or a darby. Use a devilling float to flatten the new plaster to just below the level of the surrounding hard plaster, to accommodate the thickness of the finishing coat, and provide a key for it.

Finally, apply two coats of Carlite Finish plaster, working onto the hard edge; then polish the surface when it's set, lubricating your trowel with water splashed on from a brush.

Leave the patch to dry out thoroughly for about two or three days, before decorating with paint or paper.

READY REFERENCE

PATCHING HOLES

There are basically two ways to patch a hole in a ceiling. These are:

- to replaster existing soundly-fixed laths or
- to nail a panel of plasterboard, grey side down, to the joists where the laths are missing, and to plaster the surface.

TOOLS FOR PLASTERING

To plaster a defective area of ceiling you'll need few specialist tools. These are:

- a steel laying trowel
- a wooden feather-edged rule
- a devilling float
- a comb scratcher (A)
- an aluminium darby (B).

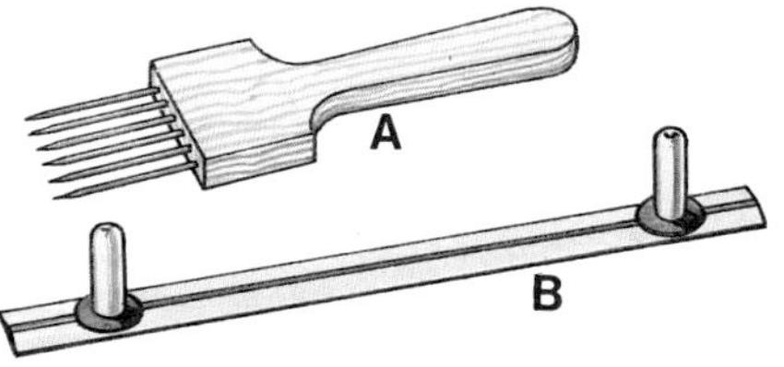

TIP: PREVENT STAINING

When you patch an area of ceiling, moisture absorbed from the fresh plaster by the existing plaster can cause serious staining. To avoid this, and greatly improve the adhesion of the new plaster, treat the laths and the plaster at the perimeter of the hole with a coat of PVA adhesive.

PROPPING SAGGING AREAS

You may be able to push back an area of ceiling that's sagged and then form new nibs. To do this:

- make a prop from 50 x 50mm (2 x 2in) softwood with a panel of chipboard screwed on top
- use the support to gently push the sagged plaster against the joists and wedge it between floor and ceiling (A)
- pour quick-setting plaster onto the laths at the top of the ceiling to form new nibs
- remove the prop when the plaster has set.

For more information on plastering see Working with plaster.
For more information on making a new ceiling see Replacing an old ceiling.

PLASTERING PLASTERBOARD

A timber stud partition makes a sturdy wall dividing up a room. You can clad it with sheets of plasterboard and finish the surface with a skim coat of plaster so it looks like an integral part of the house's structure.

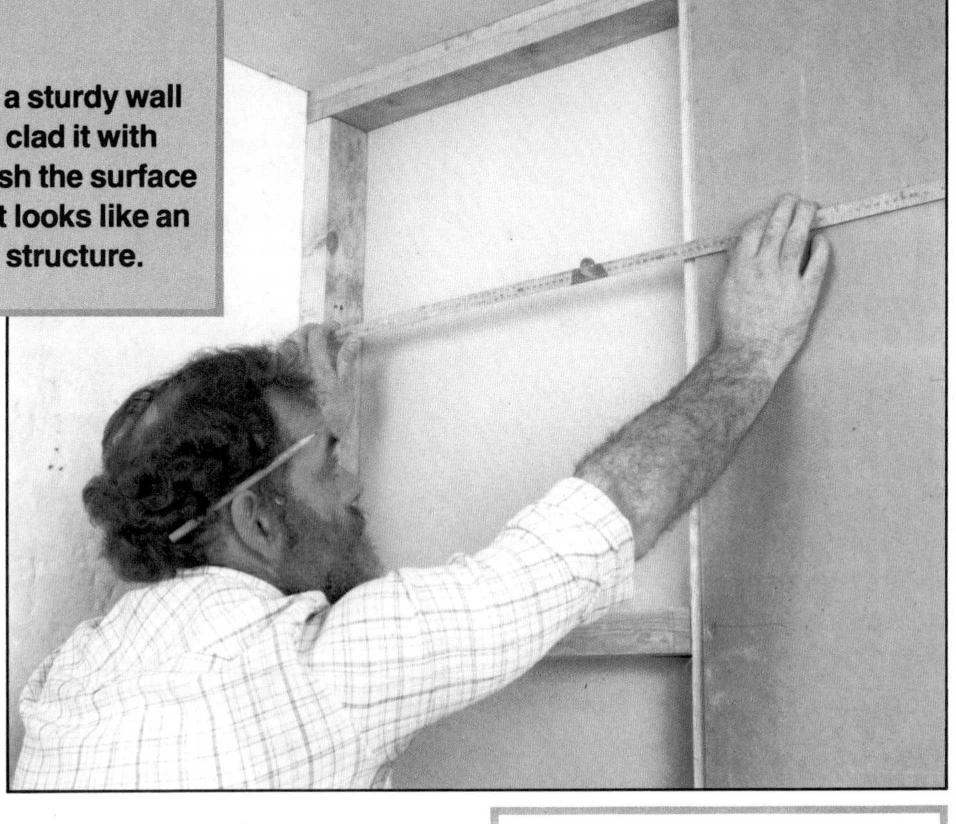

You can build a timber-frame partition to divide a room into two separate areas using only basic carpentry techniques (see Chapter 5 pages 142-146). In doing so you'll not only add areas of interest to the room but gain some extra space for putting up shelves and storing items.

Your partition can be built from floor to ceiling and wall to wall, or else simply project into the room; you can also incorporate a doorway, serving hatch or glazed areas. Timber studs form the frame of the wall and you can clad it with one of a variety of sheet materials to give a finish which can be easily decorated to match adjacent walls.

Insulating wallboards, which are made of lightweight fibre, can be used to give a surface you can paint or paper. They give good thermal and sound insulation but they're fairly soft and therefore susceptible to knocks.

You can also fix sheets of plywood and hardboard with a natural wood veneer or a plastic coating printed on one side to simulate natural wood; some even come with a decorative finish of imitation ceramic tiles. You might, however, prefer the look of real timber cladding on lengths of tongued and grooved knotty pine, for example.

Asbestos boards can also be used to clad your partition where you need some resistance to fire. They're normally used underneath other wallboards.

These materials are adequate if you want a ready-made decorative finish or a surface that you can paint or paper, but if you'd like your partition to look like a solid, integral part of the house, the best treatment is to plaster it to match adjacent walls.

To give your partition a suitable surface for plastering you'll have to nail sheets of plasterboard (see *Ready Reference*) to the studs. This is a sheet material that consists of a core of gypsum plaster sandwiched between two sheets of heavy-duty paper. There are various grades for use on ceilings or on walls that require insulation but the ones to use for a stud partition are called 'dry lining boards'.

They have a grey side intended for plastering and an ivory-coloured side specially prepared for decorating directly with paint or wallpaper.

Cladding the partition

Plasterboard cladding is nailed to the wooden framework of the partition wall so you'll have to take into account the dimensions of the sheets when spacing out the studs.

The commonest sheet size is 2440 × 1220mm (8 × 4ft), but a number of other sizes are available. Remember to space studs accurately so their centres coincide with joins between adjacent sheets; with 1220mm (4ft) wide sheets the studs should be at 610mm (2ft) centres.

If your house has very high ceilings a 2440mm (8ft) long sheet might not fit exactly from floor to ceiling height so you'll have to add a smaller panel above it. Fix extra noggins (see *Ready Reference*) to coincide with the horizontal joints in the cladding, so you can nail the boards in place.

It's best to stagger these horizontal joints in the cladding, to prevent the likelihood of the surface plaster cracking across the wall, and to stiffen the partition. The way to do this is to fix the first whole sheet at the top of the partition and fill in the gap below with a cut piece, then to fit the second whole sheet at the bottom and clad the gap at the top (see *Ready Reference*). It's a good idea to use the waste piece from the first cut sheet to fill the gap in the second row, to avoid wastage.

Leave cladding around any doorways or

READY REFERENCE

WHAT IS PLASTERBOARD?

Plasterboard is a layer of gypsum plaster sandwiched between two layers of heavy paper. There are various types available for different uses:

- dry lining boards have a grey side for plastering and an ivory-coloured side for decorating with paint, wallpaper or a textured finish

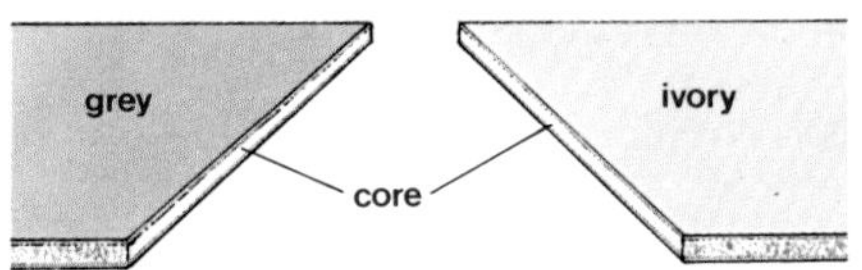

- baseboards have grey paper for plastering on both sides; they're generally used for ceilings
- insulating plasterboards have a layer of aluminium foil on one side to reflect heat back into the room and to stop moisture penetrating the partition.

PLASTER COVERAGE

One 50kg (1cwt) sack of Thistle Board Finish will make about three bucketfuls of plaster, enough to cover about 8 sq m (85 sq ft). It'll set in about 1½ hours and can be decorated after about 24 hours.

hatch openings until you've fixed all the full-width sheets you can. The best place to start fixing the boards to a half-partition is at the free end. Make sure the first sheet is flush with the end of the partition, parallel with the studs, and that the inner edge runs up the centre of an intermediate stud.

If you're cladding a wall-to-wall partition, however, you can start nailing on the boards at either end. You'll probably have to scribe and cut one edge of the first sheet to butt up with the adjacent wall accurately: few walls are truly vertical (see *Ready Reference*).

Fixing the plasterboard

You'll find that sheets of plasterboard are fairly heavy and cumbersome for one person to lift and the corners are likely to break off if knocked or dropped. When you're carrying a sheet you should grip it at each side at about shoulder height and tilt back the top so that you can walk without kicking it; if you allow it to tilt forward it'll tend to pull you over. You'll probably find it easiest to walk sideways with the board. When you reach the partition, set it down first then lean it against the wall.

To fit your first sheet of plasterboard measure the height of the partition and if it's less than the length of the board, transfer this dimension to one face and subtract about 12mm (½in). Scribe a line against a straight edge across the board using a sharp trimming knife and 'snap' back the waste piece. Run the knife up the opposite side to cut through the paper and free the waste piece.

It's important that the sheet is fixed tightly at the ceiling, so offer it up to the wall, pushing it up with a 12mm (½in) gap at the floor. There's a simple device you can use to lever the plasterboard into position. It's called a 'foot-lifter' and you can make it yourself from a small block of softwood and a thinner strip, which fits on top. It works like a seesaw, levering up the board from the floor so you can make the first fixings at both sides. When you've hammered in a few nails at each side you can remove the footlifter and hammer in the remaining nails.

Use only galvanised plasterboard nails 30mm (1¼in) long, evenly spaced about 150mm (6in) apart and no closer to the edge of the board than 12mm (½in) or there's a danger that the edge might tear away. Don't use ordinary nails that aren't galvanised or they'll rust and stain the plaster finish.

Hammer in the nail so that the head just grips the surface of the paper without tearing or punching its way through. If you don't hammer it in this far the projecting head will be visible on the finished plaster surface; if you knock it in too far so that it punctures the paper any subsequent vibrations will work the plasterboard loose.

Continue along the partition, nailing up whole sheets of plasterboard. If you find that they don't overlap the studs half-way you've made an error in the initial setting-out. This will have a cumulative effect across the wall but you won't have to remove all the boards to remedy the fault: simply cut one sheet to the required width.

Your last sheet of plasterboard will probably be narrower than a full sheet so you'll have to cut it to size. If the adjoining wall is fairly straight and vertical (test this with a spirit level) you can just measure the width of the gap, transfer this to the board and cut off the waste. Fit the board in the same way as the others. But if the wall's untrue you'll have to cut the cladding to fit the profile.

Hold the sheet against the partition, using your footlifter, and butt its edge up to the wall. Make sure that the opposite edge is parallel with the edge of the last fixed sheet or stud. There's a simple trick you can use to scribe the profile of the wall onto the face of the plasterboard: hold a small block of wood and a pencil against the wall and draw it along to mark the profile of the wall on the face of the board (see *Ready Reference*). Lie the board flat on the floor and carefully cut along the guideline with a sharp knife. Return the board to the wall and butt the cut edge up to the wall. Mark the opposite side where it falls halfway between the last stud. Cut off the waste and nail the sheet in place.

When you've covered one side of the partition you can move to the opposite side and clad that in the same way. Now's the time to add some form of thermal or sound insulation to the cavity between the two skins. You can cut strips of insulating fibreboard or sheets of expanded polystyrene to fit between the studs before you fix the second skin.

Don't forget that now is also the time to lay in any electric wiring or pipe runs that traverse the wall.

Plastering the plasterboard

You can apply a finishing coat of plaster to the partition as soon as you've nailed all the

CUTTING PLASTERBOARD

1 *Mark the height of the partition on the plasterboard, less 12mm (½in) for fitting, and scribe across the sheet with a trimming knife.*

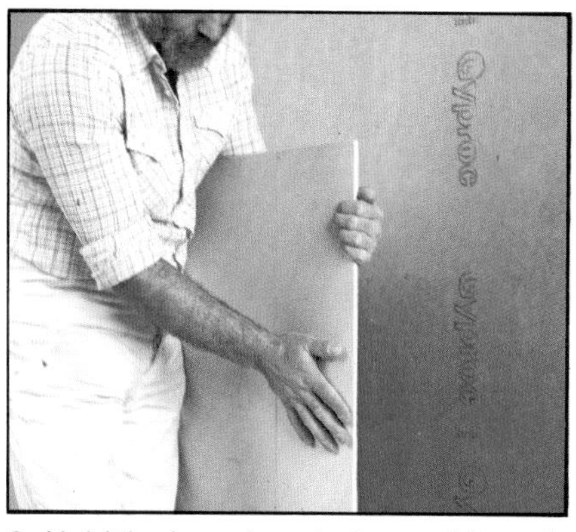

2 *Hold the board against your side and grip the top end. Slap your other hand sharply on the waste piece, pulling it back gently as you do so.*

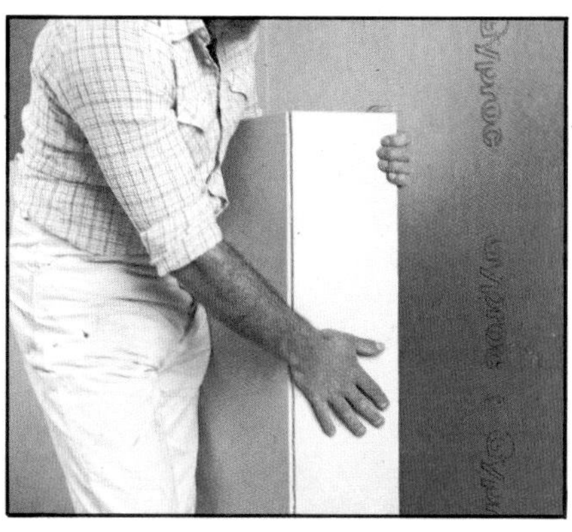

3 *The plaster core within the board should snap cleanly along the line of your cut, but the waste piece will be held by the paper lining at the back.*

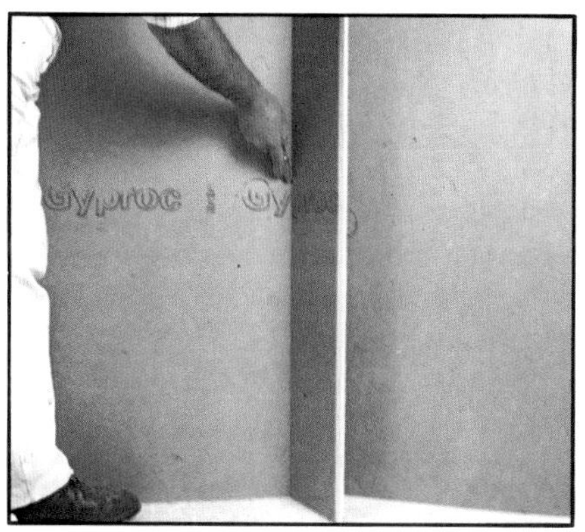

4 *Go around to the other side of the sheet and, with the waste piece bent back, run your knife up the fold in the paper backing to remove it.*

5 *You'll probably have to cut a sheet to width at the end of the partition. Measure the space at the top, centre and bottom of the partition.*

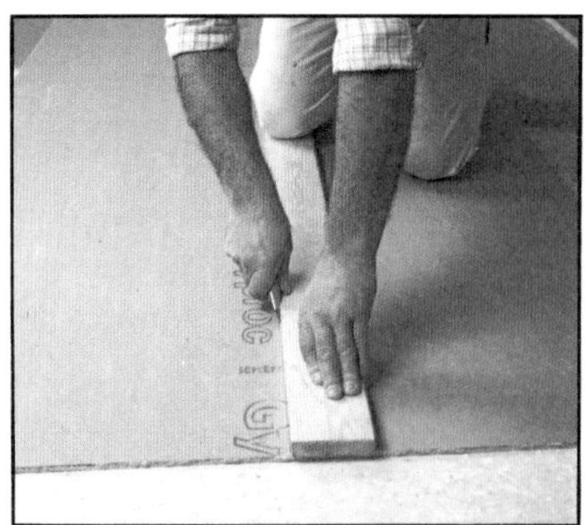

6 *If the wall is fairly even, simply transfer your measurements to the board, scribe along the side of a timber straight edge and cut off the waste.*

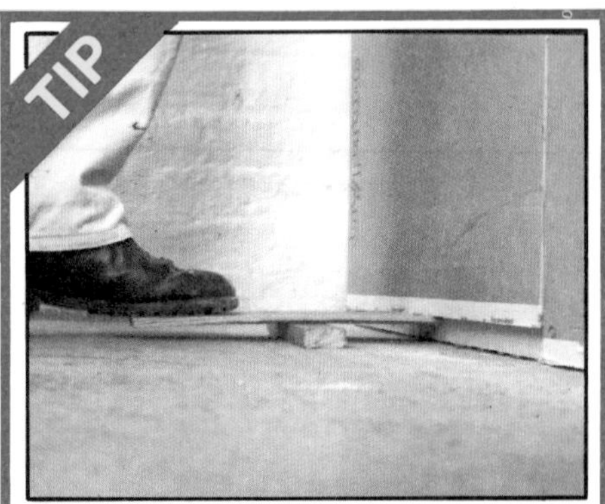

7 *To fit the board tightly into the ceiling angle you'll need a 'footlifter', made from two offcuts of wood, which you use to lever up the sheet.*

8 *You'll now have both hands free to nail the plasterboard to the studs, using 30mm (1¼in) galvanised plasterboard nails 12mm (½in) in from the edge.*

plasterboard sheets in place. When the plaster is set you can then decorate the surface with wallcovering, tiles, or just emulsion paint, to match adjacent walls.

You'll probably find plastering dry lining boards much easier than plastering a wall made of bricks or blocks because you don't need to worry about how absorbent the surface is (see Working with plaster, pages 148 -153): the heavy grey paper that lines the board does away with the need to apply a backing coat of plaster.

The correct plaster to use on an internal stud partition is a ready-mixed gypsum plaster called Thistle Board Finish. It's usually spread on to the wall in two parts; the first is intended to cover the joints between the boards and the second is a flat, finishing coat.

Using the plasterer's tools

Plastering a large area of wall requires some skill in handling the various tools (see *Ready Reference*) and the only way to acquire this is to practise. First mix the plaster; you can do this in a 14 litre (3 gallon) bucket. Half-fill the bucket with perfectly clean water – any dirt that gets into the mix could affect the setting time and mar the finish – and add handfuls of plaster until you've almost filled the bucket. Leave the mix to soak until all of the dry plaster has been absorbed by the water, then stir vigorously with a stout stick. The plaster should have a smooth, creamy consistency – like melting ice cream.

The main tools you'll need are a steel trowel to apply the plaster and a hawk to hold the plaster close to the wall. If you're right-handed, hold the hawk in your left hand (resting on your thumb and index finger) and the trowel in your right. Your index finger should be against the shank of the trowel, with the toe of the trowel pointing left. Your knuckles should be uppermost.

It's a good idea to practise loading the trowel and spreading the plaster (see photographs on pages 150 -151) on a spare

READY REFERENCE

TOOLS AND EQUIPMENT

You'll need the following tools for plastering your stud partition:

- steel trowel for spreading on and smoothing the plaster (A)
- hawk of wood or aluminium to hold small amounts of plaster (B)
- spot board to hold large amounts of plaster

- two 14 litre (3 gallon) plastic buckets – one for mixing plaster, the other to hold clean water for mixing.

WHAT IS A STUD PARTITION?

A timber stud partition consists of:

- a timber sole plate, which runs the length of the partition at floor level
- a timber top plate, which runs the length of the partition at ceiling level
- vertical timber studs at intervals along the partition, their centres coinciding with joins in the plasterboard
- horizontal timber noggins nailed between the studs to strengthen the structure and provide support for the plasterboard cladding.

CLADDING A TALL ROOM

If your house has tall rooms a standard-sized sheet of plasterboard might not fit from floor to ceiling. You'll have to cut a piece to fill the gap. Avoid a continuous line across the partition, which could cause the plaster surface to crack, by staggering the joins. Proceed as follows:

- nail the first full sheet at floor level to the studs and noggins and fill the gap above with a cut piece
- fix the second full sheet at ceiling level and fill the gap below with a cut piece; use the waste from your first cut sheet
- continue across the partition alternating full sheets and cut sheets.

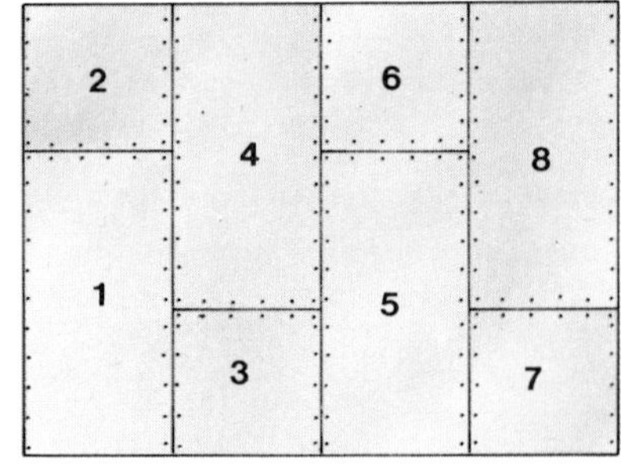

SCRIMMING THE JOINTS

1 *When you've nailed up all of the cladding, cut lengths of narrow tape (called 'scrim') to cover each joint from floor to ceiling.*

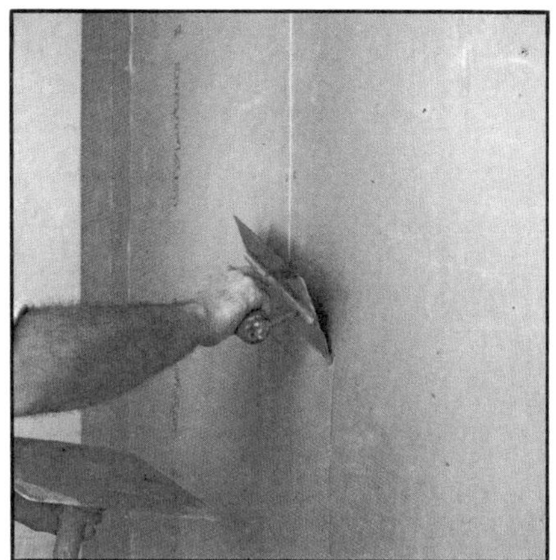
2 *Mix up some plaster and take a small amount on your hawk to the partition. Spread a thin 50mm (2in) wide layer along the joint from floor to ceiling.*

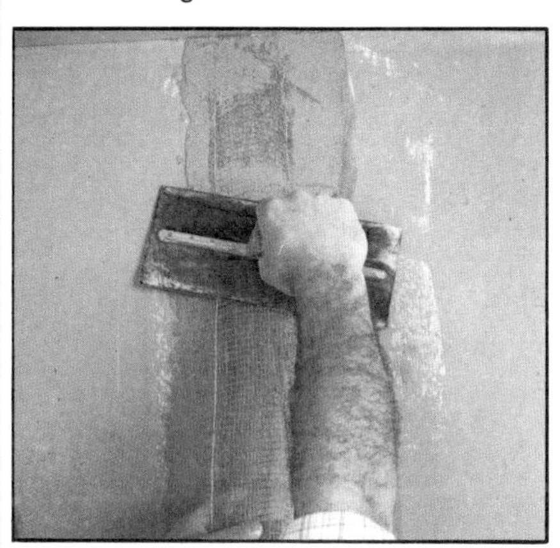
3 *Drape a length of scrim over your trowel and press it into the strip of wet plaster at the top of the partition. Then draw the trowel down the joint.*

4 *When you've laid the first length of scrim, trowel lightly over the surface from the bottom to the top to bed the tape completely in plaster.*

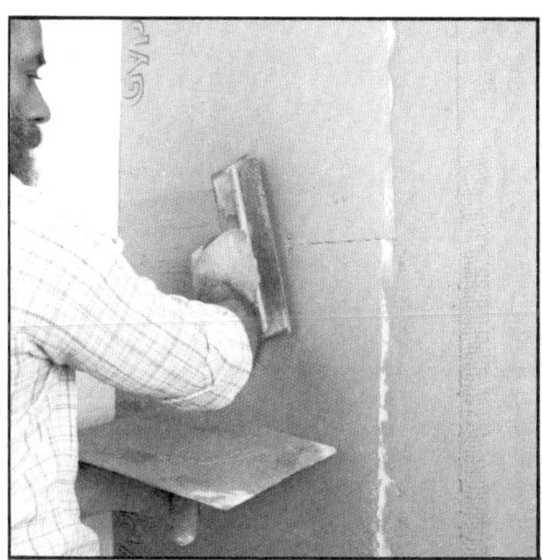
5 *You'll also have to seal any horizontal joints with scrim. Spread on the plaster as before, with your trowel blade held at 30° to the wall.*

6 *Press the scrim into the plaster and trowel it smooth. You mustn't allow an overlap of scrim where a horizontal joint meets a vertical joint.*

sheet of plasterboard until you're confident you can get the plaster to stick before attempting to plaster the partition.

Sealing the joints

Before you can apply the plaster you must seal the joints between the sheets so they won't show through on the finished surface. This is done by embedding a strip of hessian called 'scrim' in a thin layer of plaster covering each joint. Next you apply a thin coat of plaster to the board between the scrimmed joints to make the surface level again. The finishing coat is applied after this to conceal the joints, and you can polish it to a smooth, matt surface.

Scrim is sold by builder's merchants in rolls 100m (330ft) long × 75mm (3in) wide. It's best to cut your scrim to the length of the joints before you mix your plaster. If there are any horizontal joints in your wall you'll have to scrim these also. You mustn't fold or overlap the scrim as double thicknesses will prevent the plaster from sticking properly and will cause unsightly bulges on the finished surface. Where a horizontal joint meets a vertical joint you'll have to butt up the strips.

To stick the scrim to the wall, spread a thin 100mm (4in) wide strip of Thistle Board Finish along the first joint with a steel trowel. Drape one end of the scrim over the top of your trowel and position it on the screed at the top of the wall: you might need a stepladder to reach the top. Keep the trowel blade at about 30° to the surface of the wall and draw it down the joint, feeding the scrim on to the plaster strip with your free hand. Don't press too hard or you'll drag the scrim down the wall and might even tear it. When you've positioned the first strip, pass the trowel lightly over the joint from the bottom upwards to embed it in the plaster. Scrim the second joint in the same way, forming a 'bay' between the two joints. When the plaster has begun to set it'll turn from dark pink to light pink in colour and when this happens you should spread a thin layer of plaster over the bay, working from the bottom left hand side of the wall. This will bring the whole plastered surface to the level of the scrimmed joints. Scrim the remaining joints and plaster the bays between them.

By the time you've applied the first coat of plaster to the entire wall the surface will be set hard enough to acept the second, finishing coat. Apply an even layer of plaster 4mm (just over ⅛in) thick to the entire surface of the wall, again working from the bottom left, but this time make your strokes long, light, and sweeping to avoid ridges in the plaster.

When the finishing coat has almost set, go back over the area with your trowel – without any plaster – to give a smooth finish. When it's completely set, trowel again but splash a little clean water onto the wall from a brush to lubricate the trowel and create a polished and perfectly smooth surface.

PLASTERING THE PARTITION

1 *Spread on the plaster in the bays between the scrimmed joints. Work from the bottom left, spreading a very thin layer of plaster.*

2 *You'll have to spread the plaster from the ceiling angle downwards to make a clean, neat edge and to avoid smearing the ceiling with plaster.*

3 *Complete the bay, applying a very thin coat of plaster, if required, from bottom to top. Try to avoid creating a build-up of plaster over the scrim.*

4 *Plaster the subsequent bays with a thin coat; by the time you've finished the plaster will have set enough to accept the finishing coat.*

5 *Spread on the finishing coat over the entire surface of the wall, working with long, sweeping strokes to remove any ridges in the plaster.*

6 *When the plaster has set trowel over it to polish and flatten the surface. Repeat this several times, lubricating the wall with a little water.*

READY REFERENCE

FITTING TO AN UNEVEN WALL

If the wall which your timber stud partition meets is uneven or not truly vertical you'll have to cut the sheet of plasterboard to fit its profile. To do this:

- lightly tack the plasterboard against the partition, one edge butted up to the uneven wall, the other vertical
- place a small block of wood and a pencil against the wall, resting on the plasterboard
- draw the block and pencil up the angle to mark the profile of the uneven wall on the face of the plasterboard
- lie the board on the floor and cut along the guideline using your trimming knife
- return the board to the partition and butt up the cut edge to the uneven wall
- mark on the other edge of the board the position of the stud centre
- cut off the waste and nail up the board.

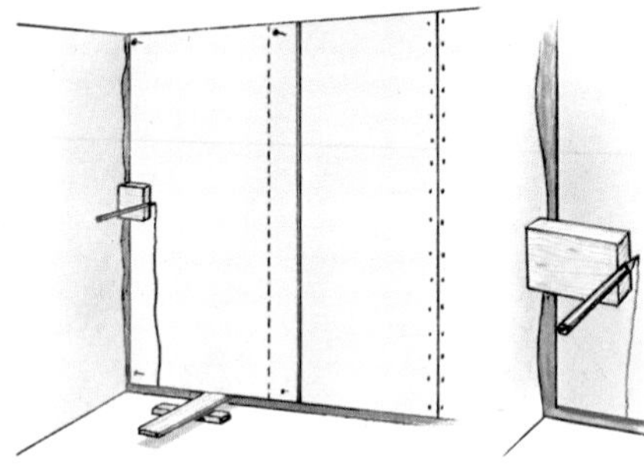

FINISHING HALF-PARTITIONS

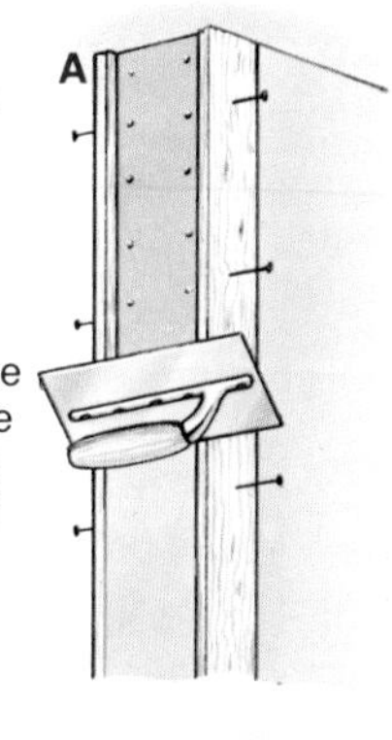

To plaster the end of a half-partition:

- nail a plasterboard strip to the end stud
- scrim the external corners
- tack thin battens to the end, projecting over the corners by 3mm (1/8in)
- plaster both faces of the partition using the battens as thickness guides
- remove the battens when the plaster's set. Refix them to the faces of the wall, projecting 3mm (1/8in) over the end
- plaster the end (A)
- or bed 'thin coat metal angle bead' in plaster on each corner and use its 'nose' as a thickness guide (B).

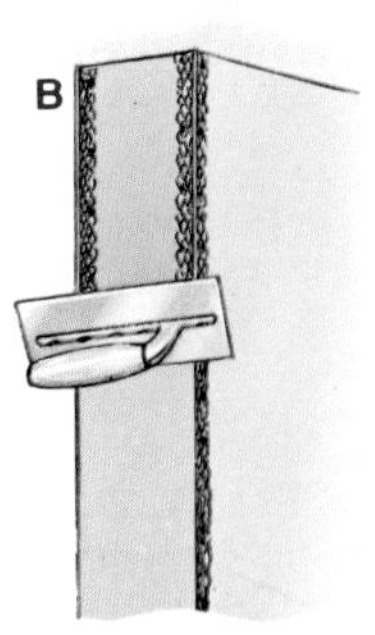

CHAPTER 7

MATERIALS, TOOLS, RULES AND REGULATIONS

There's no doubt that the right materials and equipment contribute a great deal to the success of any building job. In this chapter you will find everything you need to know about the ingredients for mortar and concrete, how to choose bricks, blocks and paving slabs, how to create safe working platforms for all sorts of building work, what tools to select and how to protect yourself from accidents. Lastly, there's vital information about all the rules and regulations you need to follow to ensure that your work is built to last...and legal.

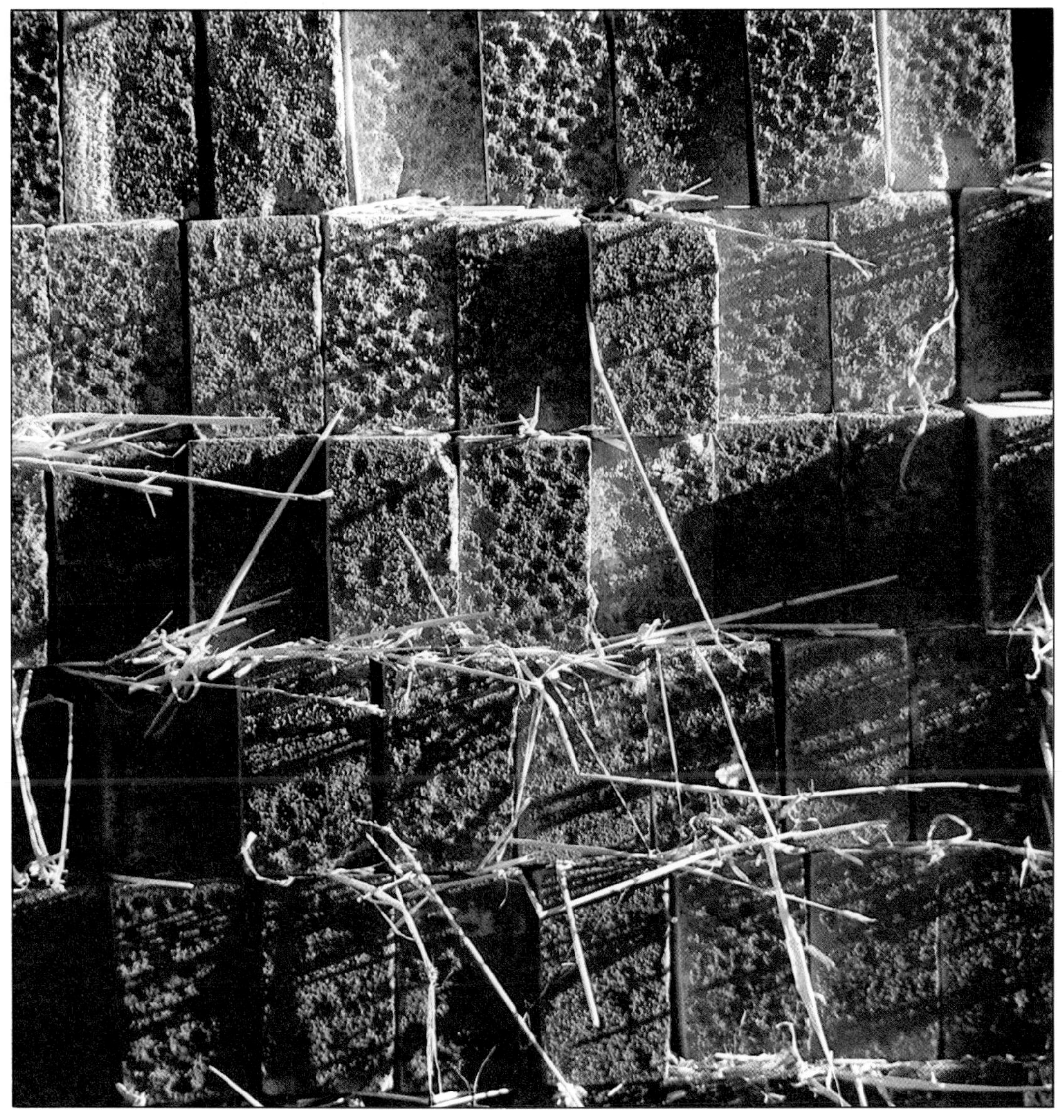

BRICKS

You may think bricks are all the same. They're not. There are thousands of different bricks available, and it really does pay to choose the bricks that are right for the job you're doing.

Bricks are one of the most versatile of all building materials. To make the most of their qualities though, you must choose the right one for the job. Basically, two things affect your choice: how the brick is to be used and whether the final appearance is important.

For most do-it-yourself jobs, strength can be ignored. Even 'weak' bricks are more than adequate for, say, a garden wall, or a small outhouse. Similarly it's not important if bricks have slots, holes, or 'frogs' in them — unless the top face is visible. These are there to make the brick lighter and key with mortar better.

What you must not ignore if you're laying bricks outside is their weather resistance. Bricks are divided into various 'qualities' according to their ability to resist extremes of temperature.

Internal quality bricks have no weather resistance at all. As their name implies, they can only be used indoors — outside they would quickly disintegrate.

Ordinary quality bricks are suitable for exterior use, but will not stand severe exposure to the elements. This means they can be used to form the bulk of a free-standing wall, but not its coping. Ordinary quality bricks are also unsuitable for retaining walls and for brickwork underground. Here, the almost complete weather resistance of *special quality* bricks is required. These are very dense and very durable.

In terms of looks, if the brickwork is to be covered with rendering, plaster, or some form of cladding, it doesn't matter what they look like. Where appearance is important, you'll find a host of 'facing' or 'faced' bricks for use indoors or out.

Engineering bricks

Technically, engineering bricks are all special quality, but they are really more than that. They are extra special: very hard, very strong, regular in size and colour, and almost completely impervious to water. Their major drawback is that they are expensive, so reserve them for situations where their virtues are really needed (eg, lining manholes, building an indestructible damp proof course into a wall, and so on). They come in two classes — A and B — which describe their exact strength and water resistance, but there are few DIY jobs where the choice is critical. You may also find bricks described as semi-engineering, but since there is no recognised definition of this type, you need to ask to find out what they are recommended for.

Commons

Bricks described as commons aren't meant to be beautiful. They are rough looking and vary considerably in colour, but they are relatively cheap and are normally either internal or ordinary quality. Some are sufficiently attractive to be left exposed (in which case they need protection by a coping). But in most cases they're used in situations where they will be covered up with some form of cladding which shields the bricks.

Faced and facing bricks

Both faced and facing bricks are designed to be put on display. The difference between them is that while faced bricks have only one or two sides that are presentable, facing bricks are attractive no matter which way you look at them.

Facing bricks made from clay are available in a variety of colours — reds, yellows, greys, blues, blue/greens, etc — and a combination of colours (called multi-coloured bricks). They also come in several textured finishes; some with commonsense descriptive names ('sand faced', 'rustic' and so on); some with names that indicate the methods of making ('hand mades', 'wire cut' etc). They offer the widest range and the best way to decide which you want is to go and look at the bricks themselves — they are available in ordinary and special qualities.

Standard 'specials'

A wide range of these bricks is made to coordinate with standard bricks. They're designed to give either protection (eg, copings) or a finishing touch to the top of a wall or an end (eg, single, double or left and right hand bullnose). Radial headers and stretchers allow you to create a curve without cutting bricks or making the mortar joints thicker. These plus plinth headers and stretchers are the ones most useful in DIY work, and if not stocked can be ordered at builders' merchants. Most are available in ordinary or special quality; copings should always be special quality.

Calcium silicate bricks

These are made from either a mixture of sand and lime, or flint and lime. They're the same size and used in exactly the same way as clay bricks. Because of the way they're made, they are much more uniform in colour and size than clay bricks. There are a total of 6 classes available, but, as with classes of engineering bricks, most of them are to do with the bricks' strength. Basically, they fall into two categories: those designed to be on show (facing bricks), and those that aren't. Sandlime and flintlime bricks contain no soluble salts and efflorescence can only occur if salts come from the ground or materials stacked against them. For severely exposed places, specify class 3 or 4.

Concrete bricks

These are similar to calcium silicate bricks but are made from either portland cement or sulphate resisting cement. Concrete bricks are as uniform as calcium silicate bricks and are classified in exactly the same way. Good quality concrete bricks closely resemble the clay bricks they are designed to imitate.

Faced bricks

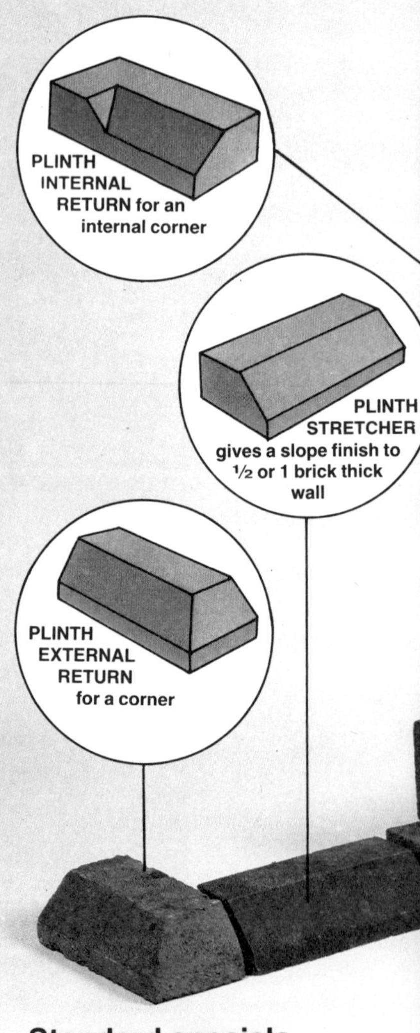

Standard specials

Standard bricks

Various brick types and colours are manufactured in different areas. If you select those from a local range they'll be more likely to blend with the colour and character of buildings in the area, and your delivery cost will also be lower.

Commons **Facing bricks** **Engineering bricks** **Calcium silicate bricks** **Concrete bricks**

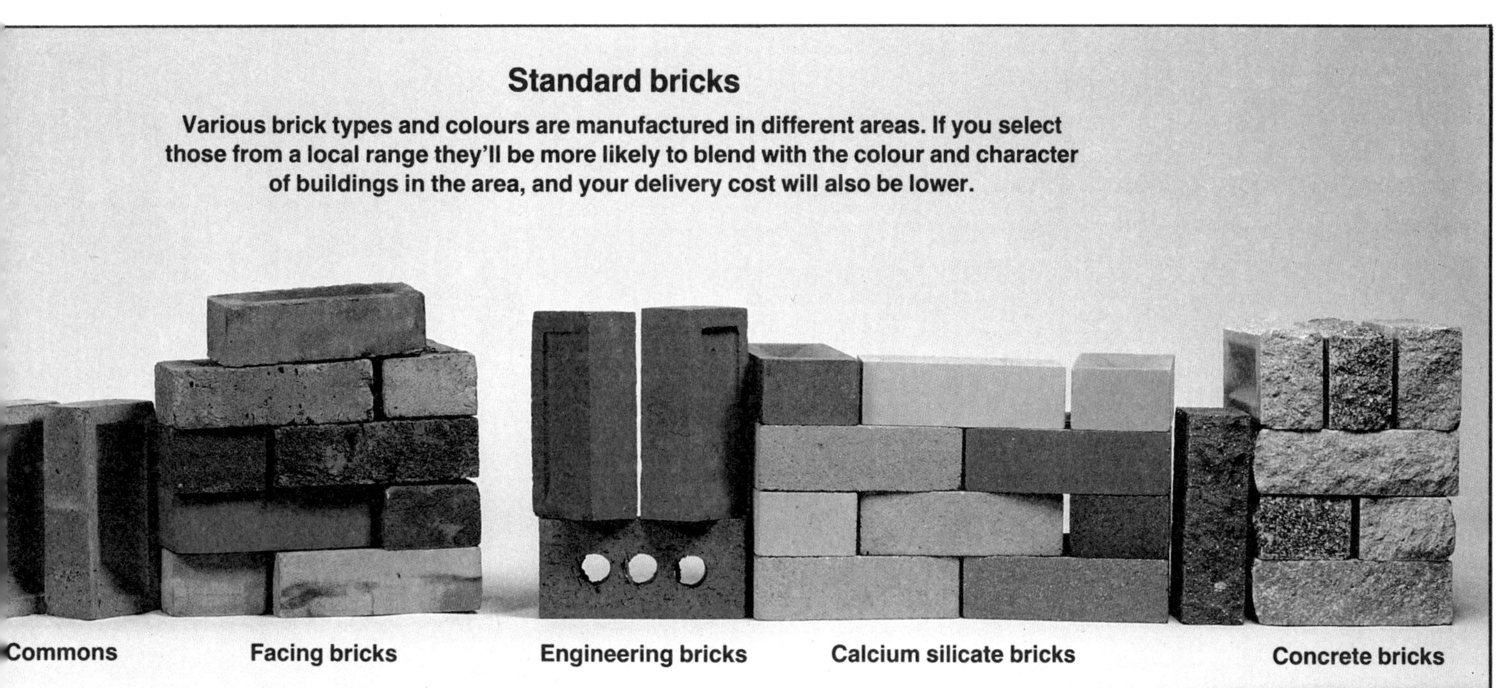

DOUBLE BULLNOSE used as a capping for 1 brick thick wall
CONCRETE COPING to protect ½ brick thick wall from rain
CLAY COPING protects 1 brick thick wall
BULLNOSE HEADER ON FLAT same as plinth bricks
PLINTH SHORT INTERNAL RETURN used instead of internal return
BULLNOSE EXTERNAL RETURN ON EDGE for a stopped end
AIR BRICK placed in wall to allow through ventilation
HALF ROUND COPING used for rounded capping on 1 brick thick walls
SADDLEBACK COPING for a pointed capping on 1 brick thick wall
PLINTH EXTERNAL RETURN for a right-angle turn at a corner
COWNOSE (BULLNOSE ON END) used to end a wall where sharp edges would be dangerous

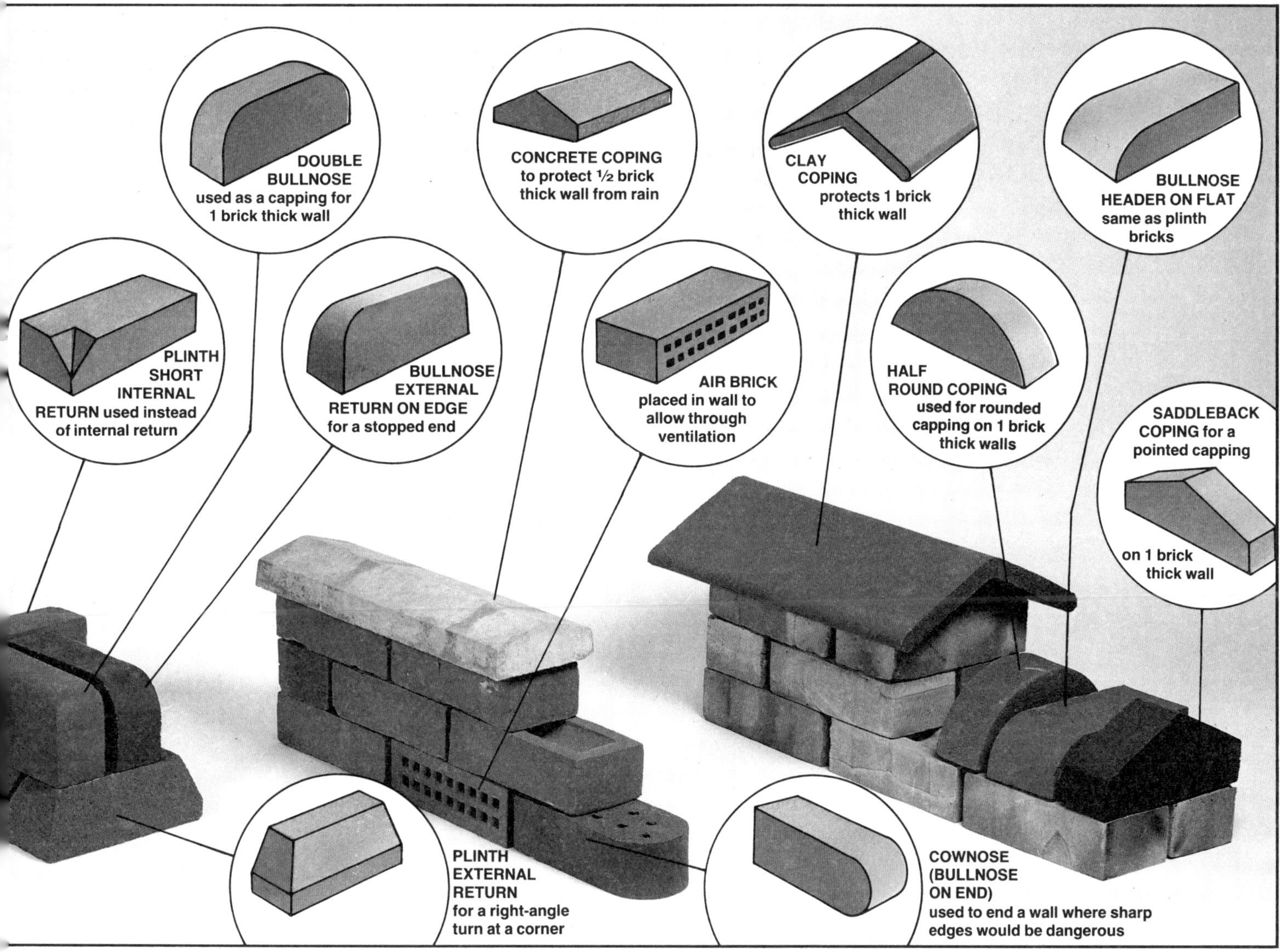

CEMENT, SAND AND AGGREGATE

Mortars and concrete are precise mixtures of cement, sand, aggregates and other additives. It's important to use the right ingredients for the best results.

You can't do much in the way of building work without having to mix up some mortar or lay some concrete, and if you're going to get good results you have to know the difference between the various types of cement, sand, aggregates and other additives you'll need. First, cement.

Cement

Cements are used to bind the sands and aggregates of a mortar or concrete mix together to give it strength. They set by the action of water, and for this reason they must be kept dry until they are used. Cement that has been stored in damp conditions and has partially hardened into lumps will not set properly and should not be used. Cements are usually sold in 50kg (1cwt) bags, but smaller sizes are available.

Portland cement is not a brand but a type of cement, and is made by several manufacturers. There are two commonly-used types.

Ordinary Portland cement (commonly abbreviated OPC) is the least expensive and most widely-used cement and is suitable for all normal purposes.

ordinary Portland cement

White Portland cement is made from white raw materials. It is used for making white mortars for pointing brick or stone walling, and for making white concrete for special decorative work. It is comparatively expensive.

white Portland cement

Masonry cements are used for making bricklaying mortars, for bedding tiles and for backing renders for decorative wall finishes, including roughcast and pebbledash. They are not suitable for making concrete.

A masonry cement is not all cement; there are other materials added to improve the mix. It needs only the addition of sand to give a good mortar, and therefore it is simple to use. You can also add colouring pigments to change the mortar colour, but nothing else. Masonry cement is sold in 50kg (1cwt) bags.

Lime

Mortars and rendering mixes made with Portland cement and sand only are not ideal because they are too strong for general use and can be difficult to work with. Lime can be added to reduce the strength and to greatly improve the workability of the mix. It retains water well, which assists the proper hardening of the cement. The addition of lime to a mix also gives a degree of flexibility to the hardened material, making it less likely to shrink and crack.

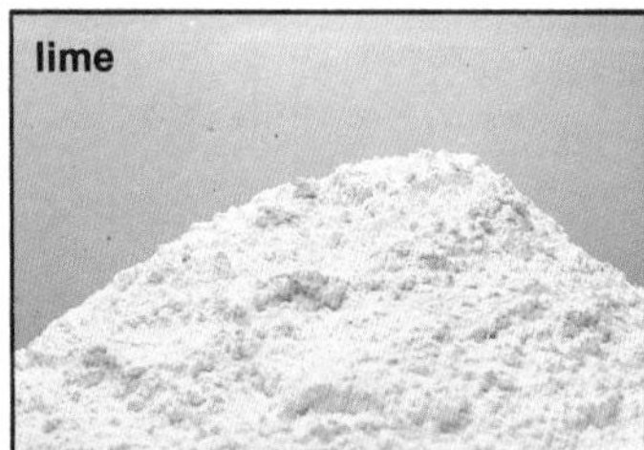
lime

Various mixes are recommended for different jobs, but in all cases the lime is similar. Builder's merchants sell 'white lime' for work where a light colour is required, or the cheaper 'grey lime' for all other work. Lime is sold in 25kg (½cwt) bags.

Plasticisers

Instead of adding lime to a mortar mix to make it workable, another additive, called a plasticiser, can be mixed in to produce minute air bubbles which act as a 'frictionless aggregate' in the mix.

Good-quality plasticisers are based on a resin that limits the amount of air bubbles to a certain proportion. This improves the resistance of the hardened mix to frost damage.

Plasticisers are sold in liquid or powder form under various brand names. Liquids are easier to measure out and mix in than powders. They are available in 1 litre cans – enough to mix with four 50kg bags of cement.

Sands

Sand is the bulk material of mortars, renderings and floor screeds. It should be clean and well-graded. Sand is cleaned either by washing with water or by dry screening. Dry-screened sands are often better than washed ones, as they retain their finest particles and these can help to produce a more workable mix.

A well-graded sand is one that has particles of various sizes. The smaller particles fill the gaps between the larger ones, and so less cement is needed to bind them all together to make a strong material. Poorly-graded sands include a lot more air space and extra cement would have to be used.

Sharp sands are coarse and are generally washed. They are best suited for floor screeds and fine

sharp sand

concrete mixes. They can produce strong, durable mortars but they are slightly more difficult to work with. *Soft sands* (often called builders' or bricklayers' sands) contain large proportions of fine particles, including clay, which makes them unsuitable for good mortars. They are used for binding or filling hard-core sub-bases to floor slabs, drives and paths.

soft sand

Silver sand is a naturally-occurring and extremely pure white sand. It is used in work where appearance is important – for example, with white Portland cement to make white or light-toned pointing mortars and white concrete. Silver sands are fine and not well graded. Additional cement is needed to make a mortar with good strength and durability.

silver sand

Bulking is a factor that must be allowed for when using sand. The proportion of sand in a mix is given by volume, and this figure always refers to dry sand. In practice, all the sand you buy will be damp to some extent. Damp sand increases in volume – sometimes by up to 40 per cent – so you should measure out extra sand to allow for this.

Coarse aggregates

Coarse aggregates are mixtures of strong stones, not chippings, and are the main material forming concrete. They are usually graded by sieving to diameters of 5 to 20mm (about ¼ to ¾in). Grading of the particle sizes throughout the material is needed for strong concrete, but the range of sizes varies for different jobs. For this reason, always tell your supplier what you require the aggregate for.

All-in aggregates (commonly called ballast) combine sand and fine coarse aggregates, and are usually used where the precise strength of the concrete is unimportant.

Bulk-buying sand and aggregates

Large quantities of sands and aggregates are traditionally sold loose by volume, measured in cubic metres or parts of cubic metres. In some areas, builders' merchants now sell sand and aggregate by weight in 25kg (½cwt) bags, which you could collect yourself, and also in ½ tonne (approx 10cwt) and 1 tonne quantities which are delivered to the customer either

all-in aggregate

loose or in canvas bags off-loaded by a crane on the special delivery vehicle.

Buying small quantities

DIY shops now offer small bagged quantities of most of the commonly used sands and aggregates as well as ordinary Portland cement and lime. Sizes vary from 6kg (approx. 13lb) to 50kg (1cwt).

fine concrete mix

The range often also includes various general-purpose mixes for popular uses, including:
sand & cement mix – for floor screeds, laying crazy paving and repair work in damp locations;
bricklaying mortar mix – for general bricklaying and internal rendering;
coloured mortar mix – for bricklaying with coloured mortars;
fine concrete mix – for paths, steps, kerb surrounds, etc, where a fine finish is needed;
coarse concrete mix – for foundations, setting fence posts and clothes line posts etc.

Most of the blended products do not state the exact composition of the mix. Read the application list printed on the bag carefully to make sure you have chosen the correct one for your job.

sand & cement mix

The range of mixes and pack sizes has been carefully worked out to provide a choice to suit commonly-occurring jobs around the home and garden. They save wastage and eliminate the trouble of buying and mixing together the separate ingredients. However, for some jobs it may be preferable to mix your own materials to the particular specification you require for the work.

coarse concrete mix

Special-purpose additives

The basic properties of standard mixes for mortars, renders and concretes can be modified by the inclusion of special-purpose additives. Here is a selection of the most common.

Colouring pigments are used to add colour to mortars and concrete. The pigments are stirred into the wet mix, and must be inert and colour-fast. Adding too much can weaken a mortar and reduce its durability, so follow the instructions provided with the pigment.

Pigments are sold in powder form in 1kg and 5kg (2.2 and 11lb) tins. The maximum amount usually recommended is 5kg per 50kg bag of cement. The stockist will have colour charts indicating the effect of various proportions of the pigment. When using pigments, keep a record of the amount added to the mix and work with accuracy to keep the colour constant.

Waterproofing additives are used to make mortars and concretes waterproof. They come as powders or liquids, and can be used with Portland cement mixes, but some are not suitable for use with masonry cements. Check the manufacturer's instructions carefully when using these additives.

Well-compacted concrete is essential for waterproof work. Do not expect the additive to do the job at the expense of good workmanship.

Frost-proofers are additives used when working in cold weather. The frost resistance of a hardened mortar, render or concrete is dependent on the strength of the basic mix; select a suitable specification and make sure that it is properly made up and used, and the work will be frost-resistant once is has set.

Frost-proofing additives are intended to give protection to work only during construction. Avoid working with mortars, renders and concrete when there is a risk of frost. Never start work in freezing weather, but if there is a risk of frost on new work or incomplete work, protect it with insulating covers such as hessian sacks, old blankets, or straw and the like, covered with polythene sheet or tarpaulins.

There is no effective frost-proofer to allow mortars and renders to be laid in near freezing conditions. However, concrete can be protected in low temperatures by the addition of an accelerator/hardener to the mix. This causes heat to be generated in the new-laid concrete, which should be covered as soon as practical in order to keep in the heat.

CONCRETE PAVING SLABS

Gone are the days when the local garden centre could offer you only paving slabs in dull grey or off-white. Today, a wide range is available in many colours and textures.

With slabs now made in an ever-increasing range of colours, textures and patterns, there's no longer an excuse for creating a concrete jungle in your garden. Your patio or path can have the mellow finish of old stone, the warm appearance of brick or cobbles, or the look of exposed aggregates.

What are concrete slabs?

Essentially concrete slabs, whatever appearance they might assume, are all made of the same material – cement. This reacts with water to form a hard, solid material. Crushed stone or gravel (coarse aggregate) and sand (fine aggregate) are added to form the body of the mix and provide strength and bulk. The cement-and-water paste coats the particles of aggregate and binds them all together into the dense mass known as concrete.

The cheapest slabs are simply cast in a mould of the required shape. A finish is then applied to the top face.

Pressed concrete slabs are stronger than cast slabs – although more expensive. They're hydraulically pressed during manufacture using many tonnes of pressure to compact and consolidate the fresh concrete. The result is a strong, double product.

Cast concrete slabs for garden use are generally 50mm (2in) thick, whereas pressed slabs are about 40mm (1½in) thick and a bit easier to lift and lay.

You might see concrete slabs described as made from 'reconstituted stone', the implication being that they are different from (and better than) concrete paving. All concrete is, in a sense, reconstituted stone. Manufacturers of this product, however, set out to simulate in concrete the appearance of a particular natural stone. Authentic stone is used for the aggregates, pigments are added and the product is made in moulds; themselves cast from an actual piece of stone in order to capture the correct texture.

Shapes and sizes

Most slabs are square or oblong, and come in a wide range of sizes, based on a 225mm (9in) or 300mm (12in) module. The most common sizes are:
225 x 225mm (9 x 9in)
300 x 300mm (12 x 12in)
450 x 225mm (18 x 9in)
450 x 300mm (18 x 12in)
450 x 450mm (18 x 18in)
600 x 300mm (24 x 12in)
600 x 450mm (24 x 18in)
600 x 600mm (24 x 24in)
675 x 450mm (27 x 18in).

Not all manufacturers make all of these sizes. The ones based on a 225mm (9in) module are the most useful, as they're appropriate in scale for most gardens. None of these sizes is too arduous to work with but normal precautions against back injury should always be taken. Bend the knees rather than the spine when lifting them, and watch your fingers and toes when stacking them: a 450 x 450mm (18 x 18in) slab weighs about 16.6kg (37lbs).

In addition to rectangular slabs you'll also find a wide range of other shapes – such as hexagons, and circles – although some are available only from specialist suppliers.

Hexagons, for instance, are particularly useful for making winding paths since you can change direction at 60° as well as at right angles. The commonest size measures 400mm (16in) in width between parallel sides. To obtain straight sides for a patio or path of hexagonal slabs half-hexagons are available.

Circular slabs come in various diameters and you can make a path of attractive stepping stones set in a lawn.

Estimating numbers

Estimating how many slabs you'll need to pave a given area is a matter of simple arithmetic with square and rectangular types. As a general rule, plan the size of your paved areas so that whole slabs are used when possible.

The easiest and most accurate way to work out how many slabs you'll need is to draw out the area you want to pave on squared paper, using one square to represent one slab.

Estimating the numbers of hexagonal slabs can be more difficult, although most manufacturers provide guides in their leaflets. When you're ordering your slabs it's wise to allow a few extra for breakages.

Colours and textures

Colour and texture must really be considered together. Relatively smooth, flat slabs can generally be found in the widest range of colours – from off-white, buff, yellow, brown and red to green, grey and dark slate.

With slabs designed to look like natural stone manufacturers naturally use stone colours. These include York stone in both buff and grey, Cotswold and a neutral stone grey. The most authentic looking slabs, however, are composed of two shades – a mixture of grey over buff to give a York – stone look and red with overtones of greys to simulate red sandstone.

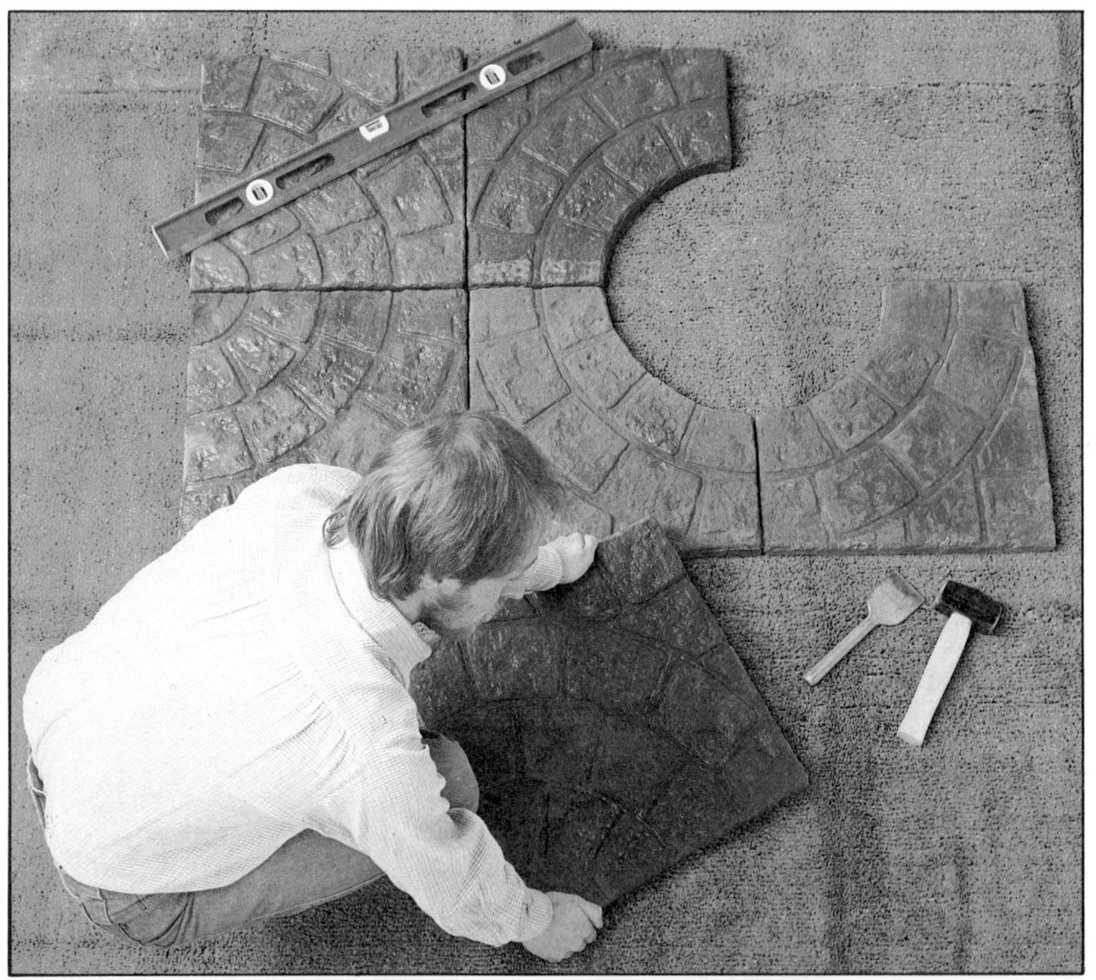

Left: You can buy paving slabs with special cut-outs for paving around trees and all sorts of textured patterns that you can lay in swirling, interwoven designs.

Not all slabs are made to simulate natural stone; there's a wide range of other surface textures and patterns available. These range from a cobbled effect to more uniformly-patterned designs.

Slabs with relief patterns such as these are made in conventional sizes and shapes but you can also buy radius slabs to make a circular pattern contained within a square. Alternatively, some can be laid in such a way that they form interwoven swirling patterns or an overall herringbone effect.

A different approach to colour and texture is adopted with 'exposed aggregate' slabs.

Right: Surface finishes for square and rectangular slabs vary from smooth and riven to exposed aggregate.

Below: With pressed slabs you get patterns resembling brick, stone or mosaic in various shapes and sizes. With hexagonal slabs, two different edging slabs are available.

hints

Stacking slabs correctly
To prevent accidentally chipping the corners and edges of your slabs, and marking their faces, stack them on edge in pairs, face to face, against a wall, with their bottom edges on timber battens.

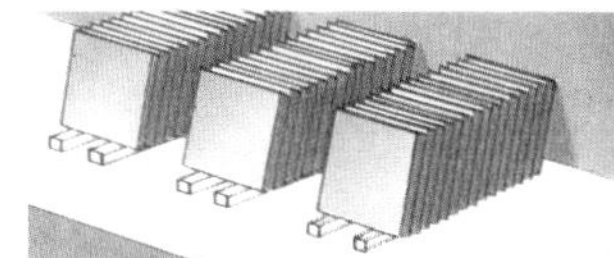

Handling slabs
Slabs are heavy and cumbersome to carry. The best precautions you can take are to wear heavy-duty shoes to protect your feet in case you should drop a slab on them, and gloves to protect your hands, especially if the slab has a rough surface. Grip the edges of the slab firmly and lift them with your knees bent and your back straight. Alternatively, you can hold the top corners of the slab and 'walk' it on its bottom corners.

Laying slabs in patterns
You've a lot of freedom in the way you lay your slabs. You can lay only one shape, size and colour or you can mix them to give various patterns.

GARDEN WALLING BLOCKS

If you're building walls in your garden, you don't have to stick to brickwork. There is a wide range of walling stone, natural or man-made, to choose from.

Many projects in the garden involve building free-standing walls, either for their decorative effect or to give shelter and privacy or to act as earth-retaining walls on sloping or banked sites.

In some areas, natural stone is still readily available at a price that makes it a worthwhile consideration for wall construction, but otherwise there is a good selection of reconstituted stone and high-quality precast concrete masonry blocks which come in a range of sizes and which offer an attractive alternative. In most areas the larger garden and DIY centres and builders' merchants keep good stocks. Stone merchants too, should not be overlooked as many offer quite an extensive range of reconstituted stone and precast concrete garden walling products as well as quantities of natural stone.

Reconstituted stone

Reconstituted stone blocks are made from concrete in which the aggregate is crushed natural stone and the sand and cement content is carefully selected so that the finished product closely resembles natural stone. They can be smooth, but most blocks have a texture intended to simulate traditional split stone walling.

These blocks are made in a co-ordinated range of sizes that can be laid in traditional stonework bonding patterns – either coursed or random. Precise sizes vary between makers and different brands are not generally interchangeable if regular bonding is to be maintained. But any size can be used on its own or together with others from the same co-ordinated range to produce the desired bonding pattern. The block sizes allow for joints 10mm (3/8in) thick. There is a choice of about half a dozen natural stone colours; yellows, reds, greys and greens predominate. They can be used in mixtures to produce a multi-coloured walling.

Not all faces of the blocks are textured; often only one long face and one end is intended to be exposed, while the others are flat and smooth for neat bonding. Check this point when you are planning your wall and estimating quantities. If you want the wall to look good on both sides you may have to use two blocks back-to-back if the blocks themselves have only one textured face. Some ranges also include special blocks for corners and the ends, and also coping stones for finishing off the top of the wall.

Reconstituted stone can be used outside for the walls of buildings, extensions, garages and greenhouse bases, for boundary walls, retaining walls and barbecues; it is also ideal for use inside the house for fireplaces and decorative feature walls.

Simulated dry-stone walling

An interesting variation among reconstituted stone walling blocks is one which is moulded to give the appearance of eight or nine individual 'stones'. The false joints are deeply recessed, and for the best effect the 'real' joints should also be deeply recessed, and should be made with a mortar to match the colours of the block.

These blocks are faced on one long side and on both ends. They can be used to turn corners, but if

SCREEN WALLING BLOCKS

KEY
***1** to **4**: pierced screen walling blocks; **5** to **7**: hollow pilaster blocks; **8** and **9**: pilaster caps; **10** to **12**: coping stones.*

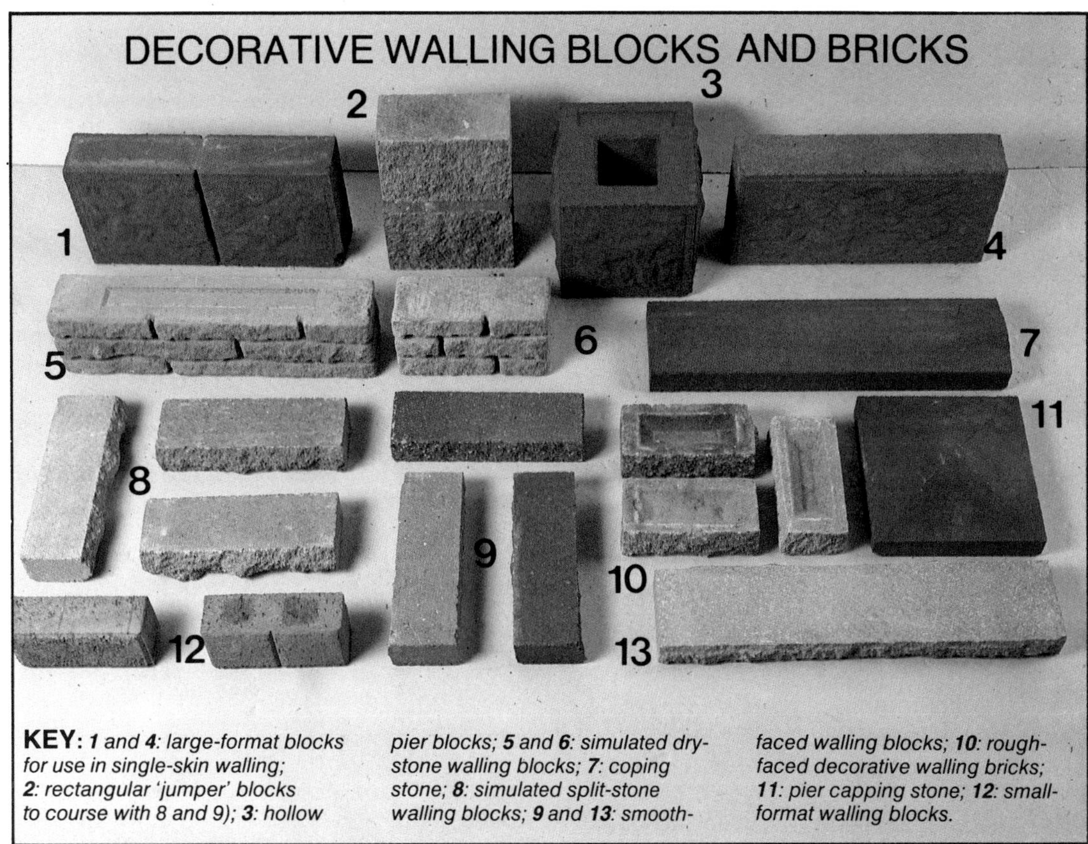

DECORATIVE WALLING BLOCKS AND BRICKS

KEY: *1* and *4*: *large-format blocks for use in single-skin walling;* *2*: *rectangular 'jumper' blocks to course with 8 and 9);* *3*: *hollow pier blocks;* *5* and *6*: *simulated dry-stone walling blocks;* *7*: *coping stone;* *8*: *simulated split-stone walling blocks;* *9* and *13*: *smooth-faced walling blocks;* *10*: *rough-faced decorative walling bricks;* *11*: *pier capping stone;* *12*: *small-format walling blocks.*

you want both sides of the wall to appear the same, then a double thickness wall will be required. The standard block size is 527 x 145 x 102mm (20¾ x 5¾ x 4in) which, allowing for a 6mm (¼in) thick joint, gives a work size of 533 x 152 x 102mm (21 x 6 x 4in). Half blocks are available for the half lap at the ends of the wall so that a soundly bonded wall can be built without cutting blocks. It is worth planning the work to suit the sizes of the block, as cutting will spoil the continuity of the pattern of false jointing. Special coping stones are made to suit single and double-leaf walls and square piers constructed from standard blocks.

These blocks are particularly suitable in the garden for retaining walls, boundary walls, piers for pergolas or pierced screen wall blocks, raised planting beds, seats and barbecues. As both ends are properly faced, a pierced screen or 'honeycomb' wall can be built with them as a design variation.

Concrete facing blocks

The lightweight concrete blocks used in houses for partitions and the inner leaf of cavity walls are not suitable for use in the garden, as they are not weather-resistant. Special concrete facing blocks should always be chosen for any outdoor work. They are walling blocks with a high-quality face intended to be left exposed. A wide variety of colours and textures are available, including sculptured, exposed-aggregate and split stone finishes. Usually they are made with dense concrete, but some types of lightweight blocks are made with dense weather-proof facing finishes. They are made in the standard work sizes of 450 x 225mm (18 x 9in) and 400 x 200mm (16 x 8in). Both sizes allow for 10mm (⅜in) thick mortar joints and they are made in several other thicknesses. Very thick ones are generally hollow to save weight. Other sizes are made to co-ordinate with these standard blocks for use at the ends or corners of walls. There are also other block sizes for special effects, including random course walling.

Facing blocks must be laid bonded – with vertical joints staggered – to give the wall strength. They can be cut with a bolster and club hammer, but it is best to set out the work to use whole block sizes as far as possible.

Decorative bricks

As an alternative to walling blocks of reconstituted stone, you could consider using specially coloured bricks – not the ordinary stocks or facing bricks used for house-building, although these would often be suitable, but special types resembling dressed stone on their exposed faces and available in yellows, reds, greens and greys. The advantage these have over walling blocks is the presence of a frog (or indent) that helps to increase the bond strength – a useful bonus on exposed walls or earth-retaining structures. They are the same size as an ordinary brick.

Pierced screen wall blocks

Open screen walls made from pierced blocks can provide an attractive feature to give partial shelter or as a background for plants. They can be used to screen a patio, to build a carport wall or a porch screen, or to hide an unsightly area such as a compost heap or the dustbin.

The blocks are made from dense concrete, often white or near white, although some makers offer them in grey or other colours. Designs are generally based on geometric patterns. Some are laid in groups of four to give a larger interlocking design. Many manufacturers also make co-ordinated blocks which are not pierced; these can be incorporated in the pattern for special effects or they can be used where full screening is required in the general run of the decorative pierced walling.

Designs vary between manufacturers and different brands may not be interchangeable. Precise sizes may vary too, but they are generally a nominal 300mm (12in) square (including an allowance of 10mm (⅜in) for joints) and build a wall approximately 90mm (3½in) thick.

Pay attention to choosing the correct mortar mix for use with these blocks. White blocks should be laid either with a mortar made with white cement and silver sand to match, or else with a darker and deliberately contrasting mortar. The light grey colour of plain mortar made with ordinary Portland cement does not offer sufficient contrast and tends merely to look dirty. Take care to keep the mortar off the face of the blocks when laying them, so as to avoid the risk of staining. Align and level the blocks carefully and finish the joints neatly, as the square grid pattern will accentuate any irregularities.

Screen wall blocks are not laid with a lapped bond, but are 'stack-bonded', one on top of the other with the mortared joints aligning throughout. Stack-bonding is not as strong as lapped bonding, so a wall that is more than a couple of blocks high needs supporting piers. These piers should usually be every 3m (10ft), but follow the manufacturer's instructions for high walls or exposed locations. For extra strength, reinforcing mesh can be bedded in the horizontal mortar joints.

Piers can be built of bricks, or reconstituted stone or concrete walling blocks, but perhaps the simplest method is to use tailor-made screen wall pilaster blocks: these are made to suit end, intermediate, corner and intersecting positions and have holes through the centre so that a steel rod can be threaded through, extending into the concrete foundation, to provide extra reinforcement for high walls; the hole is then filled with a weak concrete mix.

The tops of the pilasters are finished off with special caps, and copings are made to lay along the tops of the screen blocks. These strengthen the top of the wall, as the cappings each cover two or three blocks; they also help to shed rainwater, so minimising staining of the wall surface.

It is essential that this walling be laid on a flat base, so if you have a sloping site you will have to level it up in steps by building a dwarf wall in solid blocks or bricks. Even on a flat site this is often preferable, as two or three courses of bricks or stone walling is more practical next to soil than pierced blocks.

Mortars for garden walls

Never use mortars consisting of straight Portland cement and sand for building garden walls, since cracking is likely to result. The textbook mix is 1:1:6 cement:lime: sand, but other suitable mixes for this work include 1:4 masonry cement:sand and a 1:5 cement: sand mix with added proprietary plasticiser. In all cases the mix should be workable but not sloppy.

IRONMONGERY FOR BUILDING & REPAIR WORK

There is a range of ironmongery designed for specific jobs, and using the right product can make your work go a lot more smoothly. Here are some of the things you're likely to find most useful.

For most jobs around the house, buying what you need is usually quite straightforward. But some jobs need a special bit of hardware, and this may be hard to find unless you know exactly what you want. This list gives you an idea of what's available and gives you exact sizes so you can plan your work with confidence.

Ironmongery for building work

Concrete reinforcing mesh is used to strengthen concrete floors, paths, foundations, drives and so on. It is made from steel rods or 'wires' between 2 and 12mm (1/16 and 1/2in) in diameter, welded together to give a rigid mesh. The mesh sizes are 100mm sq (4in sq), 200mm sq (8in sq), 100x200mm (4x8in) or 100x400mm (4x16in). It is sold in sheets 2.4x6m (8x20ft), with some light meshes in 2.4m wide rolls up to 20m (60ft) long.

Brickwork reinforcing mesh is bedded in the mortar joins between courses of bricks or blocks to strengthen the bonding of the wall. It is made of 0.5mm (1/64in) thick galvanised steel and comes in widths ranging from 64 to 305mm (2 1/2 to 12in) and in rolls either 22.8 or 82.4m (75 or 270ft) long.

Choose the width that leaves a 25mm (1in) gap between the edges of the mesh and the faces of the wall. Thus, for a half brick thick wall you'll need 64mm (2 1/2in) wide mesh, and for a whole brick thick wall you'll need 178mm (7in) wide mesh.

Plaster laths and beads are used as a base when plastering across a gap, round a corner, or as reinforcement in positions where unprotected plaster is likely to crack.

Laths are made from 0.5mm (1/64in) thick galvanised steel mesh with a mesh size of 6mm (1/4in), and sold in sheets up to size 675x2440mm (2ft 3inx8ft).

Angle bead is used to reinforce corners, and there are two types available. The standard type is designed for 12mm (1/2in) thick plaster and comes in 2.4, 2.7 and 3m (8, 9 and 10ft) lengths. Thin-coat bead reinforces a 3mm (1/8in) thick skim coat of plaster and comes in 2.3, 2.4m and 3m (7ft 6in, 8ft and 10ft) lengths.

Stop bead is used to finish and reinforce edges of plaster at openings, and there are types suitable for external use to provide a clean finish to the bottom edge of rendering on an upper floor. They come in 2.4 or 3m (8 or 10ft) lengths.

You can also buy prefabricated mesh arch formers, which you fix under the head of a rectangular opening to form the arch profile and then plaster over. They are made in a range of arch shapes, for openings of various widths.

Wall ties are used to bond together the two leaves of a cavity wall. There are four types available, though the first two types are the most common – and the cheapest.

The galvanised wire butterfly pattern is 203mm (8in) long and is made from 3mm (1/8in) wire twisted together to form a double loop.

The galvanised steel twisted pattern is 200mm (8in) long, by 20mm (3/4in) wide and 3mm (1/8in) thick.

Stainless steel ties look a bit like a propeller blade without the twist. The fins are embedded in mortar and are dimpled to improve the grip. They are available with insulation-retaining discs.

Finally, there are plastic ties. They are similar to the twist pattern and are available with insulation-retaining discs.

Door frame ties are used to provide fixing points for door or window frames in brickwork. They are built into the wall between courses in the same way as wall ties. There are two types.

The fishtail pattern is similar to the twist pattern wall tie, but with a 50mm (2in) pre-drilled and countersunk upstand for fixing to the wood. They are available in various sizes.

The second type is called a patent door and window frame holdfast. They have a sharp fishtail end which is driven into the wood so screws aren't required.

Profiles made of galvanised steel, are used to tie in a new wall without having to cut out joints. They are fixed to the existing wall with coach bolts and butterfly ties are clipped in to tie in to the new wall. They come in 2310mm (7ft 6in) lengths to suit 75 to 215mm (3 to 8 1/2in) bricks or blocks.

Joist hangers are used to support the ends of the joists as an alternative to setting the ends in the masonry. Again there are two types.

One type of joist hanger has a fishtail lug which is keyed into mortar joints between courses of bricks, or set into concrete. It is made of galvanised steel and is available for 150 to 230mm (6 to 9in) deep joists, all 50mm (2in) thick. Other sizes (and different coloured finishes) are available if ordered specially.

The second type has two pre-drilled straight straps that you bend as necessary in order to hang one joist from another. It too is made of galvanised steel and is designed for nail fixing. It comes in a variety of widths from 38 to 100mm (1 1/2 to 4in) and depths from 150 to 230mm (6 to 9in).

Herringbone joist struts are available, made of galvanised steel, and can be used as an alternative to constructing the struts from timber. They are designed for joists at 400mm (16in) centres, and various sizes are available according to the width and depth of the joists.

BUILDING & PLASTERING

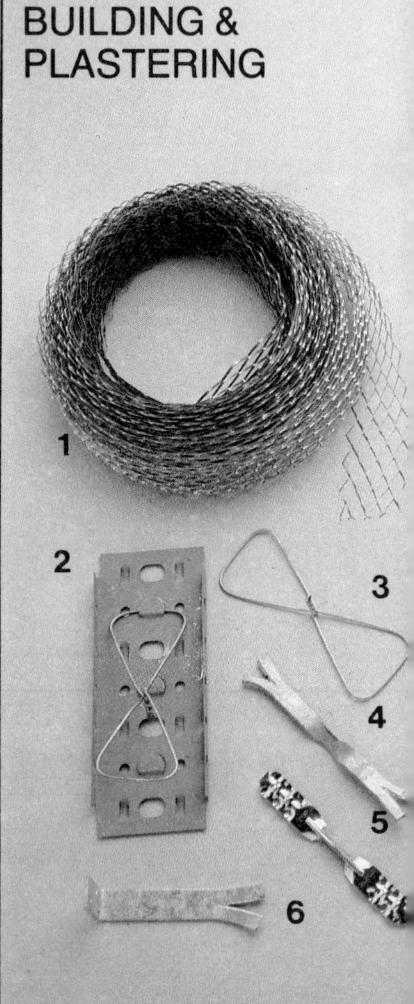

GUTTERS & FENCES

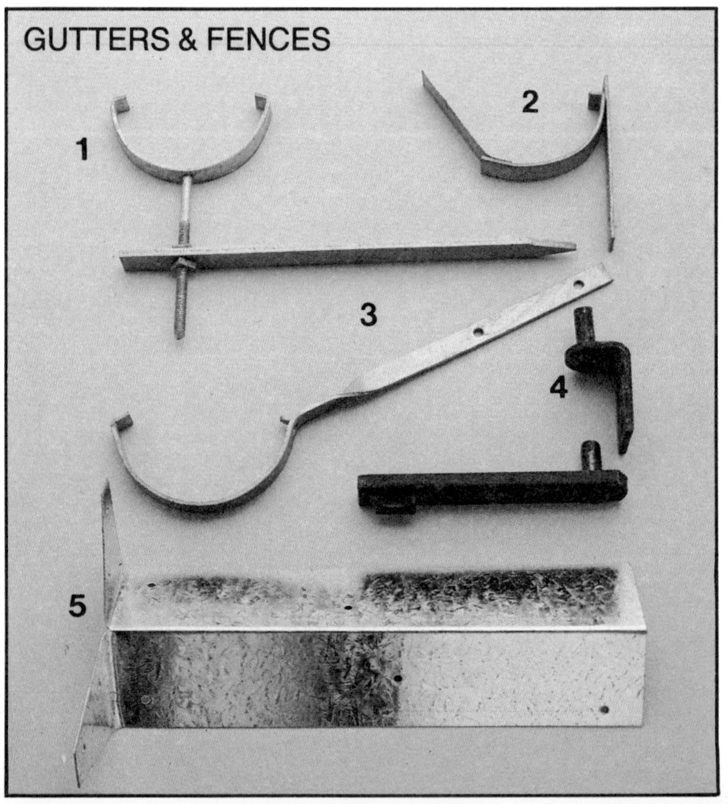

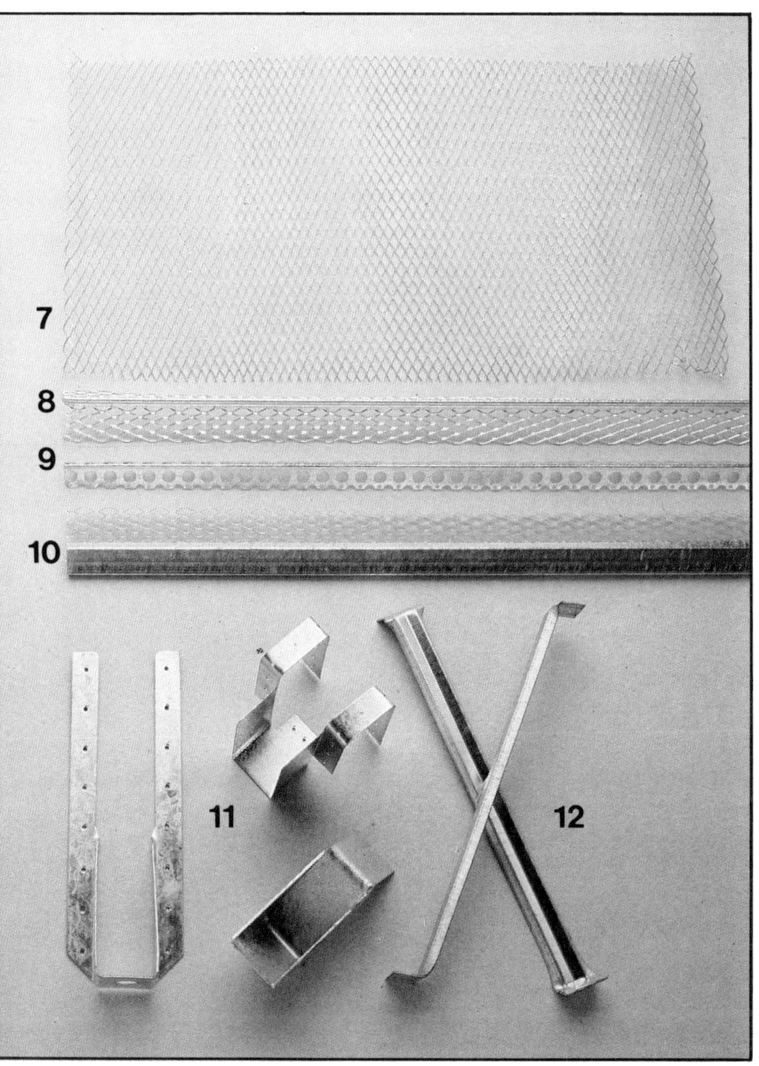

KEY

Building and plastering
1 *Brickwork reinforcing mesh*
2 *Profile wall tie*
3 *Butterfly wall tie*
4 *Twisted wall tie*
5 *Stainless steel wall tie*
6 *Door frame tie*
7 *Plaster lath*
8 *Plaster-depth angle bead*
9 *Thin-coat angle bead*
10 *Stop bead*
11 *Joist hangers*
12 *Herringbone joist struts*

Gutters and fences
1 *Rise and fall bracket*
2 *Fascia bracket*
3 *Rafter bracket*
4 *Gate hangings*
5 *Arris repair bracket*

Roofing and flashing
1 *Lead flashing*
2 *Zinc flashing*
3 *Self-adhesive flashing*
4 *Copper crampons*
5 *'Jenny Twin' slate clips*

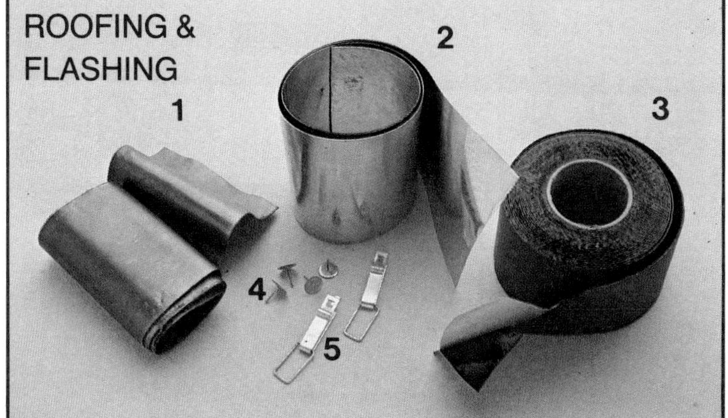

Ironmongery for carpentry and home repairs

Angle repair irons are used to reinforce butt joints in timber. They are available in four shapes – X, T, L and straight – and are screwed to the face of the timber or fitted into shallow recesses. They are made of bright steel or brass and are pre-drilled. Sizing is by leg length, which is normally between 50 and 75mm (2 and 3in).

Corner braces are also used to reinforce timber butt joints but are screwed to the inside of the angle. They are made from bright steel, with two pre-drilled countersunk screw holes per leg, with leg sizes of 50, 75 or 100mm (2, 3 or 4in).

Corner brackets are used to strengthen and square corners of cabinets and frames, etc. They are made from galvanised steel or plastic and there are two main types. One is for fixing to the face of the joint and the other for fixing in the internal angle. Both are triangular in shape and the latter has upstands on the sides to provide the fixing points.

Shrinkage plates are useful for reinforcing butt joints in natural timber where there is a possibility of shrinkage or swelling in the future. They are made from galvanised steel and have holes in one leg and slots in the other. The slot allows the timber to move without straining the joint. There are three main types – the angle bracket and the straight plates with either horizontal or vertical slots.

Fence repair brackets are used to reinforce damaged joints in fence posts. Normally made from galvanised steel, the most common type is used for joining arris rails to posts where the rail end has begun to rot away. There is an open socket to receive the end of the rail and a flange providing a screw fixing to the post.

Gate hangings are the large gate's equivalent of a hinge and consist of a bracket with an upright pin that locates in the hinge eye of the gate. There are four types – a bolt fitting for pre-drilled timber and concrete posts; a flange fitting for screwing to timber posts; a spike fitting for driving into timber posts; and lastly, a plain tail for bonding into masonry.

Gutter brackets support heavy cast iron guttering and are therefore handy for repair work. There are five main types, all made from galvanised steel. For screw fixing there are fascia brackets which fix to the surface of a wall or fascia, and rafter brackets that fix to either the top or side of a rafter. Other brackets have a long spike for bonding into masonry – 'drive' brackets have a horizontal spike and 'rise and fall' brackets have a vertical spike. The latter allows the gutter to be adjusted up or down by means of nuts and bolts.

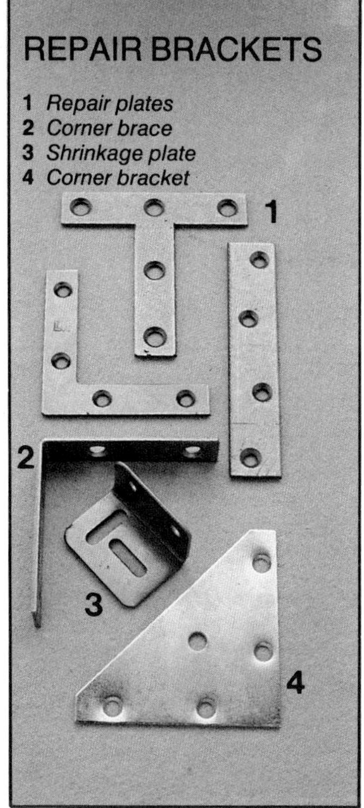

Sheet flashing is used to provide a waterproof join at junctions in a roof. It can be made from lead, zinc or aluminium.

Zinc is usually sold in 2440x915mm (8x3ft) sheets of various thicknesses and in strips 150mm (6in) wide. Lead is sold in 6m (20ft) rolls or by the metre, as a strip between 150 and 780mm (6 and 30in) wide. It comes in a variety of thicknesses, quoted as a code number from 3 to 8. Code 3 is for lightweight applications such as chimney flashings, but for general flashing code 4 is usually considered to be the minimum.

Aluminium flashing is soft, and easy to shape round any profile. It is also lightweight and much cheaper than lead or zinc. It is sold in 8m (26ft) rolls, 600 or 900mm (2 or 3ft) wide.

Self-adhesive flashing strip has a thin aluminium foil bonded to a bitumen base. It is useful for general flashing work as well as for repairing leaking gutters or old flashing. It comes in 100m (33ft) rolls with widths of 50 to 460mm (2 to 18in).

Slate clips, also known as tingles, are used to retain replacement slates. They are made of copper, 150mm (6in) long by 15mm (⅝in) wide. They are pre-drilled with a single hole to take the fixing nail.

A clever modern slate fixer called the Jenny Twin is also available. It's linked to the slate by a folding tab, and locks it onto the roofing batten as the slate is slid back into place. **Crampons** made of copper are for securing asbestos-cement tiles.

PLATFORM TOWERS

Whether you are decorating the outside of your house or carrying out vital repairs, you need to be able to reach the heights comfortably and safely. A platform tower does the job perfectly.

The main use of a platform tower is for gaining safe access to the upper storey of a house – for example when painting the upstairs walls and woodwork, or when you need to be able to get at gutters, chimney stacks or the roof. Most can also be used indoors, either as a platform for painting ceilings or, more usefully, for creating a working platform for decorating the stairwell.

The most common type of platform tower is made from steel sections that are simply slotted together. A typical height for such a tower would be 4.8m (16ft) from the ground to the platform, allowing heights of up to 6.5m (22ft) to be reached.

The tower components

For this type of tower, the main components are the *frame sections*. These are made from tubular steel, welded together to form an H-shape with two cross pieces, and slot together in parallel pairs to form the tower. The steel is either painted or galvanized to protect it from corrosion.

Most towers are about 1.25m (4ft) square, but some manufacturers also make half-width frame sections for a rectangular tower – useful for access in confined spaces.

As well as steel towers, there are also aluminium alloy towers. These have main frames about 2m (6ft 6in) high, linked by fixed platforms and fitted with an internal staircase that acts as a structural brace (see photograph, page 840).

There are three kinds of *feet* that you can fit to a platform tower. The cheapest is a simple fixed baseplate. Its disadvantage is that all four feet have to be at the same level, a problem overcome by using adjustable feet instead. Mobility can be achieved by fitting the tower with lockable castors.

All kinds of tower have diagonal *braces* or tie bars, which keep the tower rigid and, incidentally, also provide a handy grab rail when you are climbing the inside of the tower.

The *platform* itself usually consists of a number of stout boards. Some systems offer prefabricated platform sections instead of planks. Most towers also have toe boards which slot in on edge round the platform to stop paint tins and tools cascading to the ground or on to someone's head.

To top off the tower, a *guard rail* is fitted all round. It is absolutely vital to fit guard rails properly and to use them at all times.

When building towers above a certain height against a building, *outriggers* can be fitted to the two outside corners of the tower to increase its stability. These avoid the need to tie the tower to the building.

Optional extras

Many tower manufacturers offer other accessories and extras, usually intended to widen the usefulness of the tower. These include ladder sections, hinged platforms, staircase frames, trestle feet and workbench attachments.

Buy or hire?

Because of the cost, most people will hire a tower rather than buying one. Always order well in advance, and check on delivery that all the components are present.

SAFETY RULES

When building a tower:

- ensure that the feet are resting on firm ground
- on soft ground, set the feet on stout planks
- check that the tower is standing level, and adjust the feet if necessary
- check that the height recommendations are not exceeded; fit outriggers or tie the tower to the building if they are
- beware overhead power lines
- fit guard rails and toe boards round the platform.

When using a tower:

- lock castors, if fitted, before starting to climb
- climb the inside, NOT the outside, of the tower
- don't lean ladders against towers
- don't attempt to move a tower if anyone is on the platform.

When dismantling the tower:

- don't throw components to the ground; pass them to a helper
- don't force components apart; tap them with a hammer and a block of wood if they are stiff.

Interlocking toe boards are a vital safety feature, stopping you from slipping beneath the guard rails or knocking tools to the ground.

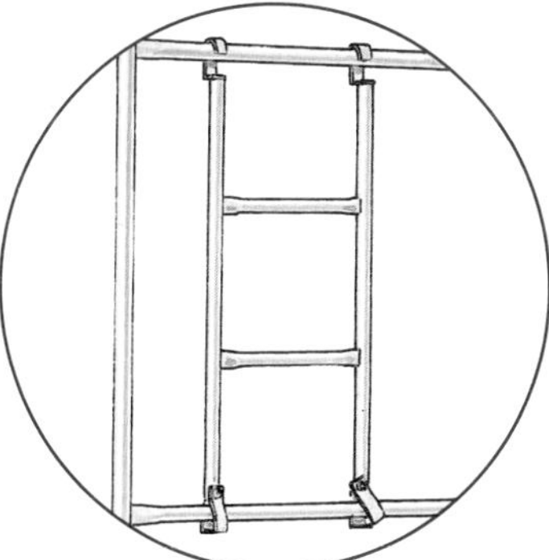

You can fit a hinged platform section alongside a fixed landing panel for easy access.

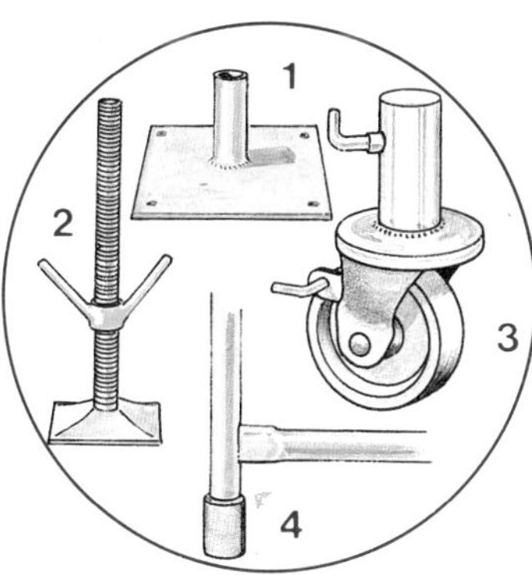

Small clip-on ladder sections bridge successive H-frames to make climbing the tower easy.

You can fit fixed base plates (1), adjustable ones (2), lockable castors (3) or rubber feet (4 – for indoor use only).

hints

- *it's often difficult to reach upstairs walls with a ladder when a single-storey flat roof gets in the way. Build a cantilevered arrangement like this, resting the feet of the cantilevered part on boards to protect the roof surface.*

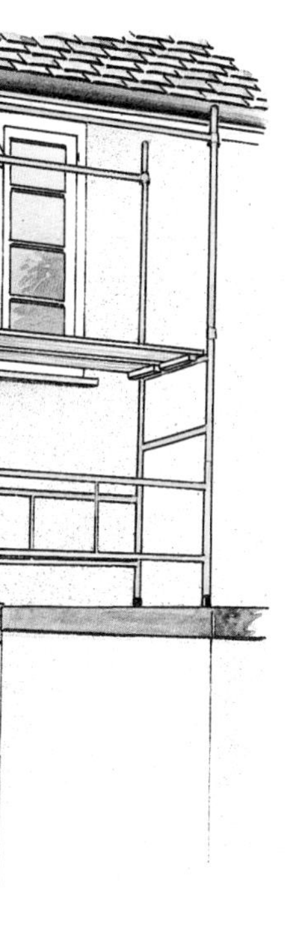

- *where the cantilever aims to bridge a pitched roof, protect the tiles with sandbags and again use boards to help spread the load. Use narrow H-frames to reach over a shallow roof.*

- *if you want access to the area above a door that is in constant use, build a tower at each side of the door and link them together with two frames.*

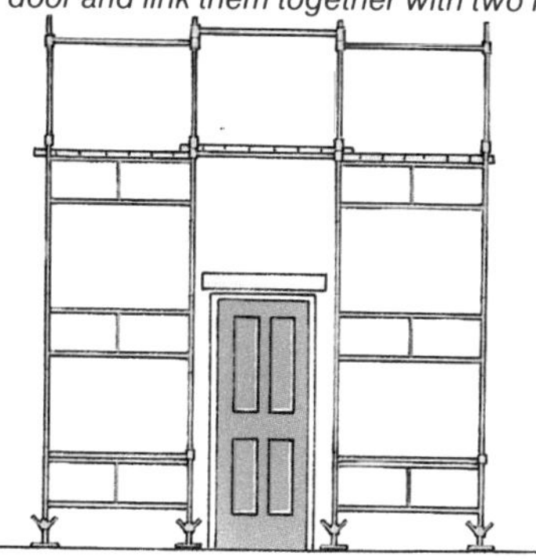

- *to work on a chimney stack in complete safety, use the tower components to form a working platform as shown. Rest the feet of the frames on boards laid on sandbags, and use a roof ladder for access.*

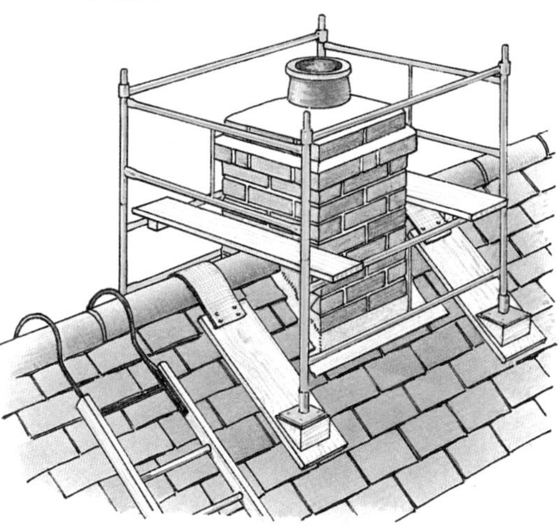

- *you can use tower components indoors as well – to make a low mobile work platform for painting ceilings (left), or a stepped platform for decorating in stairwells (right). A special staircase frame is usually used for this arrangement.*

SAFETY GEAR

Lots of jobs around the home can be dangerous, dirty, or just plain unpleasant. There's a wide range of protective clothing and accessories designed to help you get on with the job while keeping clean and reducing the risk of accidents.

There are times when you simply cannot avoid working in hazardous situations or with potentially dangerous materials and machines. And accidents will happen. The trick is to minimise the risk by following the safety rules appropriate to the job and by wearing the right gear.

This need not be anything elaborate. It all depends on what you are doing. If you are working with machinery, the 'right gear' may mean nothing more than buttoning your shirt sleeves and taking off your tie so they will not get tangled in the moving parts. If you are standing on a ladder for long periods, it may simply mean wearing comfortable 'sensible' shoes to reduce 'foot fatigue'. The right gear can even mean clothes which protect you from the weather, including warm gloves and sweaters, waterproofs or a sun hat. If you are too cold or too hot, your concentration will go, leaving you prone to errors. If you are extremely cold, you may actually injure yourself without knowing it.

But of course there are also times when only specialist equipment will give you the protection you need. The range available is detailed below.

Safety glasses/goggles

These will protect your eyes from dust, harmful liquids, and also from the flying debris produced by such jobs as chiselling masonry.

The cheapest are like ordinary spectacles, with impact-resistant plastic lenses. To stop things getting past them, more expensive types also have protective side screens, or curved main lenses. They are light and comfortable to wear, even for extended periods, but, like ordinary spectacles, have a tendency to slip if you are doing anything really active.

For greater protection, particularly against dust and liquids, safety goggles are better. To ease the problem of perspiration, the body of the goggles contains ventilation holes. Many can be worn over ordinary spectacles.

Protective masks/respirators

To stop you inhaling harmful dust, sprayed paint and so on you will need one of these.

The cheapest and most widely available consist of a simple replaceable cotton gauze pad mounted in an aluminium frame contoured to fit over the nose and mouth. For more demanding work, there are also moulded rubber and plastic masks (normally referred to as 'respirators') which accept more efficient replaceable cartridge filters. These filters are generally designed to filter specific substances, and it's important that you choose the right type. For example, some work against dust; others against the vapours produced by spray painting.

Heavy-duty gloves

If you want to avoid blisters, gardening gloves will probably do the trick. For some jobs though, you need far greater protection against abrasion on rough surfaces, cuts from sharp edges, and/or corrosive chemicals, and should wear purpose-made heavy duty gloves.

There are three common types. One is a chrome leather version, either with double thickness palm, or armoured with metal staples, to give good protection against cuts and abrasions.

Natural rubber gloves are the usual choice against corrosive chemicals. Unlike their domestic washing-up counterparts, these are not only proof against chemicals, but also have good resistance to tears, snags and abrasion. There are ordinary medium weight versions and also heavy gauntlets with extended cuffs to protect the forearms.

Then there are PVC gloves. Most are equivalent to medium weight gloves; some perform like chrome leather but give a degree of chemical resistance as a bonus.

Overalls

When tackling a dusty, dirty or potentially messy job, these simply stop your ordinary clothes getting in too much of a state. They don't offer much more than splash protection though, particularly against oil, grease, paint and so on. If it's cold you will still have to wear old clothes underneath (if its warm, for many jobs you can just wear the overalls).

Work boots

If you have ever dropped a brick on your foot, or can imagine how it feels, the advantages of wearing a good strong pair of boots with steel-reinforced toe caps when carrying out general maintenance and building work are immediately obvious. They are not expensive, and since most have hard-wearing leather uppers and oil and alkali resistant rubber soles, they do represent very good value for money. Protective shoes made to the same standards are also available. The only thing they are not much good at is coping with mud and deep water, so, if you are digging trenches, wear your wellingtons.

Safety helmets

More commonly known as 'hard hats', these are really worth considering only if you are engaged in substantial demolition work, or intend to go clambering up professional type scaffolding. They can be hired. They are normally made from glass fibre or high density polythene and should have a fully adjustable harness allowing you to make them fit well enough not to fall off. The gap between harness and helmet also cushions any impact, so never wear a helmet without a properly adjusted harness. As an added safety precaution, make sure the helmet conforms to British Standards (BS5240 to be precise).

Ear protectors

These are designed to protect your ears from the damage that can result from prolonged exposure to high levels of noise.

Most are foam plastic filled plastic muffs mounted on a plastic covered sprung-steel headband. For the sake of general safety, when you are wearing the protectors you should still be able to hear moderately loud noises in your vicinity (so you can hear warning shouts for example) which is why they are preferable to ordinary ear plugs.

Knee pads

You will really appreciate these when tackling a job which means spending hours crawling about the floor. Some are simple cushioned rubber mats, others are strapped onto your knees. If you can't buy them, improvise with pads of cloth, or some old cushions.

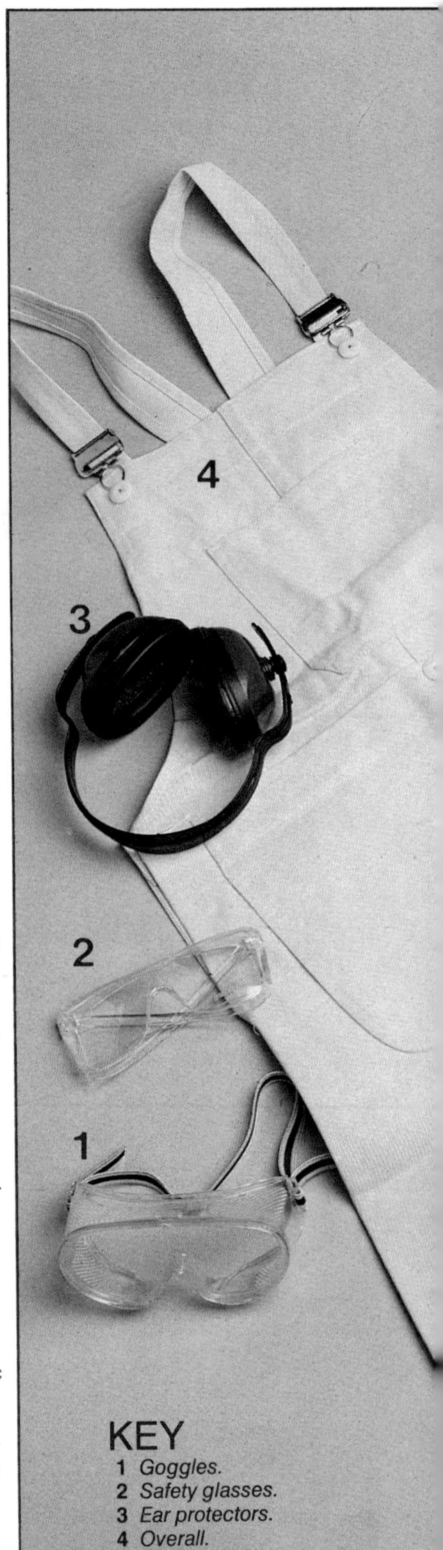

KEY
1 *Goggles.*
2 *Safety glasses.*
3 *Ear protectors.*
4 *Overall.*
5 *PVC gloves.*
6 *Boiler suit.*
7 *Helmet.*
8 *Respirator.*
9 *Face mask.*
10 *Steel toe-capped boots.*
11 *Knee pads.*
12 *Thick cloth gloves.*

6
7
5
8
12
9
10
11

MASONRY TOOLS

The term 'masonry' can mean anything from a brick or piece of stone to a complete wall, so it is important to be specific when choosing the right tool for a cutting or shaping job.

The hardness of the material to be worked is, perhaps, the most important consideration when choosing masonry tools. Some stone-cutting tools are designed for use only on relatively soft stones such as sandstone, and to use them on a hard stone like granite could cause damage to the tool or workpiece. Most 'brick' tools are intended for use on ordinary, relatively soft bricks and tend not to cope with hard engineering bricks, which have to be treated more like hard stone. Some are tempered just to cut brick and nothing else, whereas others may cut a variety of materials; it is important to check. When dealing with mixtures of materials, the general rule is to pick a tool that will handle the hardest element in the mixture.

Cutting and shaping tools

The commonest tools for cutting and shaping masonry of all types are cold chisels. There are several general-purpose and specific versions.

The **flat-cut cold chisel** is frequently used on masonry for splitting, chopping out, cutting chases and, occasionally, rough shaping. Like the rest of the cold chisel family, it is a hexagonal steel bar with a cutting tip formed at one end – in this case a straightforward wedge-shaped tip a little wider than the bar. The other end has chamfered edges to prevent chipping when struck with a heavy hammer.

The **cross-cut cold chisel** also known as the Cape chisel, has a cutting edge very much narrower than the bar from which it is made, allowing it to cut slots and grooves with great accuracy.

The **half-round cold chisel** is a variation of the cross-cut chisel and may also be known as the round-nosed chisel. It has a single cutting bevel ground into the tip to produce a semi-circular cutting edge. Used mainly for cutting grooves, it can produce rounded internal corners as well.

The **diamond-point cold chisel** is yet another variation of the cross-cut chisel and is sometimes known as the diamond-cut chisel. It has a diagonally-ground single cutting bevel. Use it for making V-shaped grooves and neatly angled internal corners.

The **plugging chisel** is otherwise known as the seaming chisel or seam drill. It has a curious slanting head, and is used for removing the mortar pointing in brickwork. Two types are available: one with a plain head and the other with a flute cut into the side to help clear waste material.

The **concrete point** is a fairly rare cold chisel that tapers to a point rather than a normal cutting edge and is used for shattering concrete or brickwork in areas previously outlined with a flat chisel.

The **dooking iron** is intended for cutting holes through brickwork and stone. This extra-long, flat-cut cold chisel has a narrow 'waist' let into the bar just behind the head to help prevent waste from jamming it in the masonry.

The **brick bolster** has a extra-wide, spade-shaped head and is designed to cut bricks cleanly in two. Most have a 100mm (4in) wide cutting edge, but other widths can be found, and care should be taken not to confuse these with other types of bolster chisel such as the mason's bolster (see below) or even the floorboard chisel. They are not tempered in the same way and may be damaged if used incorrectly.

The **mason's chisel** comes in two varieties. Narrow versions look like ordinary cold chisels, but the wider versions are more like brick bolsters. They are intended for general shaping and smoothing of stonework. The very narrow types (sometimes called edging-in chisels) are used to make a starting groove for a bolster when splitting large blocks and slabs.

The **mason's bolster** is a much tougher tool than the brick bolster, being designed for use on stone or concrete. Use it to split blocks and slabs, or to smooth off broad, flat surfaces.

Breaking tools

When it comes to breaking up solid masonry, heavier-duty tools are needed. The **pickaxe**, usually known simply as a pick, is the tool most people think of for breaking up masonry. It has a pointed tip for hacking into hard material and a chisel or spade tip for use on softer material such as ashphalt. In practice, though, you may find it easier merely to crack the masonry with a sledge hammer (see below) and then use the spade tip of the pickaxe to grub out the debris. Neither tool is worth buying, hire them instead.

The **club hammer** is a double-faced hammer used for breaking up masonry and for driving chisels and bolsters. It is also known as the lump hammer.

The **brick hammer** is designed specifically for driving cold chisels or bolsters when cutting bricks. It has a head that incorporates a chisel end for trimming the brick after it has been cut.

The **sledge hammer** is used for directing heavy blows at masonry in order to break it up. For light work, it should be allowed to fall under its own weight, but for more solid material, it can be swung like an axe.

Electric hammers and hydraulic breakers

If you have a lot of demolition work to do or have to break through thick concrete, it is possible to hire an electric hammer to do the job. This will come with a variety of points and chisels.

If no electricity supply is available, then you can hire a hydraulic breaker which is powered by a small petrol driven compressor.

Saws for masonry

For cutting blocks and slabs, or even masonry walls, a hand or power tool can help to achieve a neat finish.

Although the **masonry saw** resembles a normal woodworking saw, its extra-hard tungsten carbide teeth and friction-reducing PTFE coating are capable of slicing through brick, building blocks and most types of stone.

A large two-man version, which has a detachable handle at one end so that an assistant can help pull the saw through, will even cut through walls. Unfortunately, with the exception of small **chasing saws** used to cut electric cable channels in walls, using masonry saws is very hard work.

The **cut-off saw** is just another name for a heavy-duty circular saw which may be electrically or petrol driven. The key to cutting masonry, though, is not so much the power of the saw as the special cutting wheel – a rigid disc of tough abrasive that grinds its way through the stone. Various grades are available to match the material being cut.

Such saws are professional tools that can be hired if there is sufficient work to warrant it, or if a particularly deep cut is required. However, if you already own a circular saw, you should be able to buy a masonry cutting disc for it. Take care, though, to get the right grade for the job.

The **angle grinder** is usually used to cut and grind all types of metal – pipes, rods and sheets. However, fitted with the appropriate stone-cutting disc it can also be used to make cuts and shallow channels in brick, stone and concrete. It is extremely useful for cutting earthenware drainpipe sections.

Tools for drilling holes

When drilling holes in masonry of any type, it's vital to use a drill bit or other tool that is specially hardened to cope with the task. Never attempt to use ordinary twist drills.

Masonry drill bits allow you to drill into brick, stone, mortar and plaster with an ordinary hand or electric drill. Each has two small 'ears' at the end to help break up the waste, and is tipped with tungsten carbide.

Special long versions and extension sleeves are available for drilling right through walls.

Core drill bits are used for boring large-diameter holes – up to 50mm (2in). The bit is a hollow tube that cuts out a 'plug' of material much like a woodworking hole saw. A reduced shank allows it to be fitted into a normal drill chuck.

One thing that masonry drill bits cannot cope with is hard aggregate, such as that found in concrete. One solution is to break up the aggregate particles with a **jumping bit** as they are met. This is a hole boring tool that is driven in with a hammer and twisted by hand at the same time. Sometimes the bit is mounted in a special holder and is interchangeable with bits of different sizes. The **star drill** is a heavier-duty one-piece relation of the jumping bit, and has four tapered flutes to clear debris as the drill is hammered into the masonry.

Electric percussion drills look like normal electric drills and are used in the same way, but there is an important difference. While the drill bit rotates it is hammered in and out, offering the benefits of the twist drill and jumping bit in one.

If you intend buying such a tool, make sure the hammer action can be switched on and off as required and that it has a strong steadying handle. Also, check that it is powerful enough for your needs. This may be indicated by the wattage of the motor or by the chuck size; generally, the larger the chuck capacity, the more powerful the drill.

Make sure that any bits you use with the hammer action are designed for such use, since otherwise they may shatter.

Finally, remember that safety goggles or glasses are essential when using masonry tools.

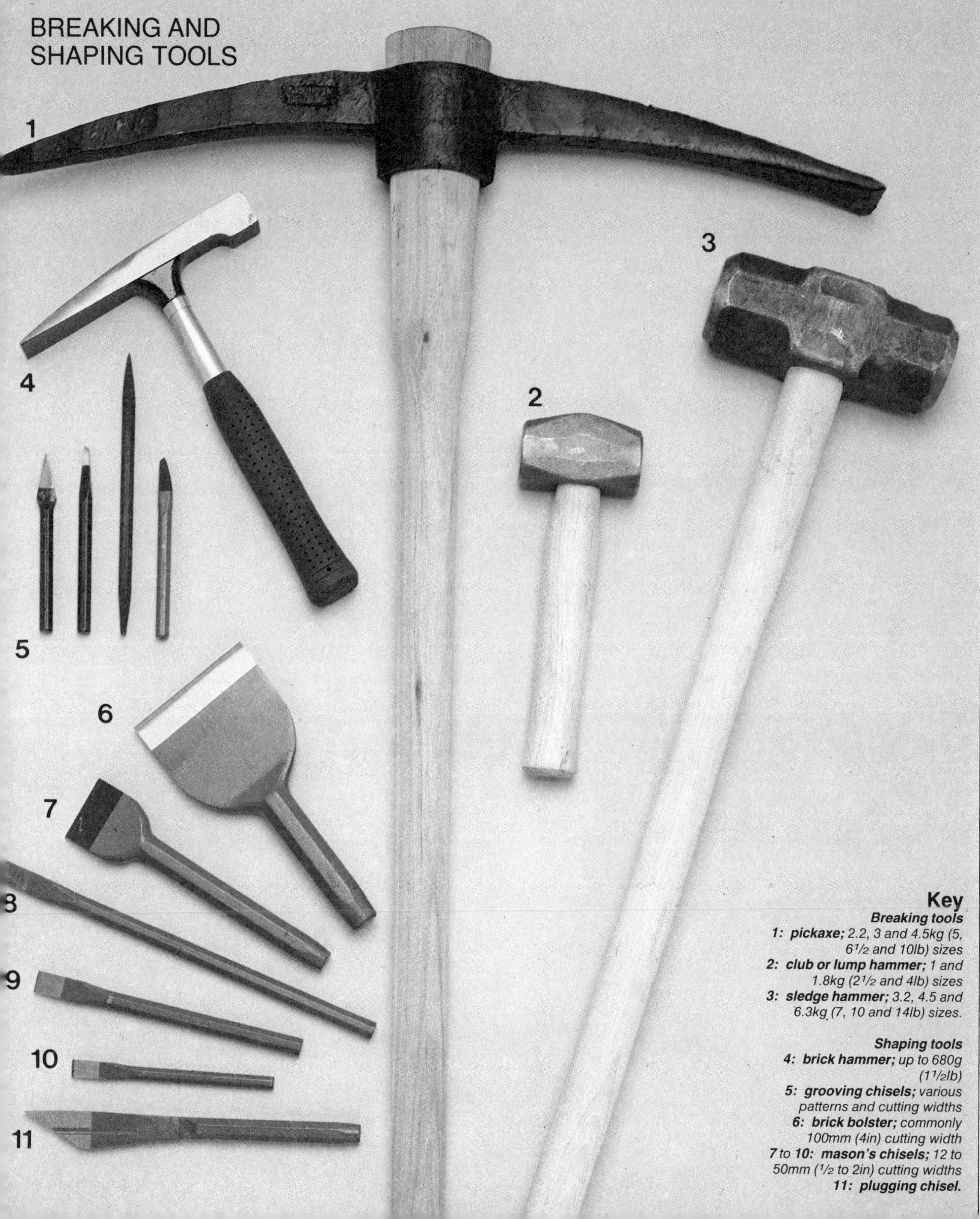

Key

Breaking tools

1: ***pickaxe;*** *2.2, 3 and 4.5kg (5, 6½ and 10lb) sizes*

2: ***club or lump hammer;*** *1 and 1.8kg (2½ and 4lb) sizes*

3: ***sledge hammer;*** *3.2, 4.5 and 6.3kg (7, 10 and 14lb) sizes.*

Shaping tools

4: ***brick hammer;*** *up to 680g (1½lb)*

5: ***grooving chisels;*** *various patterns and cutting widths*

6: ***brick bolster;*** *commonly 100mm (4in) cutting width*

7 to 10: ***mason's chisels;*** *12 to 50mm (½ to 2in) cutting widths*

11: ***plugging chisel.***

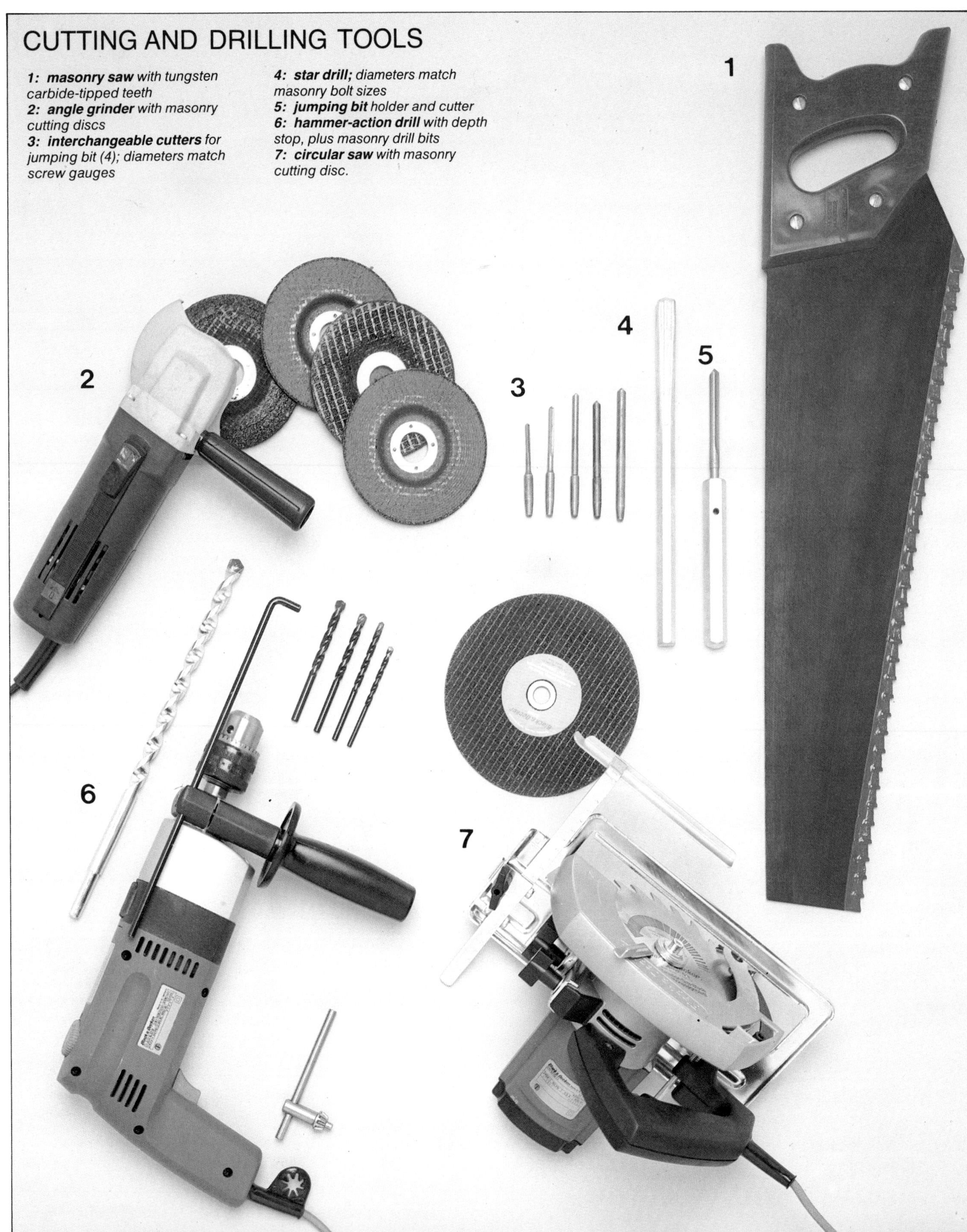

CUTTING AND DRILLING TOOLS

1: masonry saw *with tungsten carbide-tipped teeth*
2: angle grinder *with masonry cutting discs*
3: interchangeable cutters *for jumping bit (4); diameters match screw gauges*
4: star drill; *diameters match masonry bolt sizes*
5: jumping bit *holder and cutter*
6: hammer-action drill *with depth stop, plus masonry drill bits*
7: circular saw *with masonry cutting disc.*

BUILDING RULES AND REGULATIONS

If you propose to carry out any work to improve, enlarge or modify a building, you must obey certain rules and regulations. These can be rather confusing, but here's an idea of what's involved.

In Britain there are two main sets of rules governing building work. These are the Town and Country Planning Acts and the Building Regulations. These rules are quite separate and cover different sorts of things. In Inner London the London Building Acts replace the Building Regulations, and Scotland and Northern Ireland have their own rules too.

The Planning Acts govern the way in which land is developed, stipulating what kind of building can be built on that particular spot. The Building Regulations, broadly speaking, lay down methods of construction and materials, and cover certain other internal works.

You may, therefore, need Building Regulations approval for certain work – knocking down an internal wall for instance – but not planning permission. Other work, such as putting up a garden fence, may need planning permission but not Building Regulations approval. In general, though, most building work will require approval under the Building Regulations.

Where to get help

The whole system of rules and regulations is very complicated and because of this there is often a temptation not to bother about getting permission. But you ignore the need for it at your own peril. For the local authority can issue an order compelling you to undo the work – and put the building back in its original state. If you refuse, it can impose a hefty fine for every day of your non-compliance. It even has the right to do the work itself and charge you for it, although it would only do that in extreme cases. But if a year has passed since you did the work, the authority has to get an injunction before acting.

So if any job you want to tackle does require planning consent or Building Regulations approval, you will, in many cases, require the help of an expert, for example an architect or surveyor. With many Building Regulations applications you'll need engineering calculations to show that the construction will be strong enough. Even architects get the help of an engineer for these calculations.

Moreover, some of the rules are rather vague. Often they are saying little more than that a construction or component must be fit for its purpose. That leaves a lot of leeway for local authorities to put their own interpretation on the Regulations.

To find out if the project you have in mind does require official approval, the simplest thing to do is to approach your local authority. The department that supervises the Building Regulations is called the Building Control Department in England and Wales, and the Building Authority in Scotland; in Inner London you should get in touch with the District Surveyor. To ask about planning you should get in touch with the Planning Department.

When contacting them, just say you have a query about the Regulations or the planning rules and you will be directed to the right office. Tell the officials what you would like to do and ask if permission is needed and if it is likely to be granted. Architects often take this way out and there's no reason why you shouldn't do the same. Most officials are only too anxious to be of assistance. Their main concern is to see that the rules – which after all were devised for your health, safety and comfort – are followed. And they will often give you valuable advice.

However, to save you – and them – unnecessary bother, it's worth knowing a little about what the rules and Regulations do cover, and what work is exempted.

Planning permission

The planning laws cover the way in which land is developed. In particular they cover: building work such as home extensions; change of use such as converting a garage to a living room or dividing a house into flats; and other visible exterior work such as altering the access to your property or erecting a garden wall. However, a certain amount of work is classified as 'permitted development' and can be carried out without planning permission provided it conforms to certain restrictions of size and location. But you will still need Building Regulations approval of course.

For example, you can build an extension without having to get planning permission as long as its volume is less than 70 cu m (2472 cu ft), or, alternatively, as long as it is less than 15 per cent of the volume of the original house (up to a maximum of 115 cu m/ 4060 cu ft), whichever is the greater. For terraced houses, the figures are 50 cu m (1766 cu ft) or 10 per cent. For this rule, the term 'original house' means the house as it was first built or as it stood on 1 July 1948. The volume is calculated from the external dimensions and includes the volume of the roof. However, you would need planning permission if the extension projects beyond the 'building line' (usually the front wall of the house) or above the roof line, if it obstructs the view of drivers on bends and corners, or if it needs new access from a class A or B road.

Despite these restrictions there is still enormous scope for building a garage or extension, or both, without bothering about planning permission. Moreover, with one of the most popular forms of home extension, the loft conversion, you are not adding significantly to the volume of the house, but simply making use of space that would otherwise be lying idle. Dormer windows do of course count as extra space, but their volume is unlikely to be very large.

One point to note is that the volume of permitted development is the total volume allowed. If your home already has an extension this must be taken into account. The volume of porches, sheds, greenhouses, dog kennels and other outbuildings, do not count towards your extension limit and you don't need planning permission to erect them as long as they conform to certain restrictions of size and location. See the drawing on this page for more details.

Other works which are exempt from planning permission are general repairs and maintenance which do not affect the external appearance of the building – repointing brickwork for example. Nor is permission needed for internal work, though you would need it if this involved 'change of use' of the property such as converting a house into flats, or a non-habitable room such as a garage into a habitable room or starting to run a business from your house. Converting flats back into a house, however, does not need permission.

How to apply

If the work you intend to do is not part of the permitted development then you'll need to apply for planning permission. It's worth going along to your local planning department to discuss the work with them, as they will be able to help you fill out the application forms and give you advice on the drawings and documents you'll need to provide. In most cases you will have to fill in four copies of the application form, and make detailed scale drawings. If you are getting a builder to do the work or you're employing an architect, then they can make the application on your behalf.

You can, instead, apply for 'outline planning permission'. This is useful where you are considering buying a house but want to know in advance whether you are likely to get permission for what you want to do. Only simple drawings are needed for this so you should be able to make them yourself. Of course, you would still need full permission before you started work.

The planning department will send you a decision notice after about five to eight weeks to let you know if the work has their approval or not. You may be given full permission or only be allowed to do the work subject to certain conditions such as using materials to match the rest of the house. If the application is rejected the authority should give you their reasons and you may then be able to alter your plans to make them more acceptable. If you consider their refusal to be unreasonable you can appeal to the Department of the Environment within six months of receiving the decision notice.

DO YOU NEED PLANNING PERMISSION?

There are three cases where you will always need planning permission regardless of any of the other rules. These are:

- *if the work will cause danger by obstructing the view of people using a public highway*
- *if the work requires new or wider access to a classified road, and*
- *if the work is restricted or prohibited by the original planning permission for your house.*

A garage (1) *built in a conservation area or, in other areas, built within 5m (16ft) of your house, is treated as a house extension. If it is more than 5m away it is treated as an outbuilding. See the relevant section for each case.*

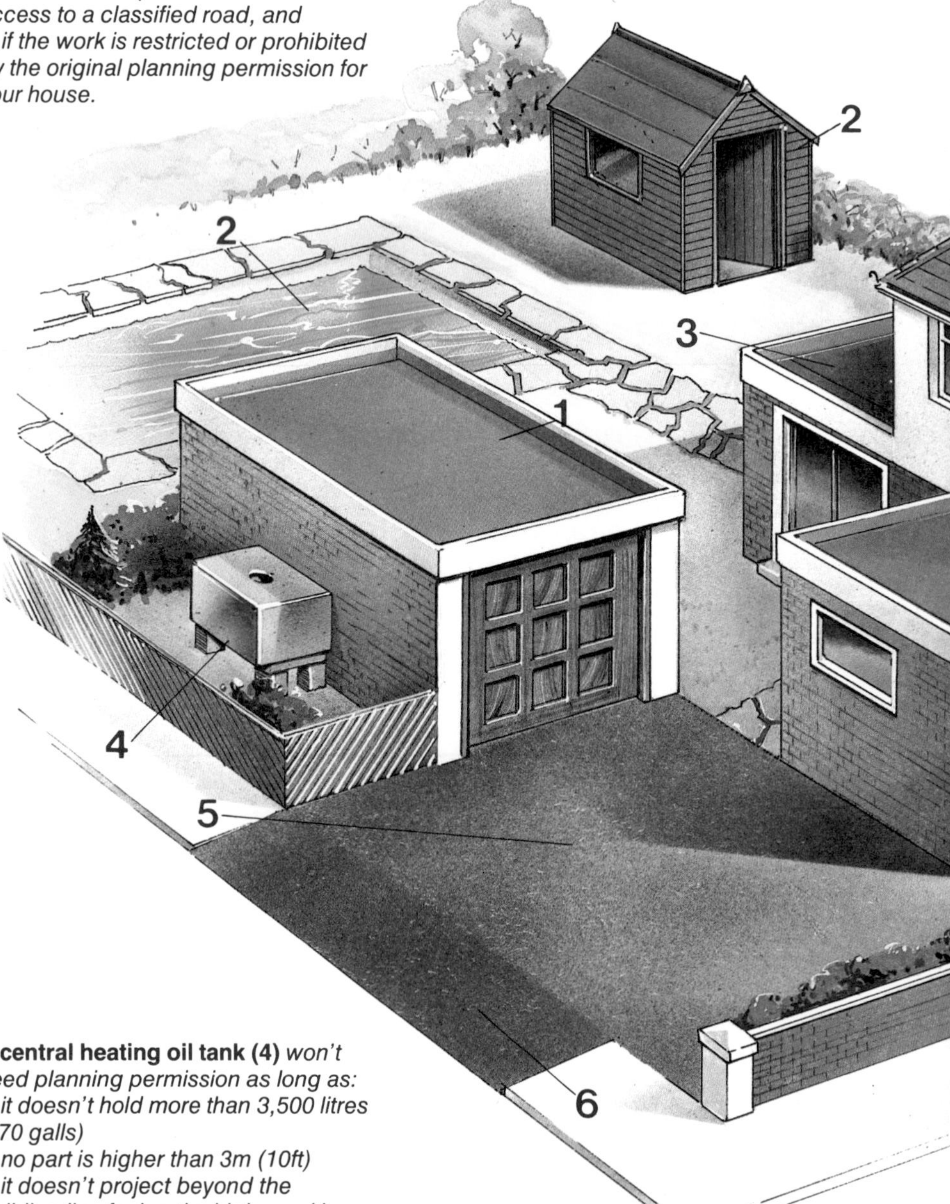

A central heating oil tank (4) *won't need planning permission as long as:*

- *it doesn't hold more than 3,500 litres (770 galls)*
- *no part is higher than 3m (10ft)*
- *it doesn't project beyond the building line facing the highway (the tank in the picture* will *need permission)*

Headstanding for a car (5) *does not need permission as long as it is constructed within your garden and the car is used mainly as a private vehicle.*

A new access (6) *will usually need permission unless it's to an unclassified road and is required for a development such as a garage, which does not itself need permission.*

An outbuilding or other structure (2) *– shed, rabbit hutch, swimming pool, etc – does not need permission if:*
- *it won't be used for business*
- *no part projects beyond the building line of the house facing a highway*
- *it is not more than 4m (13ft) high if it has a ridged roof, or 3m (10ft) otherwise*
- *not more than half the original garden will be covered by building structures.*

An extension or loft conversion (3) *won't need permission as long as:*
- *the extra volume is within the permitted allowance (see page 185)*
- *it doesn't project above the highest part of the house*
- *it doesn't project beyond the building line which faces a highway*
- *no part that comes within 2m (6ft 6in) of the boundary is more than 4m (13ft) above the ground*
- *the extension doesn't result in more than half the original garden being covered with buildings (these last two points don't apply to loft conversions)*
- *it is not a separate dwelling.*

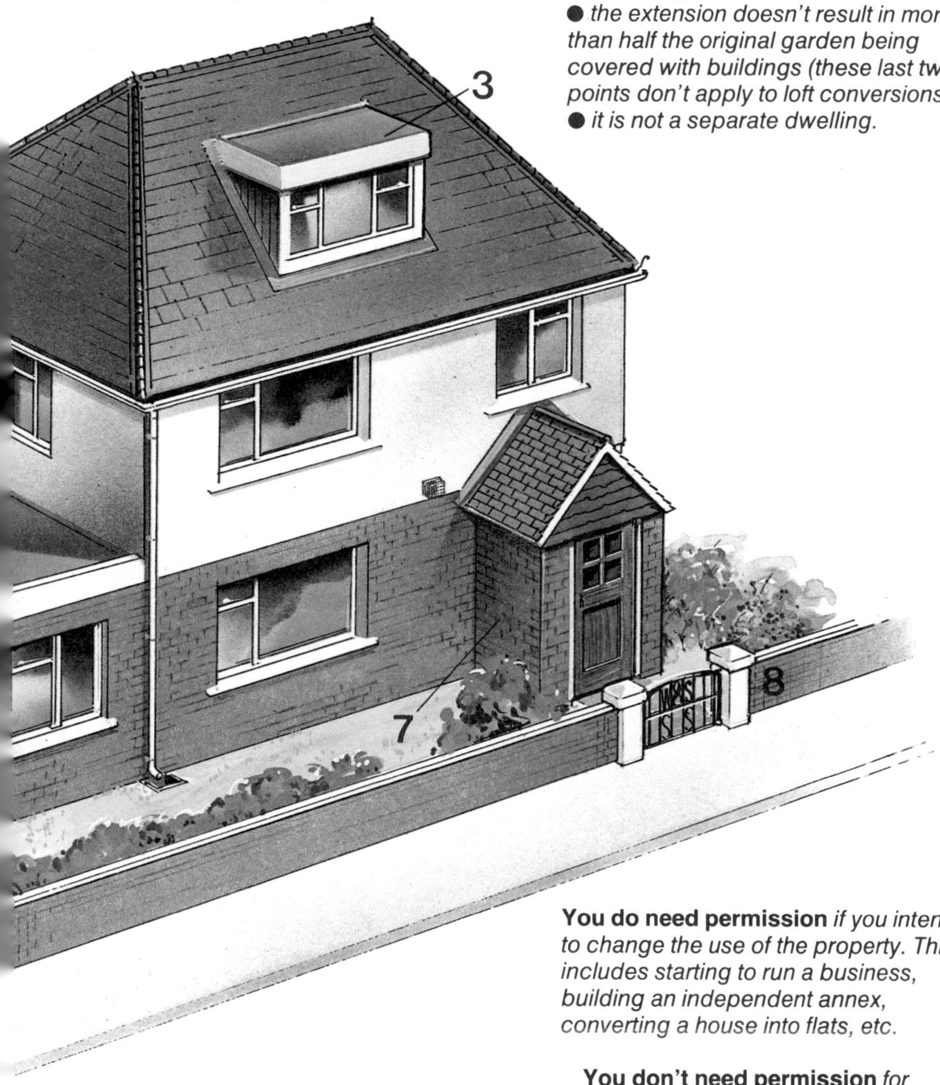

3

7

8

You do need permission *if you intend to change the use of the property. This includes starting to run a business, building an independent annex, converting a house into flats, etc.*

You don't need permission *for general repairs and internal alterations, nor to erect a TV aerial (though you would for a flagpole or radio mast), nor to demolish part of the house or any outbuilding. But remember your property may be covered by other conservation or listed building regulations. And you may need Building Regulations approval for some of these jobs.*

Porches (7) *don't need permission if*
- *the floor area is 2sq m (22sq ft) or less*
- *no part is higher than 3m (10ft)*
- *no part is less than 2m (6ft 6in) from a boundary.*

Fences (8) *need permission if they are more than 1m (3ft 3in) high along a boundary with a highway, or 2m (6ft 6in) elsewhere. Hedges are exempt.*

Building Regulations

With the planning laws just described, the do-it-yourselfer is most interested in the exceptions to the rules, but this is not the case with the Building Regulations. They affect almost every aspect of house construction and cover both the materials and the methods used. In particular the Regulations are concerned with:
- materials used and site preparation
- structural stability
- fire resistance
- damp resistance
- thermal and sound insulation
- ventilation of rooms
- open space beyond windows
- room heights
- construction of stairways
- drainage and sewage disposal
- WCs and water sources
- chimneys, flues and fireplaces
- electrical installation (Scotland).

Repairs and replacements

General repairs and decorations are about the only things that are not covered by the Regulations. You can, therefore, decorate your house as you wish and lay down any floorcovering. You can also make minor repairs such as patching plaster, repointing brickwork, replacing a rotten floorboard, etc. Even where the regulations do make stipulations – about, say, gutters and downpipes – no one is likely to bother if you take down rusty old cast iron ones and replace them with new plastic ones of the same size.

But large-scale repairs may be classed as replacements and you would need approval for the work. For instance, if there were a few tiles missing from the roof you could replace them without anyone worrying about the size of battens to which the tiles are fixed or the fire resistance of the roofing felt below. But, if you're replacing the whole roof, the council will certainly insist that it is built to conform to the Regulations. The same is true of floors. However, in both these cases, it may well be that your council would skip the formality of a proper application, but you should still inform them of the work you intend to do.

Replacement windows are not normally a problem, except that in conservation areas and on listed buildings, they should not alter the external appearance of the house.

New look

You must get approval for any structural alterations or any work involving interference with the drains. One of the most obvious examples of this would be the taking down of a load-bearing wall. Here, the local authority must ensure that the work meets the requirements as the consequences could be disastrous. The same applies to the removal of chimney breasts. In fact in London, you need a

BUILDING REGULATIONS REQUIREMENTS

You must get Building Regulations approval for almost all building work you tackle, except for general repairs and re decoration. It is impossible to give all the requirements in detail, but here are some examples of the more common ones.

A loft room *must comply with all the fire and structural regulations. This will usually mean strengthening the floor and fitting fire doors.*

A habitable room (1 and 5) *must be at least 2.3m (7ft 6in) high, but in a loft this rule is relaxed slightly. It must be 2.3m high only over part of the floor area, not over the whole floor. That is, the floor area below A must be at least half the area below B.*

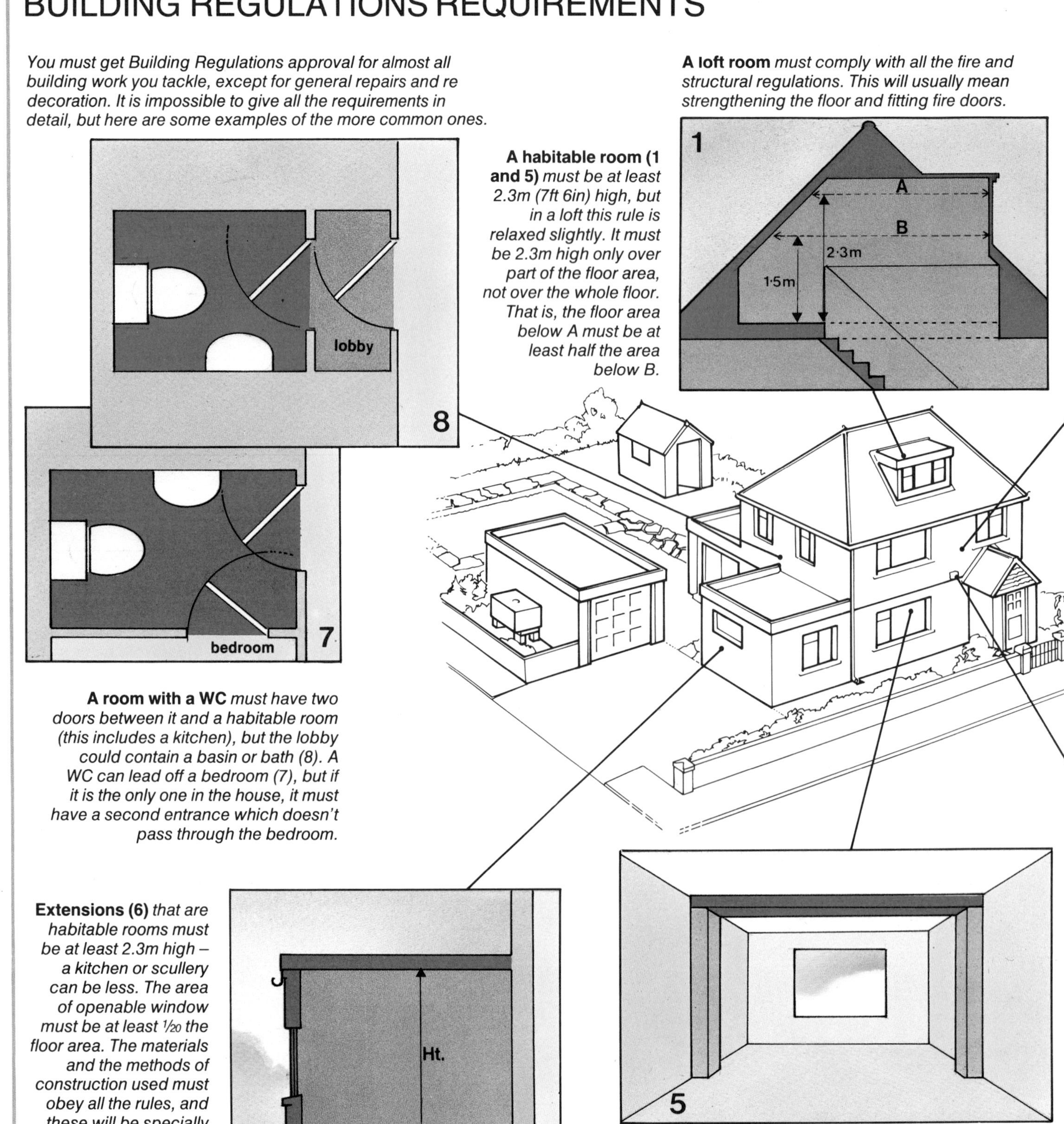

A room with a WC *must have two doors between it and a habitable room (this includes a kitchen), but the lobby could contain a basin or bath (8). A WC can lead off a bedroom (7), but if it is the only one in the house, it must have a second entrance which doesn't pass through the bedroom.*

Extensions (6) *that are habitable rooms must be at least 2.3m high – a kitchen or scullery can be less. The area of openable window must be at least 1/20 the floor area. The materials and the methods of construction used must obey all the rules, and these will be specially concerned with foundations, dpms and drainage.*

Knocking two rooms together (5) *will mean consulting a specialist to calculate the correct size of beam to comply with the rules.*

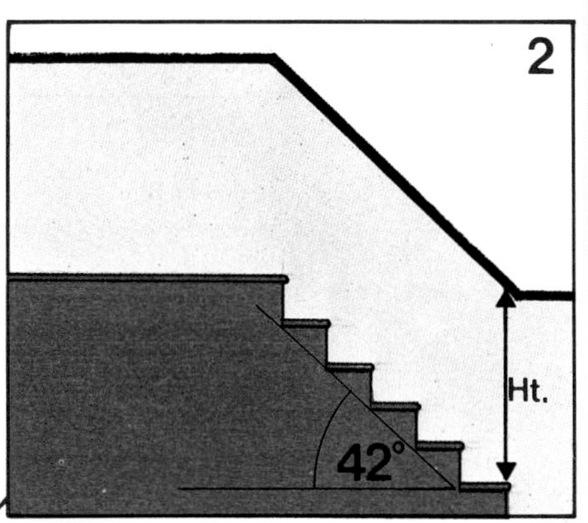

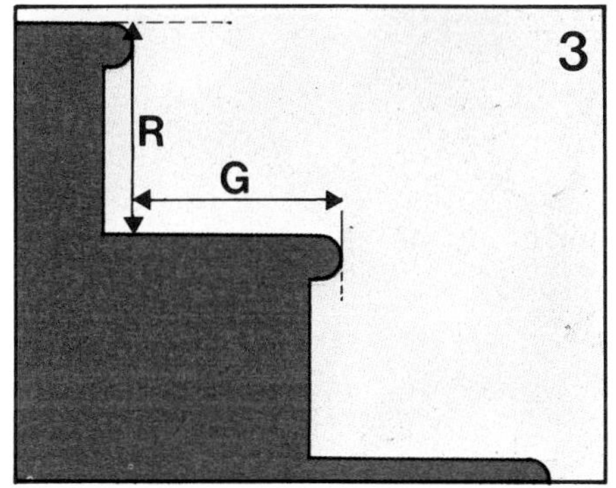

A new staircase (2) *must have at least 2m (6ft 7in) headroom, and a handrail at a height of 840 to 1000mm (33 to 40in). The maximum pitch allowed is 42°. Each 'going' (G) must be equal and so must each 'rise' (R). There are regulations governing their dimensions, but an off-the-shelf stairway will conform to these.*

With flue pipes (4), *the rules determine how near windows they can be and how high above the roof they should discharge. Balanced flue outlets must have a guard if they are below 2m (6ft 6in) from the ground.*

'chimney breast certificate' before you remove a chimney breast from a party wall, in case undue strain is put on the remaining chimney breast on the other side.

Fireplaces of course must be installed to conform with the Building Regulations, otherwise they could be dangerous. And if you move a WC or put in an extra one then you must also get approval as it involves alterations to the drains.

Extensions

When it comes to large-scale new work such as building an extension – whether it takes the form of extra living accommodation, a garage, a porch or a conservatory – then it must be constructed in line with the Regulations. (You may need planning permission, too.) Nowadays, of course, many porches and extensions are bought as prefabricated buildings. Provided they have been manufactured by a reputable company they will be all right but you must still get approval before you start. You must get special approval if the extension will cover a drain or sewer.

Loft conversions, too, must satisfy all the Regulations and this will mean providing the correct type of staircase, proper structural support for the floor, and the correct fire insulation. You will certainly need the help of an architect for this.

Getting approval

The first thing to do, as always, is to discuss your plans with the Building Control Officer. Take along a rough scale drawing if possible. He may suggest how the work should be constructed and what materials you should use. He will also explain how to make the application. The number of drawings and other details you have to provide depends on the type of job you're doing. You will usually have to submit an estimate of the cost of the work, and pay the relevant fee. The fee you have to pay depends on the cost of the work covered by the Building Regulations, and for small jobs there's no fee.

The local authority must approve or reject your plans within five weeks unless the time is extended by written agreement. Once you have approval, you must give 24 hours' notice before starting work and before commencing various stages of the work such as covering up foundations, damp courses, drains, etc. You may be given a set of postcards that you can send to the Building Control Officer to give the notification. If you don't give the notice, the local authority can require you to pull down or cut into the work so that it can be inspected.

In Inner London the situation is slightly different. You don't have to get approval but you must give the District Surveyor at least 48 hours' written notice before you start work. You may also have to provide scale drawings of the work, and you'll have to pay a fee. The District Surveyor will then inspect the work at various stages to make sure the work complies with the Regulations.

Other rules and regulations

Apart from the Planning Acts and the Building Regulations there are certain other rules affecting the development of your property and the work you can do.

If you live in a conservation area then the permitted development is restricted and you can do nothing to alter the exterior appearance of your house without special planning permission. And if you live in a listed building you will need listed building consent for *any* alterations.

The original planning permission for the house may have included some restrictions on the type of work you can do. The original deeds to the property may contain a restrictive covenant preventing you doing certain work even though you may have planning permission; this is particularly common on large estates. If you have a mortgage you should consult the lender as well.

There are certain other regulations concerning the work you can do. For instance, water authorities have their own rules. There is no obligation in law for you to get permission from them for the work you have in mind, but anything you do must be in accordance with their rules and they can, if they wish, inspect the work and demand alterations. You can replace a tap or install a shower without any worries, but if you propose ambitious new plumbing schemes you should discuss your plans with the authority – their address will be on your water rate demand. They will have firm views on the size of tanks and cisterns, and will want to ensure there is no risk of contamination of the water supply.

In Scotland, electrical installation work is governed by the Building Authority and the regulations must be followed. But in England and Wales there is no law requiring you to get permission from anyone. However, for your own safety, you should ensure that all wiring jobs are carried out according to the standards laid down by the Institute of Electrical Engineers (IEE). In fact, you may only get approval for an extension, say, on condition that all wiring is done in accordance with these rules. And, of course, the electricity board can refuse to supply houses where they feel the wiring has not been done properly.

Finally, there are tree preservation orders that must be obeyed, and restrictions on advertising material that can be displayed. It's hardly likely, though, that anyone will bother you if you want to stick a notice advertising the local jumble sale in your window.

INDEX

Additives, special purpose 171
Aggregates 171
Air bricks 189
Angle beads, 151, 167, 176
Angle grinder 182, 184
Arches, brick 83-7
Asbestos boards 162
Ashlar jointing 92

Ballast 171
Bats, brick 14, 18, 20
Beads, metal 91, 151, 154-5, 157, 174
Bed face, 9
Block partitions
  see Partitions, block
Blocks,
  cutting 141
  facing 175
  garden walling 174-5
  pierced screen wall 174-5
  types of 137
Blockwork laying 141
Bonding pockets 141
Bonds, bricklaying 9, 14, 19-20
Boots, work 180
Breaking tools 183
Brick bolster 182
Brick hammer 182
Brick joints 11, 13
Brick pavers 50
Brick trowel 11
Bricklaying 12
  glossary 9
  mortar for 8-9
  tools for 13
  use of line 23
  use of trowel 10, 11
Bricks,
  cutting 18, 21, 53
  finishing 23
  for paths 50-53
  sizes of 8, 9
  storing 8
  types of 8, 168-9
Brickwork,
  arches 83-7
  bonding 14, 19-22
  corners 14-17, 20-21
  fireplace blocking 131
  piers 15, 17-18, 21
  pointing 11
  reinforcing mesh 176
  setting-out 12
  wall ends, 15, 17
Builder's recess 127
Builder's square 16, 17, 37
Building Control Department 185
  Officer 189
Building line 185
Building Regulations 97, 185, 187-9
Bullnose bricks 169
Buttering 13

Calcium silicate bricks 8, 168
Carlite plasters 148
Casement windows 114
Cavity wall closing 107, 117
Ceilings,
  finishing 136
  patching plastered 158-61
  replacing 132-6
  sagging 158, 161
  temporary board support for 133
Cements, types of 170
Centering, arch 87
Chimney capping 128
Chisels, masonry 182
Club Hammer 182
Colouring pigments 171
Comb scratcher 159, 160
Commons (bricks) 8, 168
Concrete 26-9
  compacting 42-3
  curing 37, 39, 44
  drainage guide 29, 40, 43
  finishing 29, 38, 44
  laying 31-2, 38
  laying slabs 35, 40-41
  levelling 28, 36
  marking-out for 31-2
  material quantities 26, 37, 39
  mixes 26, 29, 30, 39
  mixing of 27, 37
  ready-mix 37, 41
    laying 40, 42-4
    quantities 43-4
  reinforcing mesh for 176
Concrete block pavers 50, 68-9
Concrete bricks 168
Concrete point (tool) 182
Concrete slabs 47-9, 67, 172-3
Concrete subsill construction 122
Coping bricks 169
Corners,
  blockwork 139
  brickwork 16-17
Course, brickwork 9
Cownose bricks 169
Crampons 177
Crazy paving 61-3
Cutting tools 184

'Deadmen' (stabilisers) 81
Devilling float 150-155
District Surveyor 185, 189
Dooking iron 182
Door frame ties 176
Doorway framing 139
Doorway making 95-9, 100-103
Drills, masonry 182, 184
Dry lining boards 162
Dry stone walling, simulated 174, 175

Ear protectors 180
Efflorescence 8, 34
Electric drill 182, 184
Electric hammer 182
Engineering bricks 8, 168
English bond 12, 19, 20-21
Expansion joints,
  in blockwork 138, 141
  in concrete 29, 37, 43
  in walls 80, 82
Extensions,
  permission for 187
  regulations for 188, 189

Faced bricks 168
Facing bricks 8, 168
Fencing ironmongery 176-7
Finishing coat 153, 157
Fireback construction 125
Fireback replacement 123-6
Fireplace,
  opening blocking 130-31
  parts 124, 127
  removal 127-31
Flashings 177
Flaunchings 125
Flemish bond 19, 22
Flemish Garden Wall bond 20
Floating coat 152, 156
Floor plate 138
Footlifter 163, 164
Former, arch 85-6
Formwork 29, 36-7, 41, 42
  for slabs 35-7
Foundations for garden walls 30-34
Frame-ties 176
Frog, brick 8, 11
Frost-proofers 171

Garden Wall bonds 19-20
Gauge rod 13, 18
Glazing 119
Gloves, heavy duty 180
Grounds (plastering) 148, 151
Gutter ironmongery 176-7
Gyproc wall board 132-3

Half-partitions 166
Hammers, types of 182
'Hard hats' 180
Hawk 150, 155
Headplate
  see Top plate
Header bond 9, 14, 19
Hearth removal 128-9
Herringbone joist struts 177

Horns, frame 118
Hydraulic breaker 182

Ironmongery,
building 176-7
carpentary 177
roofing 177

Jenny Twin fixer 177
Joist hangers 176
Joist struts 176
Jumping bit 182

Keystone 84, 87
Knee pads 180

Lath and plaster ceilings 132, 158
Laths, plaster 176
Laying to a line 16, 23
Levels, checking 27
Lime 170
Lintels, fixing 97-9, 115
Locks, safety 112
Loft conversions,
permission for 188
regulations for 188, 189
Lump hammer
see Club hammer

Manhole covers, re-setting 44
Masonry saws 182, 184
Masonry tools 182-4
Metal frames, fitting 121-2
Mortar,
for bricklaying 8-10, 11
for rendering 89, 90
mixing 9, 91
quantities 11
Movement joints
see Expansion joints

Needles (supports) 95, 96
Nibs, plaster 158
Noggins 134, 142, 145

Open bonding 20

Overalls 180

Partitions,
block 137-41
finishing 141
marking-out 140
planning 137-9, 142
stud 99, 142-6, 164
Paths 46-7, 50
brick and block 50-53
concrete block pavers 68-72
concrete slab 46-9
crazy paving 61
Patio construction 64-7
Patio doors 104-107, 108-112
Paving materials 51, 172-3
Paving patterns 53
Pebbledash 91, 92, 93
Pickaxe 182
Piers, brick 15, 17-18, 21, 84-5
Pilaster blocks 74, 75
Planning laws 185-7
Planning permission 186-7
Plaster,
haired 160-61
mixing 149, 151
types of 148
Plaster beads 174
Plasterboard 132, 134, 162
fixing 163-4
joint sealing 165
plastering over 133-6, 163-6
Plastering 148-53
angles and reveals 154-7
finishing coat 153, 157
floating coat 152, 156
patching ceilings 135, 158-61
plasterboard 133-6, 163-6
scratch coat 160
scrimming 165
tools 149, 155, 161, 164
Plasticisers,
for concrete 170
for mortar 9
Platform towers 178-9
Pointing, brickwork 11

Portland cement 170
Profile boards 15
Protective masks 180

Queen closers 18, 20

Ready-mix concrete,
see Concrete, ready-mix
Reconstituted stone blocks 174
Regulations 185-9
Reinforcing mesh 174, 176
Rendering walls 88-93
Replacements, permission for 187
Re-plastering 159
Resin-based coatings 90
Respirators 180
Retaining walls 33-4, 78-82
Reveal guide 157
Reveals, plastering 154-7
Roofing ironmongery 177
Roughcast finish 91, 92, 93

Safety gear 180-81
Sands, types of 170-171
Saws, masonry 182, 184
Scraped finish 93
Scratch coat 90-91, 160
Screed beads 151
Screeds, plastering 148
Screen block walls 74-7
Scrimming 165
Shaping tools 183
Sheet flashing 177
Shrinkage plates 177
Sill construction 119
Site preparation 28, 30, 36, 46, 48-9, 65, 70
Slabs, cutting 48-9
Slate clips 177
Sole plate 142, 143
Spatterdash coat 91
Spot board 149
Springing point 84
Stack bonding 74
Staircases, regulations for 189
Steps, door 103

garden 54
built-in 58
drainage 57
free-standing 54
sizes for 55
Stipple coat 89, 91
Stop beads, 151, 176
Stopped end 14, 15, 17
Stretcher bond, 9, 14, 19
Stud partitions,
see Partitions, stud
Studs 142, 144, 145
Subsoil problems 34

Tamping beam 37
Thistle Board Finish 164
Throat restrictor 124, 125
Timber frames, fitting 121
Tingles 177
Toothing-in 57
Top plate 142, 143
Towers, platform 178-9
Trowel, use of 10-11
Tyrolean finish 90, 92, 93

Ventilators, fireplace 130-31

Wall profile 140
Wall ties 176
Wallboards 132
Walling blocks 174-5
Walls,
foundations for 30-34
marking-out for 33
reinforcing 82
rendering exterior 88-93
screen block 74-7
Weep holes 82
Window boards 121
Windows,
construction of 119
enlarging 116-17
installation of 113-17, 118-22
types of 113-14
weatherproofing 121